PLEASE STAMP DATE DUE, BOTH BELOW AND ON CARD

DATE DUE	DATE DUE	DATE DUE	DATE DUE

GL-15

Monitoring Underground Nuclear Explosions

Color picture on cover:
First moment of an atmospheric French nuclear explosion
(Courtesy of C.E.A. — D.A.M., Commissariat à l'Energie Atomique, Montrouge, France)

Monitoring Underground Nuclear Explosions

by

OLA DAHLMAN
and
HANS ISRAELSON

*National Defense Research Institute,
Stockholm, Sweden*

ELSEVIER SCIENTIFIC PUBLISHING COMPANY
Amsterdam — Oxford — New York   1977

ELSEVIER SCIENTIFIC PUBLISHING COMPANY
335 Jan van Galenstraat
P.O. Box 211, Amsterdam, The Netherlands

Distributors for the United States and Canada:

ELSEVIER NORTH-HOLLAND INC.
52, Vanderbilt Avenue
New York, N.Y. 10017

Library of Congress Cataloging in Publication Data

Dahlman, Ola.
  Monitoring underground nuclear explosions.

  Bibliography: p.
  Includes index.
  1. Underground nuclear explosions--Detection.
2. Atomic weapons--Testing--Detection. 3. Seismology.
I. Israelson, Hans, joint author. II. Title.
UG465.5.D33   623'.37   77-8619
ISBN 0-444-41604-8

© Elsevier Scientific Publishing Company, 1977.
All rights reserved. No part of this publication may be reproduced, stored in a retrieval system or transmitted in any form or by any means, electronic, mechanical, photocopying, recording or otherwise, without the prior written permission of the publisher, Elsevier Scientific Publishing Company, P.O. Box 330, Amsterdam, The Netherlands

Printed in The Netherlands

PREFACE

Since more than a decade the achievement of a Comprehensive Test Ban Treaty (CTB) banning all nuclear test explosions in all environments has been one of the key issues in the arms-control and disarmament discussions. The fact that a CTB, despite its political importance and the lengthy negotiations, has not been achieved so far is generally attributed to the lack of the necessary political will of the nuclear-weapon powers. The difficulty in adequately verifying such a treaty has also by some countries been put forward as a main obstacle.

Many countries, including the US and the USSR, have now expressed their strong interest and readiness to conclude a CTB. Recently draft texts to such an agreement have been presented by the USSR (CCD/523, 1977) and Sweden (CCD/526, 1977).

An *ad hoc* group of scientific experts from the two superpowers and several other countries (Eastern, Western, and nonaligned) has been established by the Conference of the Committee on Disarmament (CCD) in Geneva to investigate the technical possibilities to establish a cooperative international exchange of seismological data to facilitate the monitoring of a CTB.

As the question of adequately verifying such a treaty remains one of the most important questions in the CTB negotiations, we have found the time appropriate to summarize the various technical aspects of the monitoring of a CTB. The political aspects of this issue have recently been reviewed by Mrs. Alva Myrdal, former Swedish Delegate to the CCD (Myrdal, 1976).

Our aim with this book has been that it should be generally understandable not only by experts on seismology and disarmament, but also by politicians and by laymen interested in these questions. To this end the six first chapters give an introduction to the basic issues: the present test-ban negotiations, the nuclear test activities, aspects of seismology relevant to test-ban monitoring, and the capabilities of existing seismological stations. The key issue in the discussions of CTB monitoring—the identification of explosions and earthquakes—is given a fairly comprehensive treatment. This includes not only seismological methods, but other methods as well, e. g. on-site inspection and satellite surveillance, and further a discussion of various proposed ways of evading identification.

The estimation of nuclear explosion yields from seismological data is of interest both for estimating the capability of a monitoring network and for interpreting the actual testing activities. A summary, including yield estimates, is given of announced and presumed nuclear explosions conducted for peaceful and military purposes since the conclusion of the Partial Test Ban Treaty in 1963.

All drafts so far presented on a Comprehensive Test Ban Treaty have contained provisions for a cooperative international exchange of seismological data. We therefore present a layout for a monitoring network that in our opinion would meet the political requirements so far put forward for monitoring a CTB.

Although we have attempted to give a broad presentation, much of the material included in this book is inevitably colored by our involvement in the research program on Nuclear Explosion Detection run by the Swedish National Defense Research Institute (FOA). In particular, the seismological illustrations and examples have to a great extent been taken from data obtained at the Hagfors Observatory operated by FOA.

Ola Dahlman   Hans Israelson

Stockholm, March 1977

ACKNOWLEDGEMENTS

This work has been carried out with the generous support of the Swedish National Defense Research Institute (FOA).

We are most grateful to Dr. Olov Alvfeldt, Scientific Editor at FOA, for his comprehensive linguistic review of the manuscript and diligent editorial work. The manuscript was typed by Mrs. Rigmor Berg, and all line drawings were made by Miss Christina Engström; their patient efforts are very much appreciated.

We would also like to express our appreciation to Dr. Ulf Ericsson, Scientific Adviser to the Swedish Foreign Office, for many fruitful discussions and for constructive criticism of the manuscript.

We have benefited in various ways from the assistance of our associates Erik Åhs, Gunnel Barkeby, Karl-Erik Beckman, Nils-Olof Bergquist, Eva Elvers, Nils Gustafsson, Ingvar Jeppsson, Bryant Lindh, Ingvar Nedgård, Leif Nordgren, Ragnar Slunga, Karl-Erik Strandberg, and Britt-Marie Tygård, and from improvements of and additions to various chapters suggested by several other colleagues at FOA: Ingvar Åkersten, Lars Beckman, Kay Edvarson, Lars-Erik de Geer, Tor Larsson, Torleiv Orhaug, and Gunnar Persson.

Valuable suggestions on matters related to the test-ban negotiations have also been offered by Baron Gustaf Hamilton, Ambassador, and Captain Ulf Reinius, R.S.N., Military Adviser, of the Swedish Delegation to the CCD.

Finally, we are indebted to Dr. Milo D. Nordyke of the Lawrence Radiation Laboratory, Livermore, Cal., for comments on the chapter on Peaceful Nuclear Explosions and for making available the photo of the "Sedan" crater (Fig. 14.4), and to Dr. Bernard Massinon of the Commissariat à l'Energie Atomique in Paris for providing the photograph appearing on the cover of this book.

CONTENTS

1.	INTRODUCTION	1
2.	TEST BAN NEGOTIATIONS	5
	2.1 Review of test-ban negotiations	6
	2.2 Disarmament treaties	11
	2.3 Present positions on the test-ban issue	22
3.	NUCLEAR EXPLOSIONS	35
	3.1 A and H bombs	35
	3.2 Bomb designs	38
	3.3 Nuclear explosion phenomena	41
	3.4 Nuclear test activities	44
	3.5 Technical, military, and political significance of nuclear-weapon tests	49
4.	SEISMOLOGICAL BACKGROUND	52
	4.1 Earth structure	52
	4.2 Seismic waves	58
	4.3 Seismicity	73
5.	EXPLOSIONS AND EARTHQUAKES AS SEISMIC SOURCES	92
	5.1 Explosion-source models	92
	5.2 Earthquake-source models	97
	5.3 Comparison between explosion-source and earthquake-source models	102
	5.4 Theoretical calculation of seismic waves	104
6.	SEISMOLOGICAL STATIONS	108
	6.1 Instrumentation	109
	6.2 Station networks	121
7.	DETECTION	134
	7.1 Seismic noise	134
	7.2 Detection processes	143
	7.3 Station detection capabilities	156
	7.4 Network capabilities	163
8.	EVENT DEFINITION AND LOCATION	171
	8.1 Location by an array station	172
	8.2 Location by a network of stations	176
	8.3 Event location at short distances	186
	8.4 International data exchange	188

9.	DEPTH ESTIMATION	190
	9.1 Surface-reflected phases	192
	9.2 P-Arrival times	200
	9.3 Near distances	202
	9.4 Long-period signals	204
	9.5 Short-period signals	206
	9.6 Comparison and combinations	208
10.	IDENTIFICATION	211
	10.1 Statistical aspects	212
	10.2 The $m_b(M_s)$ discriminant	219
	10.3 Short-period discriminants	235
	10.4 Other discriminants	251
	10.5 Multistation discriminants	256
	10.6 Operative aspects of discrimination	258
11.	YIELD ESTIMATION	263
	11.1 Theoretical amplitude–yield relations	265
	11.2 Observed m_b–yield relations	267
	11.3 Observed M_s–yield relations	270
	11.4 Yield estimates of underground nuclear explosions 1963–1976	272
12.	PEACEFUL NUCLEAR EXPLOSIONS	291
	12.1 Possible applications of PNE	291
	12.2 Nuclear explosions conducted for peaceful purposes	294
	12.3 Future prospects for PNE	303
13.	EVASION	306
	13.1 Decoupling	306
	13.2 Multiple explosions	311
	13.3 Hide-in-earthquake method	315
	13.4 The feasibility of evasion	321
14.	NONSEISMOLOGICAL IDENTIFICATION	325
	14.1 On-site inspection	326
	14.2 Reconnaissance satellites	329
	14.3 Intelligence methods	334
15.	MONITORING A COMPREHENSIVE TEST-BAN TREATY	337
	15.1 Political requirements of the verification of a CTB	337
	15.2 A monitoring system	341
	15.3 General conclusions	354
	Appendix 1. Treaties related to nuclear explosions	357
	Appendix 2. Announced and presumed nuclear explosions in 1963–1976	385
	References	406
	Abbreviations and symbols	432
	Index	433

1. INTRODUCTION

Since the first nuclear explosions were carried out in 1945, the threat of nuclear weapons has been hanging over the world like the Sword of Damocles. Few kinds of weapon, if any, have, as nuclear weapons, frightened people and dominated strategic doctrines and the military balance between the two superpowers. Immediately after the second World War, negotiations started to limit the spread of nuclear weapons. These negotiations are still going on, and so are the development and testing of new nuclear weapons. More than 1000 nuclear weapon tests have so far been carried out. Some of these tests have had explosion yields more than one hundred times as large as the yield of the bombs that at the end of WWII were dropped over Hiroshima and Nagasaki. At least five countries: France, the People's Republic of China, the United Kingdom, the United States of America, and the Soviet Union possess nuclear weapons. India has also conducted a nuclear explosion, officially said to be for peaceful purposes. A few treaties with the aim of limiting the spread and testing of nuclear weapons have been concluded. The Partial Test Ban Treaty of 1963 prohibits nuclear explosions in the atmosphere, and the bilateral Threshold Treaty between the US and the USSR limits the yield of nuclear weapon tests carried out by either party as from April 1976 to 150 kilotons. The Non Proliferation Treaty prevents nuclear-weapon states from supplying nonnuclear-weapon states with nuclear weapons, material, and know-how. The Tlatelolco, Antarctic, and Outer Space Treaties prohibit the introduction of nuclear weapons into South America, Antarctica, and the outer space, respectively. These treaties are not adhered to by all states, and do not limit nuclear-weapon development in existing nuclear-weapon states, nor the proliferation of such weapons to a number of so-called near-nuclear-weapon states.

For a long time a Comprehensive Test Ban Treaty, CTB, that would prohibit all nuclear-weapon tests by all countries and in all environments, has been considered by many countries and many people to be a significant step towards nuclear-arms control and disarmament. The urgency of concluding such a treaty has also year after year been emphasized in resolutions by the UN General Assembly. The conclusion of a CTB is fundamentally a political question. The technical discussions of how to verify adherence to the treaty have, however, been much more extensive than for other arms-limitation treaties. Already in an early stage of the negotiations it was realized that seismological means would come to play a central role in the monitoring of a test-ban treaty. Seismic waves generated by explosions can be detected at distances as large as 10 000 km. These waves can also be discriminated with a high degree of confidence from similar waves from earthquakes. In the opinion of the US it is, however, necessary to have provisions for on-site inspection of areas where clandestine tests may have been conducted, to obtain an adequate verification of a CTB. Until recently, the USSR has been against such on-site inspection.

To assess, and also to improve, seismological methods to detect, locate, and identify underground nuclear explosions, research programs have been conducted in several countries. Some of the national research programs are quite extensive. The US, for example, is estimated to have allocated about 200 million $ over a period of ten years for unclassified research and development in this field. Significant research programs have also been carried out in nonnuclear countries like Canada, Norway, and Sweden. These national research programs have led to a considerable development of the science of seismology as a whole. A more or less separate branch associated with the seismological verification problem, often called *detection seismology,* has developed.

The purpose of the present volume is to summarize and interpret the state of the art of detection seismology. Attempts are made to review both the technical capabilities and the political requirements of monitoring a comprehensive test-ban treaty. We feel that the research results and data acquired in this field have reached a point where it is appropriate to make such a summary, both to estimate the present capabilities and to discuss an operative system for monitoring a CTB. Many of the problems encountered in the initial stage in the 1960's have largely been solved. Few dramatic changes are likely to occur in this field in the near future.

Reviews of different aspects of verification methods and programs have been published earlier. The first significant review was published by the UK Atomic Weapons Research Establishment in 1965, and deals mainly with the early development of methods for detecting seismic waves generated by explosions and earthquakes (UKAEA, 1965). In 1968, the Stockholm International Peace Research Institute (SIPRI) brought together a group of seismologists from several countries to assess the advances in seismic verification. The report of this group summarizes many of the early results of the work on the discrimination between explosions and earthquakes (SIPRI, 1968). Estimates of the global capabilities to detect, locate, and identify seismic events, using data from existing seismological stations, were evaluated by Basham & Whitham (1970). Results of work in the USSR are reviewed in a monograph by Pasechnik (1972). A comprehensive account of the historical development of the verification issue in the US is given by Bolt (1976). A treatise on the theoretical basis for seismological discrimination has been worked out by Archambeau (1976).

We will conclude this introduction by giving a general outline of the following presentation. Initially, a brief background to the test-ban negotiations is given, which includes a short historical review as well as a presentation of the present positions taken by some parties in the negotiations. A brief description of nuclear devices and of explosions as seismic sources is given initially. The military, technical, and political significance of conducting tests is considered. The testing activity in different countries is presented. For the reader who is unfamiliar with elementary seismology some fundamental concepts and definitions of earthquakes and seismic waves generated by earthquakes and explosions are briefly summarized. The fundamental principles and characteristics of the instrumentation used to measure seismic waves are given, together with a review of the seismological stations operating all over the world. Data acquired at several globally distributed stations are of paramount importance for seismic monitoring. Throughout the presentation much attention is given to international cooperation, with particular regard to exchange of data acquired at the various stations. Exchange of data among countries will be a key element of an international monitoring system. Methods to detect signals from seismic events and ways of improving and optimizing the detection capability are also described. Estimates of the detection capabilities of individual stations and station networks of particular significance to the verification problem are also given. Different ways of estimating the location of an explosion or an earthquake from seismic signals are reviewed. The problem of estimating the depth of earthquakes is of great importance not only for monitoring a CTB but also in a broader geophysical context. A separate chapter has been devoted to methods of depth estimation and to the accuracy obtainable by such methods. The problem of discriminating between the seismic waves from explosions and those emanating from earthquakes is the main remaining problem in detection seismology. A variety of identification methods for this purpose have been suggested. A fairly complete review of the most useful methods and their performance is made. The significance of nonseismic methods for explosion verification, such as on-site inspection and surveillance with satellites, as complements to the seismic methods is also discussed. The Threshold Test Ban Treaty, banning explosions in the US and the USSR with yields above 150 kilotons, has increased the need for methods to estimate the yield of an explosion. We present available methods and also estimated yields of explosions conducted since 1963. Various methods proposed, in particular by the US, to evade verification are discussed.

The use of nuclear explosions for peaceful purposes has been an important issue in the discussions of a CTB, and a special agreement on such explosions has been reached between the US and the USSR in connection with the Threshold Treaty. Peaceful nuclear-explosion projects carried out so far and the future prospects of such projects are discussed in a separate chapter.

The final chapter attempts to make a synthesis of the different elements presented in the book. In particular, a discussion is given of the political requirements put forward for monitoring a CTB in relation to the technical capabilities. The operational capabilities of presently existing resources are also discussed. A monitoring system is suggested consisting of a seismic station network, a communication system, and an international center for data analysis and assessment.

2. TEST-BAN NEGOTIATIONS

The view has often been expressed that a discontinuance of nuclear-weapon testing would contribute more towards arms control and general disarmament than any other measure. For well over twenty years, efforts have been spent in an attempt to bring about a Comprehensive Test Ban Treaty (CTB), i. e., a treaty banning the testing of all nuclear weapons. Today, when this is written (July 1976), these negotiations are still going on at the Conference of the Committee on Disarmament (CCD) in Geneva. Although no CTB has been agreed upon, partial solutions, in the form of various multilateral and bilateral treaties, have been achieved. There is one question that has, probably more than any one else, caused disagreement during the CTB negotiations, and that is the problem of verifying underground nuclear explosions. The United States and the Soviet Union have all along held different opinions on the technical aspects of control. The USSR considers national means to be sufficient for adequate verification, whereas the US considers such means to be inadequate and therefore requires on-site inspection. Although the control issue has been the main obstacle to the test-ban negotiations, such issues as the use of nuclear explosions for peaceful purposes (PNE) and the universality of a CTB have also been controversal. The superpowers have expressed dissent over the accommodation of nuclear explosions for peaceful purposes under a CTB. The USSR has also demanded that before a test ban becomes effective, its provisions must have been accepted by *all* nuclear countries.

In this chapter, the development of the test-ban negotiations, since the test-ban issue was raised in the middle of the 1950's, is reviewed. Other nuclear-disarmament treaties related to a CTB are presented, although only briefly. Because of the significance of the control issue in the CTB negotiations, close attention is given to this subject. More comprehensive and general accounts of disarmament measures and treaties can be found in, for example, the SIPRI yearbooks (SIPRI, 1969–1976) and a summary on disarmament issued by the United Nations (UN, 1967). The chapter is concluded with a brief summary of the positions, as of July 1976, on the test-ban issue of the nuclear countries and a few other countries participating in the CCD.

2.1 REVIEW OF TEST-BAN NEGOTIATIONS

After the first nuclear explosions in the Alamogordo desert in New Mexico and over Japan in 1945, the testing of nuclear weapons was kept at a comparatively low level until the early fifties. In 1949, the USSR became a nuclear power, and by the end of 1952 the US exploded its first thermonuclear device. This was followed by a similar USSR test only eight months later. At that time the UK also joined the club of nuclear powers. The intensified testing activity that followed gave rise to a growing concern also about the environmental effects of the radioactive products produced by large atmospheric explosions. A case in point were the sufferings of the crew of the Japanese fishing boat *Lucky Dragon,* which was seriously hit by fallout from one of the megaton shots, "Bravo", conducted by the US in the Pacific in 1954. In this year the first proposal for a test ban was put forward in a statement by the Indian Prime Minister, Mr. Nehru, who said that, among the steps to be taken towards the elimination of weapons of mass destruction, consideration would be given to:

> "... some sort of what may be called "standstill agreement" in respect, at least, of these actual explosions, even if arrangements about the discontinuance of production and stockpiling must await more substantial agreements among those principally concerned." (UN, 1954)

At that time the disarmament discussions concentrated on how to put an end to the arms race and how to bring about general and complete disarmament.

In 1955, the USSR, in its proposal for a general and complete disarmament, introduced a special clause on a test ban. Shortly afterwards, the USSR declared its preparedness to negotiate a test-ban treaty as a partial measure towards arms control. This was initially denounced by the Western Powers, the UK and the US, who considered the test-ban issue only as part of a package of disarmament measures, all closely linked together. Already at this time there was fundamental dissent about the verification of a test ban.

The USSR considered what was called "national means" adequate enough for verification, whereas the Western Powers on their part preferred to set up systems of inspection and control (UN, 1967). After having presented a variety of proposals, the parties agreed to initiate a study of the technical aspects of verification. To this end, technical experts from Canada, France, the UK, and the US met in Geneva with experts from Czechoslovakia, Poland, Romania, and the USSR during two months in the summer of 1958. It was then concluded that nuclear explosions down to the low-yield range, 1–5 kilotons (kt), could be detected and identified using a system of 160 to 170 land-based and 10 sea-based control posts (UN, 1958). The control posts should be equipped with instruments to detect radioactive debris and seismic, acoustic, hydroacoustic, and radio signals. In addition there should be options for on-site inspection of ambiguous events.

With this technical analysis as a base, negotiations between the UK, the US, and the USSR were initiated in Geneva in the autumn of 1958 in an effort to achieve a treaty on the discontinuance of nuclear weapon tests. Some time before the opening of the negotiations, the UK and the US announced their intention to suspend nuclear tests for one year, once the negotiations had begun, provided that the USSR did not resume its tests. This suspension could be extended on a year-by-year basis. By the end of 1958, all three parties had voluntarily suspended nuclear testing. At an early stage the assessment made by the Geneva expert group was criticized by the US delegation. It was maintained that the seismic verification capabilities had been overestimated, and also that the possibility of muffling or decoupling an underground explosion had not been taken into account properly. There was also disagreement over the practical implementation of on-site inspections. Various treaty proposals considering these technical problems were put forward, for example a so-called threshold treaty based on, in principle, the same concept as the more recent Threshold Test Ban Treaty between the US and the USSR.

The positions of the two sides were apparently quite close during this period. At this time, France conducted her first nuclear explosion. Then the U-2 aircraft episode occurred, dramatically changing the atmosphere of the political scene. Shortly afterwards, in 1961, nuclear testing was resumed, first by the USSR, and then by the US. This ended the voluntary suspension of testing that had been in effect since late 1958, and the notable, high-yield USSR device, estimated to 58 megatons, was detonated in the autumn of 1961 (Glasstone, 1964).

The negotiations between the UK, the US, and the USSR within the framework of the Conference on the Discontinuance of Nuclear Weapons Tests ceased early in 1962. The negotiations were resumed in March 1962 in a Sub-Committee of the *Eighteen-Nation Committee on Disarmament* (ENDC), a committee that had developed from the *UN Disarmament Commission* established in 1952. The subcommittee, which consisted of the UK, the US, and the USSR, was called upon to consider the problem of halting the nuclear weapon testing. In 1960, the Disarmament Commission, the work of which aimed at general and complete disarmament, was composed of five NATO and five Warsaw Pact countries. In 1962, eight nonaligned countries were added to this committee.

Various proposals for a comprehensive agreement on the cessation of tests were put forward and considered by the subcommittee. In October 1962, the world was shaken by the Cuban crisis. In historical and political analyses of this period it is often pointed out that the Cuban crisis created a need for reconciliation between the two superpowers (Kennedy, 1969). An analysis by SIPRI (1972) of the significance of the exchange of messages between the USSR Premier and the US President in the period right after the crisis lends support to this idea. Although the USSR at that time accepted the idea of on-site inspection, there was still disagreement over the number of inspections. When the parties were at the closest, the USSR proposed two or three, the US seven inspections a year. The impasse was never broken, and the parties instead agreed upon a partial solution by signing the so-called *Partial Test Ban Treaty* (PTB) in Moscow on August 5, 1963. The text of the PTB is attached as Appendix 1. In a following section of this chapter we pursue to discuss the PTB together with a few other nuclear disarmament treaties.

During the summer of 1969 the ENDC was enlarged with several new members, and decided to change its name to the *Conference of the Committee on Disarmament* (CCD). The number of members had gradually increased to 31 in 1975. The latest parties to take their seats at the negotiation table as of Jan. 1, 1975 were the Federal Republic of Germany, the German Democratic Republic, Iran, Peru, and Zaire. Table 2.1 lists the CCD members of 1976. All nuclear powers, except China, are members of the Committee. It should be noted, though, that France has never taken part in the negotiations. A comprehensive test

TABLE 2.1. *Members of the Conference of the Committee on Disarmament (1976).*

Argentina	F. R. of Germany	Mexico	Romania
Brazil	France	Mongolia	Sweden
Bulgaria	German D. R.	Morocco	United Kingdom
Burma	Hungary	Netherlands	United States
Canada	India	Nigeria	USSR
Czechoslovakia	Iran	Pakistan	Yugoslavia
Egypt	Italy	Peru	Zaire
Ethiopia	Japan	Poland	

ban is not the only issue at the CCD, which today is the principal forum for multilateral disarmament negotiations. A number of other issues, concerning, for example, chemical weapons and weapons of mass destruction, are also on the agenda.

More than ten years of negotiating at the Conferences in Geneva has not brought us very close to an agreement on a CTB. Up to 1976, the delegates have met more than 700 times at the conference table, and well over 400 working papers, dealing with different technical matters, have been prepared, some 50 of which deal with the verification of a CTB. There have also been a few so-called informal meetings, with participation of technical experts in the discussions in Geneva (CCD/397, 1973). Nineteen experts from eight countries, for example, attended the informal expert meeting held in 1976. This has given opportunities for an exchange of views and ideas not only between technical experts and delegates, but also among the experts from the different countries.

Very few of the proposals for a CTB have been put forward in formal treaty language. One draft treaty was tabled at the CCD in 1971 by the Swedish delegation (CCD/348, 1971). In 1975, the USSR submitted a draft treaty to be considered by the UN General Assembly at its 30th session. Pending a CTB, various measures of a partial nature have been suggested, and already in 1965 the idea of a threshold treaty prohibiting nuclear tests above a certain seismic magnitude was brought up (ENDC/PV. 224, 1965). Shortly afterwards, the UK proposed an annual quota of nuclear explosions that should be phased out over a period of five years (ENDC/232, 1968). These proposals for partial measures have been revived, with minor variations, on several occasions and by different delegations.

Even if the CCD is the main forum for test-ban discussions, nuclear explosion testing has been debated and negotiated in other contexts

TABLE 2.2 Nuclear disarmament treaties signed and ratified by nuclear countries.

Treaty	Antarctic Treaty	Partial Test Ban Treaty	Outer Space Treaty	Treaty of Tlatelolco[a]	Non-Proliferation Treaty	Sea-Bed Treaty	Threshold Test Ban Treaty	PNE Treaty
Signed	1959	1963	1967	1967	1968	1971	1974	
Effective	1961	1963	1967		1970	1972		1976
Total number of parties[b]	19	106	71	20	97	58		

Country	Signed / Ratified	Signed / Ratified	Signed / Ratified	Signed / Ratified	Signed / Ratified	Signed / Ratified	Signed / Ratified	Signed / Ratified
China				1973(P II) / 1974				
France	1959 / 1960		1967 / 1970	1973(P II) / 1974				
India		1963 / 1963	1967			1973		
UK	1959 / 1960	1963 / 1963	1967 / 1967	1967(P I, P II) / 1969	1968 / 1968	1971 / 1972		
US	1959 / 1960	1963 / 1963	1967 / 1967	1968(P II) / 1971	1968 / 1970	1971 / 1972	1974	1976
USSR	1959 / 1960	1963 / 1963	1967 / 1967		1968 / 1970	1971 / 1972	1974	1976

[a] P I and P II denote Additional Protocol I and Additional Protocol II, respectively (see text).
[b] As of December 31, 1975.

also. This is illustrated by, for example, the recent *Threshold Test Ban Treaty,* which is presented in the following section. This treaty is the result of a summit meeting between the US and the USSR in the summer of 1974. The test-ban issue has also been frequently debated and discussed in the UN General Assembly, whose concern about nuclear-weapon testing has repeatedly, for a long time, been expressed in resolution form. An excerpt from a resolution in 1975 reads (UN, 1975 a):

> "... emphasizes the urgency of reaching agreement on the conclusion of an effective comprehensive test ban ... urges the Conference of the Committee on Disarmament to give highest priority to the conclusion of a comprehensive test ban agreement."

2.2 DISARMAMENT TREATIES

In addition to the Partial Test Ban Treaty, a number of other bilateral or collateral arms-regulation and disarmament treaties, agreements, or conventions have been concluded during the sixties and seventies. It is beyond the scope of this presentation to even outline the historical and political development of all of these treaties. In this section we will therefore confine ourselves to a presentation of the PTB and a few other nuclear treaties related to the test-ban issue. The following treaties are dealt with:

>Antarctic Treaty
>Partial Test Ban Treaty
>Outer Space Treaty
>Treaty of Tlatelolco
>Non-Proliferation Treaty
>Sea-Bed Treaty
>Threshold Test Ban Treaty
>PNE Treaty

Some basic data for these treaties are listed in Table 2.2. The full texts of the Partial Test Ban Treaty, the Non-Proliferation Treaty, the Threshold Test Ban Treaty, and the PNE Treaty are given in Appendix 1. The treaties listed above are multilateral, with the exception of the Threshold Test Ban Treaty and the PNE Treaty, which are bilateral treaties between the US and the USSR. At the end of this section we also briefly discuss other bilateral agreements between the US and the USSR.

In the following the treaties are presented in the chronological order in which they were signed. For each treaty we have attempted to de-

TABLE 2.3. *Elements of verification in nuclear-disarmament treaties.*

Verification element	Antarctic Treaty	Partial Test Ban Treaty	Outer Space Treaty	Treaty of Tlatelolco	Non-Proliferation Treaty	Sea-Bed Treaty	Threshold Test Ban Treaty	PNE Treaty
1. Collection of information								
Obligatory declaration and notification	X		X	X			X	X
Ground, naval, and air observation	X					X		
Special detection and identification techniques							X	X
International exchange and reports of data				X				
Bilateral exchange and reports of data							X	X
2. Inquiry	X			X		X	X	X
3. On-site inspection by parties								
Obligatory continuing, periodic, or in a limited number								X
On the basis of free access	X		X	X				
On the basis of consultation, cooperation, or invitation			X			X		
4. International supervision and inspection								
Specially established control organization				X				
Existing organization					X			
5. National self-supervision and inspection	X		X	X		X	X	X
6. Complaint procedure								
Consultation and cooperation	X			X				
Reference to a conference of the parties	X			X				
Reference to the International Court of Justice	X			X				
Recourse to the UN Security Council				X		X		

scribe the main provisions, the means of verification, and the adherence. Table 2.3 summarizes how the question of verification is dealt with in the different treaties. Data in this table are in all essentials identical with the findings of the Swedish working paper CCD/398 (1973).

Antarctic Treaty

The *Antarctic Treaty* is sometimes referred to as a model for international cooperation. Negotiations for the treaty were initiated in the wake of the International Geophysical Year in 1957–58, and the scientific community is often considered to have played an important role in bringing about the treaty (SIPRI, 1973). The Antarctic Treaty was also the first multilateral disarmament treaty to be signed after the second world war. It became effective in 1961.

In the first article of the treaty it is ensured that Antarctica shall be used for peaceful purposes only. The treaty prohibits any activities of a military nature in Antarctica, such as establishment of military bases, military maneuvers, or testing of any kind of weapon. The treaty also prohibits exploding of nuclear charges and disposal of radioactive waste products in Antarctica. The provisions of the treaty apply to the area south of latitude 60 S, except for the high seas within this zone. The treaty has an initial duration of 30 years.

The treaty recognizes two kinds of parties: (a) the twelve contracting parties, all of which participated in the scientific exploration of Antarctica during the International Geophysical Year, and (b) parties later acceding to the treaty. At present there are five nations of the latter category. An acceding party may become a contracting party by conducting substantial scientific exploration in Antarctica.

To verify compliance with the treaty, each contracting party is entitled to designate observers to carry out any inspections with complete freedom of access at any time to any or all areas of Antarctica. All installations and equipment should be open to inspection. In addition, aerial observations of any part of Antarctica can be carried out by each contracting party. Up to 1975, six inspection operations (three by the US, one by Argentina, one by New Zealand, and one jointly by the UK and Australia) have been carried out. None of these inspections revealed any violation of the treaty. Finally it can be mentioned that the treaty provides for periodic meetings for consultation and exchange of information. This type of meeting has so far been held almost every other year.

Partial Test Ban Treaty

The *Partial Test Ban Treaty* (PTB) was signed in 1963 and became effective the same year. The text of the PTB is given in Appendix 1. The PTB prohibits nuclear-explosion tests in the atmosphere, in the outer space, and under water. Thus, underground tests are not prohibited by this treaty. The PTB does, however, ban underground explosions that generate radioactive debris outside the territory where the explosion is conducted. In the preamble to the treaty, the three depositary parties, the UK, the US, and the USSR, have undertaken to carry on negotiating in order to bring to an end all testing of nuclear weapons for all time.

In 1975, 106 states had adhered to the PTB; as can be seen from Table 2.2, this is the largest number of states having joined a nuclear-disarmament treaty. The two nuclear countries China and France are the most notable among the nonsigners. Among nations that up to and including 1975 had not ratified the PTB were also some so-called near-nuclear countries, like Argentina and Pakistan (SIPRI, 1975). China and France have since 1963 been conducting explosion tests in the atmosphere, although on a much smaller scale than did the superpowers in the late fifties and the early sixties.

The PTB does not specify how to verify compliance with the treaty, as can be seen from Table 2.3. In practice, verification has to be carried out by national means.

There have been no reports on any remarkable violation of the PTB. Up to 1975, leakage of radioactivity from at least five underground tests in the USSR and from at least two in the US had, however, been detected outside the territories of these countries.

Although it is only a partial disarmament measure and in spite of the fact that it has not been adhered to by all nations, the PTB has reduced the number of high-yield weapon tests and made it impossible to test nuclear weapons in their destined environment. The PTB has also slowed down the contamination of the atmosphere by radioactive particles.

Outer Space Treaty

The *Outer Space Treaty* entered into force in 1967 and is at present adhered to by 71 nations. The treaty, which concerns nuclear weapons as well as any other kinds of weapons of mass destruction prohibits the placing in orbit around the Earth objects carrying such weapons. Installation of mass-destruction weapons on other celestial bodies as well as stationing such weapons in outer space are also prohibited. Moreover, the treaty prohibits establishment of military bases, testing of any type of weapon, and conduct of military maneuvers on celestial bodies.

In 1974, the UN General Assembly commended the Convention on Registration of Objects launched into outer space (UN, 1974). This convention formalizes the voluntary registration system existing since the early sixties, and is in accordance with the principle of the Outer Space Treaty stating that states bear international responsibility for national activities in outer space.

Treaty of Tlatelolco

The *Treaty of Tlatelolco* is the only major disarmament treaty not being initiated by or being entirely dependent on the superpowers. Originally proposed by Mexico, it was signed in Mexico City in 1967. The name of the treaty derives from the Aztec name of the district in Mexico City where the treaty was approved.

The treaty aims at making Latin America a nuclear-free zone, and prohibits "testing, use, manufacture, production or acquisition by any means as well as receipt, storage, deployment, and any form of possession of nuclear weapons by countries in Latin America". One article of the treaty defines a nuclear weapon as "any device which is capable of releasing nuclear energy in an uncontrolled manner and which has a group of characteristics that are appropriate for use for warlike purposes". The parties should follow special safeguard agreements on their peaceful nuclear activities. These safeguards have been laid down by the International Atomic Energy Agency (IAEA). Nuclear explosions for peaceful purposes are dealt with in a separate article, on the interpretation of which there is dissent. Argentina and Brazil have claimed a right to conduct nuclear explosions for peaceful purposes. According to the view of other signatories, no nuclear-weapon device according to the definition of the treaty can be used for a nuclear explosion for peaceful purposes. The Agency for the Prohibition of Nuclear Weapons in Latin America, usually called OPANAL, an acronym based on its name in Spanish, has been set up to ensure compliance with the obligations assumed by the parties. There are also two additional protocols annexed to the treaty.

According to Additional Protocol I, continental or extracontinental states which de jure or de facto are internationally responsible for territories within the limits of the Latin American zone undertake to apply military denuclearization to such territories. According to Additional Protocol II, the nuclear-weapon states undertake to respect the military denuclearization of Latin America. Moreover, the nuclear-weapon states should not contribute to acts involving violation of the treaty, nor should they use or threaten to use nuclear weapons against the parties to the treaty.

Non-Proliferation Treaty

After seven years of negotiations, the *Non-Proliferation Treaty* (NPT) entered into force in 1970 for an initial duration period of 25 years. The text of the NPT is given in Appendix 1. Basically the NPT aims at stopping the spread of nuclear weapons and limiting the number of nuclear countries to those having nuclear weapons in 1967. The treaty prohibits the transfer of nuclear weapons to any state. Nuclear-weapon states, defined by the treaty as states having carried out a nuclear explosion prior to 1967, are prohibited from assisting nonnuclear states to aquire nuclear devices. A nonnuclear-weapon state party to the treaty is prohibited from manufacturing nuclear devices. It may, however, make all preparations necessary for the manufacture of nuclear weapons, without thereby formally violating the treaty. The nuclear-weapon states have undertaken to make available to nonnuclear-weapon states nuclear charges to be used for peaceful purposes, "pursuant to a special international agreement or agreements, through an appropriate international body ... Negotiation on this subject shall commence as soon as possible after the Treaty enters into force" (cited from the NPT text). As of early 1976, no such negotiations had, however, taken place. The treaty also prescribes that all parties shall pursue negotiations on effective measures for nuclear disarmament.

All parties to the treaty have a right to use nuclear energy for peaceful purposes. In their peaceful nuclear programs, the nonnuclear states are then obligated to follow safeguards stipulated in special agreements with the IAEA.

Among the nuclear countries, China, France, and India have not signed the NPT. This means that the Indian explosion of May 18, 1974 was not a violation of the NPT. Among the countries that have not signed the NPT, there are also a number of so-called near-nuclear states, like Argentina, Brazil, Israel, Pakistan, and South Africa. Up to July 1976, a number of states, among them Egypt, Indonesia, and Switzerland, had signed but not ratified the treaty. Article VIII of the NPT prescribes that a review conference be held five years after the treaty became effective, the purpose of the review being to ascertain that the aims of the treaty had been accomplished (NPT, 1975). The review conference was held in 1975 in Geneva. It does not appear to have had any effect in the particular respect of making any further near-nuclear states adhere to the treaty.

Sea-Bed Treaty

The *Sea-Bed Treaty* entered into force in 1972. The parties to the treaty undertake not to emplace any nuclear weapons or any other weapons of mass destruction, as well as structures, launching installations, or facilities for storing, testing or using such weapons on the sea-bed and the ocean floor and in the subsoil thereof beyond the outer limit of the twelve-mile zone referred to in the Convention on the Territorial Sea and the Contiguous Zone of 1958.

The treaty is adhered to by all the nuclear countries except France and China. At the end of 1975, a total of 55 states had ratified the treaty.

The treaty includes elaborate provisions for ensuring compliance by its parties. Each party has a right to verify through observations the activities of other parties in the region specified in the treaty, provided that the observations do not interfere with these activities. If violation is suspected, consultations between the parties concerned should be held to remove the uncertainty. If the situation persists, other parties to the treaty should be notified, and cooperation for further verification, including inspection, may be agreed upon. If doubt still remains, the matter can be referred to the UN Security Council.

Threshold Test Ban Treaty

The *Threshold Test Ban Treaty* (TTBT), which was signed in July 1974 and was due to enter into force on March 31, 1976, is a bilateral agreement between the US and the USSR. (Other bilateral agreements are discussed in the next section.)

The treaty prohibits underground nuclear-weapon tests with yields exceeding 150 kilotons. Nuclear-weapon tests must also be confined to defined test sites. To ensure compliance with the treaty, each party shall use national technical means of verification in a manner consistent with the generally recognized principles of international law. The parties also undertake not to interfere with each other's means of verification, and, if necessary, to consult with each other, make inquiries, and respond to such inquiries. A protocol, which is considered as an integral part of the treaty, specifies reciprocal exchange of data to facilitate verification. It is stipulated that data on the location and geological characteristics of the nuclear-weapon-test areas be exchanged. Moreover, the yields of two explosions in each geophysically distinct test area shall be provided for calibration purposes. The detailed geological and yield information stipulated in the treaty is important from a technical point of view as regards the monitoring of the treaty, as will become apparent from the presentation later on in this book.

Nuclear explosions for peaceful purposes (PNE) are not covered by the TTBT as it was phrased (see Appendix 1). The parties have negotiated rules for PNE separately, and an agreement on these was signed as a separate treaty. This so-called PNE Treaty is presented below.

The value of the TTBT has been much questioned—it has been noted, for example, that only a small percentage of the explosions conducted during the seventies did have yields larger than 150 kilotons. This matter is further discussed in Chapter 11.

Some positive aspects of the treaty can be found, however. It will probably increase the difficulties of developing new, high-yield nuclear explosives. The treaty can also, in principle, be considered as a comprehensive test-ban treaty effective outside the defined test sites. This condition must, of course, have an important implication for the discussion of the monitoring of a CTB also at the test sites. The reciprocal exchange of geological data stipulated in the protocol may also be a step towards a greater openness (SIPRI, 1975). If detailed information on explosion and geological conditions becomes available to seismologists all over the world, this may improve the present methods of seismic verification (CCD/PV. 647, CCD/PV. 653, 1974).

PNE Treaty

The Treaty on Underground Nuclear Explosions for Peaceful Purposes, or the *PNE Treaty*, which was signed on May 28, 1976, is a bilateral agreement between the US and the USSR. The agreement, the text of which is given in Appendix 1, consists of a treaty, a detailed protocol, and an agreed statement, the latter of which defines certain activities which do not constitute peaceful applications of nuclear explosives. The PNE Treaty, which was negotiated pursuant to Article III of the TTBT, is considered as a companion treaty to the TTBT. The interrelationship of the TTBT and the PNE treaties is recognized from the provisions that the PNE Treaty is supposed to enter into force simultaneously with the TTBT, and that neither party is entitled to terminate the PNE Treaty while the TTBT remains in force.

The PNE Treaty governs all nuclear explosions carried out at any place outside the weapon testing grounds specified under the TTBT. This means that it also governs all nuclear explosions for peaceful purposes conducted by the US or the USSR in the territory of a third state. Such cases should be conducted in accordance with the procedures of Article V of the Non-Proliferation Treaty.

The PNE Treaty prohibits any individual explosion having a yield exceeding 150 kt, a limit identical to that placed by the TTBT. The same limit has been placed on the two treaties because it is not considered possible to distinguish between explosives for peaceful purposes and for weapon-related purposes. The PNE treaty further prohibits any group explosion consisting of a number of individual explosions with an aggregate yield exceeding 1 500 kt. Although the PNE Treaty prohibits individual explosions greater than 150 kt, it should be noted that it also provides for consideration of the question of conducting individual explosions exceeding 150 kt at an unspecified future time.

Detailed restrictions described in the protocol are put on the emplacement of both individual and group explosions. There is, for example, a minimum depth requirement on any explosive. This minimizes the possibility of obtaining militarily significant information, like blast and electromagnetic effects.

To provide assurance of compliance with the provisions of the treaty not only the use of national technical means of verification, but also quite extensive bilateral procedures, are included. These procedures, which are highly technical and complicated, are described in detail in the protocol. The aim of the verification measures is essentially to ensure that no individual explosion has a yield exceeding 150 kt. The amount of verification measures which should be taken increases with the aggregate explosion yield. For yields below 100 kt certain information should be conveyed prior to and after the explosion.

The amount of information required increases with increasing yield, and for explosions having yields between 100 and 150 kilotons observers may be present on the basis of consultation between the parties. The principal function of the observers is to confirm geological and other information to assist in the determination of the explosion yield by the national technical means. For explosions with an aggregate yield exceeding 150 kt access by observers to the explosion sites is guaranteed. For certain explosions the observers have a right to bring and use instruments to estimate the yield of the explosion. Detailed procedures are laid down in the protocol to ensure the verifying side a valid set of measurements, and the other that the instruments are not being misused to obtained unwarranted information.

The provision for on-site observation is the most significant feature of the verification procedures. As can be seen from Table 2.3, the principle of on-site inspection has not been accepted in other arms-control agreements. It has therefore been said that the PNE treaty constitutes an important milestone in this respect. It should be noted, however, that on-site observation is guaranteed only for explosions exceeding 150 kt. This limit is quite high compared to the yields of explosions so far carried out for peaceful purposes (Chapter 12). Few such explosions, if any, have had yields exceeding 150 kt. It can therefore be argued whether on-site observation of a PNE will ever be conducted.

Other US and USSR agreements

Table 2.4 lists some of the agreements and treaties between the US and the USSR reached after the second World War. Here we outline very shortly the provisions of some of these agreements and discuss in brief terms the verification issue. It can be seen from Table 2.4 that most bilateral agreements made after the establishment of the hot line in 1963, in the wake of the Cuban crisis, resulted from the Strategic Arms Limitation Talks, SALT.

The number of antiballistic missile (ABM) systems is limited by the so-called SALT ABM treaties. According to the ABM treaty of 1974, the deployment of ABM systems shall be limited to a single area in each of the two countries. The components of each ABM system are also subject to limitation. The SALT Interim Agreement provides for a freeze of the aggregate number of fixed, land-based intercontinental missile launchers and of missile launchers on modern submarines. The

TABLE 2.4. *Bilateral agreements between the US and the USSR.*

	Signed	In force
Memorandum of understanding regarding the establishment of a direct communications link (*Hot Line Agreement*)	1963	1963
Agreement on measures to improve the USA–USSR direct communications link (*Hot Line Modernization Agreement*)	1971	1971
Agreement on measures to reduce the risk of outbreak of nuclear war between the USA and the USSR (*Nuclear-Accidents Agreement*)	1971	1971
Agreement on the prevention of incidents on and over the high seas	1972	1972
Treaty on the limitation of anti-ballistic missile systems (*SALT ABM Treaty*)	1972	1972
Interim agreement on certain measures with respect to the limitation of strategic offensive arms (*SALT I Interim Agreement*)	1972	1972
Protocol to the Agreement on the prevention of incidents on and over the high seas, signed on 25 May 1972	1973	1973
Agreement on the prevention of nuclear war	1973	1973
Protocol to the Treaty on the limitation of anti-ballistic missile systems	1974	
Treaty on the limitation of underground nuclear weapon tests (*Threshold Test Ban Treaty*)	1974	
Treaty between the USA and the USSR on underground nuclear explosions for peaceful purposes (*PNE Treaty*)	1976	

interim agreement does, however, cover a period of only five years. Even though the possibilities of a quantitative expansion of nuclear defence systems have been somewhat reduced, the treaties still allow a qualitative improvement of these weapon systems. There is also a possibility that the number of nuclear warheads carried by each missile will increase. This means that the development of nuclear weapons is likely to go on, with a corresponding need for continued explosion testing. That the interest in high-yield explosion testing is diminishing is, however, indicated by the 150-kiloton threshold of the Threshold Test Ban Treaty.

To provide assurance of compliance with the provisions of the various SALT agreements, national technical means of verification are used. It is noteworthy that in this context the US has refrained from insisting on on-site inspection, an issue which in the case of the test-ban negotiations has caused such a long-lasting impasse. Moreover, the use of deliberate concealment measures that impede verification and interference with the national verification means is prohibited.

The article on verification of the Interim Agreement on the Limitation of Strategic Offensive Arms reads:

> "1. For the purpose of providing assurance of compliance with the provisions of this Interim Agreement each Party shall use national technical means of verification at its disposal in a manner consistent with generally recognized principles of international law.
>
> 2. Each Party undertakes not to interfere with the national technical means of verification of the other Party operating in accordance with paragraph 1 of this article.
>
> 3. Each Party undertakes not to use deliberate concealment measures which impede verification by national technical means of compliance with the provisions of this Interim Agreement. This obligation shall not require changes in current construction, assembly, conversion, or overhaul practices."

This is in sharp contrast to the possible use of evasion methods that the US have maintained to be one of the obstacles to an adequate verification of a CTB by seismic means.

2.3 PRESENT POSITIONS ON THE TEST-BAN ISSUE

We conclude this chapter by summarizing the positions, as of 1976, on the test-ban issue of the countries possessing nuclear explosives and of a few other countries which are members of the CCD. This summary attempts to describe the general political attitudes of these countries, as expressed either at the CCD or at the UN General Assembly, and the views on the control issue and on nuclear explosions for peaceful purposes. The verification issue is also briefly treated. The countries in question are presented in alphabetical order.

Canada

Canada has been actively engaged in studies of the problems involved in the CTB issue. Pending a CTB agreement, the Canadian CCD delegation has proposed various measures of constraint, such as reducing the size or number of nuclear-weapon tests, or a moratorium on testing (CCD/PV.546, 1971). Interim proposals have also been put forward by Canada in the UN General Assembly, as is evident from the following statement of the Canadian delegate in 1975 (UN, 1975 b):

"In the view of my Government, to be effective, a comprehensive test-ban treaty must provide adequate means for the nuclear-weapon States to assure each other and the world community that they are fully complying with its provisions. It must ensure that any testing or application of nuclear explosions for peaceful purposes does not contribute to the testing or refinement of existing nuclear-weapon arsenals or to the acquisition of nuclear explosive capability by additional States. Agreement by some testing Powers to stop their tests should not, in the view of my delegation, have to await the participation of all nuclear-weapon States. We believe that the two super-Powers and as many other nuclear-weapon States as possible should enter into an interim agreement, open to all States and containing appropriate provisions to ensure its effectiveness. Parties to such an agreement would halt their nuclear-weapon tests at least for a specified time. At the end of that time the agreement could be reviewed to determine whether it could be continued or should be replaced by an agreement involving all nuclear-weapon States."

The Canadian delegation at CCD has put great efforts into examining the technical matters of seismological verification. In 1969 the UN General Assembly adopted a resolution proposed by Canada to assess international exchange of seismic data (UN, 1969). According to this resolution, all UN members were requested to supply information on seismic installations from which they were prepared to exchange data on a guaranteed basis. The returns served to assess the world-wide seismic detection and identification capabilities. The results of this assessment were presented in a working paper at CCD (CCD/305, 1970). The Canadian attention to international cooperation for exchange of seismic data has also resulted in working papers presented jointly with Japan and Sweden (CCD/380, 1972; CCD/457, 1975).

China

China is the only nuclear power that is not a member of the CCD. Having not signed the PTB, China has linked the discontinuance of nuclear tests to the question of prohibition and destruction of nuclear weapons, as stated by the Chinese delegate in the UN General Assembly (UN, 1971):

"The two super-Powers have been working on their nuclear weapons for decades. They have conducted innumerable nuclear tests of all kinds and their nuclear arsenals have swollen immensely. In these circumstances, the partial or complete halting of nuclear tests will not inhibit the continuation of the production and use of nuclear weapons. Therefore the prohibition of nuclear tests will be of no positive significance if not linked with the prohibition and the destruction of nuclear weapons. It can only serve to consolidate the super-Powers' nuclear monopoly, deprive the other countries of their just right to develop nuclear weapons and resist nuclear threats posed by the super-Powers; it can only spread a false feeling of security and weaken the struggle of the peoples of all countries for the complete prohibition and the thorough destruction of nuclear weapons. The world can not gain peace and security from the prohibition of nuclear tests which, on the contrary, can only increase the nuclear threat and the nuclear blackmail of the two super-Powers and increase the danger of a nuclear war."

Although this statement was made already in 1971, it was still representative of the Chinese position in 1975, when the Chinese UN delegate made the following statement about the draft resolution on a CTB proposed by the USSR (UN, 1975 c):

"With regard to the Soviet draft resolution on the conclusion of a treaty on the complete and general prohibition of nuclear-weapon tests, the Chinese delegation has solemnly pointed out on many occasions that this is another fraud of sham disarmament, which is solely aimed at preserving the nuclear monopoly of the super-Powers. Therefore, the Chinese delegation will vote against the said draft resolution, and states that it will never enter into the so-called "negotiations" as such, nor will it be bound in any way by the result of such "negotiations"."

France

France is formally a member of the CCD, but has so far not taken part in the negotiations in Geneva. According to the French position, the cessation of nuclear tests is not disarmament, as expressed by a French delegate to the UN General Assembly in relation to a resolution on the conclusion of a treaty prohibiting nuclear-weapon tests (UN, 1975 b):

"It is certainly not for the French delegation to praise the work of the CCD; more authoritative voices than ours should be heard on the matter. I should like to say, however, that on this issue the prohibition of nuclear-weapon tests, working documents of great technical value have been submitted to the CCD, with a view to clarifying the conditions for such a prohibition. Despite those studies, despite the discussion in depth to which they have been giving rise for four consecutive years, we can see that so far the three nuclear Powers participating in the CCD have not been able to arrive at an agreement.

In these circumstances, what usefulness can the draft resolution before us have? Can we reasonably believe that the presence of China and France in these discussions, in a new framework, will enable us to do away with the obstacles that so far have been insurmountable for those Powers which first began the exploration of the atomic domain and which have arrived at a high degree of technical knowledge after having carried out numerous and varied tests?

Frankly, we do not believe so, and the draft resolution before us seems to us, politically speaking, to be unrealistic. France has repeatedly stated that it was ready to study with the parties concerned all aspects of the nuclear problem. Thus, from the very beginning we agreed to a conference of the five nuclear Powers. Since the conference was not held, for reasons beyond our control, we supported the idea of a world disarmament conference to deal with these problems. We truly regret that so far this idea has not been realized. We are still ready to undertake the study of effective disarmament measures whenever any real opportunity of doing so emerges.

But having thus made manifest our goodwill, we cannot agree to a draft resolution which isolates, in the area of nuclear disarmament, a specific point which does not affect the substance of the problem. The prohibition of tests is the same type of enterprise as the Non-Proliferation Treaty; that is, it is not of a nature to put an end in any way to the production of nuclear weapons. If an agreement were reached on this point, the tremendous privilege of nuclear Powers would still be maintained without their undertaking obligations regarding real disarmament: that is, the limitation and the reduction of nuclear weapons.

We believe that partial measures are only a palliative and can only be a false security measure. We repeat this now. This is not how we want to see the problem of nuclear disarmament approached. Rather, what we want to see is negotiations concerning the real elimination of these weapons.

Only then would an end to nuclear tests be an episode in the process, without which it has no real significance."

India

India conducted its first explosion in May 1974 and has officially announced that the explosion was carried out as part of the program of study of peaceful uses of nuclear explosions (CCD/PV. 637, 1974).

During the discussions of a test ban, India has also distinguished between nuclear-weapon tests and explosions for peaceful purposes. India's position on the role of a CTB and on the Non-Proliferation Treaty is clear from the following excerpt from a statement by India's CCD delegate (CCD/PV. 664, 1975):

"India has been among the first countries to work relentlessly in various international forums for the total elimination of nuclear weapons. It is for this reason that India is opposed to all moves and suggestions which might shift the focus of the international community from the over-riding objective of nuclear disarmament. India has also objected to the Non-Proliferation Treaty because, among other reasons, it is not a treaty which will lead to arms limitation and disarmament.

For nuclear disarmament the first requirement is stoppage of the production of nuclear weapons and a cut-off of the production of fissile material for weapons purpose. It will then be easy to devise a universal non-discriminatory system of safeguards. A step on which the CCD should concentrate immediately in order to control the nuclear arms race is a comprehensive agreement to ban all nuclear weapon tests, an agreement which will find universal acceptance. The Indian delegation has always been of the opinion that there is no justification whatsoever for continuing with nuclear weapon testing. Strategic superiority in nuclear weapons has ceased to be a relevant factor because of the over-kill capacity of the two super-Powers."

The position on nuclear explosions for peaceful purposes is expressed as follows in the same statement:

"In connexion with peaceful nuclear explosion technology, we have heard the argument that intentions do not matter but that what matters is the technology of conducting nuclear explosions. The argument goes further, that a country should therefore be restricted from developing explosion technology. While one can understand the appeal that a country should not go in for nuclear weapons, it is difficult to accept the principle that a technology should be restricted to some because it may be used for weapons purposes by others. This is a strange argument. We are being asked to fight the wrong enemy. We cannot stop the proliferation of nuclear weapons by controlling the development of peaceful explosion technology."

"...To sum up, we feel that only nuclear-weapon tests are relevant to the question of nuclear-arms development and proliferation. As far as the question of regulating PNE's is concerned, it can only be taken up after achieving a comprehensive test ban."

Japan

Japan joined the CCD in 1969 and has during the years taken several initiatives to break the impasse on the CTB. Japan considers the conclusion of a comprehensive test ban to be an important issue, which might be achieved by partial solutions (CCD/PV.692, 1976):

"There is no need to stress that the most urgent issue at present in the field of nuclear disarmament is the realization of a comprehensive test ban. Although the Partial Test Ban Treaty of 1963 about a nuclear test ban already exists, the nuclear-weapon States which are not yet Parties to the Treaty ought still to be urged to accede to it and in this way protect the atmosphere completely from radioactive contamination. Therefore, we urge again the nuclear-weapon States which have not yet become Parties to the Partial Test Ban Treaty to accede to the Treaty and assume the treaty obligation not to conduct nuclear tests in the atmosphere. In this content, it is with much regret that we note the atmospheric nuclear test conducted by the People's Republic of China in January this year: and we appeal to that country to discontinue atmospheric nuclear tests as early as possible.

As nuclear tests in the atmosphere, outer space and under water are already banned by the Partial Test Ban Treaty, the realization of a comprehensive test ban of nuclear weapons in the form of an international instrument would only require the conclusion of a treaty banning nuclear weapon tests underground, which is the remaining environment. In concluding a treaty banning underground nuclear weapon tests, two major problems are cited—namely, (1) verification and (2) how to deal with peaceful nuclear explosions. So I shall now address myself to these two problems.

As to the first problem, as a result of discussions over many years on an underground nuclear weapons test ban, it has become clear that one of the greatest obstacles to the conclusion of the treaty is the difference between the super-Powers about verification. So we have stressed time and again that, if a comprehensive test ban of nuclear weapons cannot be expected in the near future because of the differences of position on verification, both the United States and the Soviet Union should show their sincerity towards this question by banning underground nuclear weapon tests from wherever verification is possible, and at the earliest possible date, as an intermediate measure leading to a comprehensive test ban. As a concrete measure along these lines, we have suggested the banning of underground nuclear explosions above a certain level which can be detected and identified by present seismological means, and then the gradual lowering of the threshold of underground nuclear tests, which would eventually lead to a comprehensive test ban."
...

"As to the next question of peaceful nuclear explosions, I need to reiterate that it is necessary to ensure that these peaceful nuclear explosions should not be used for military purposes. The fact that negotiations between the United States of America and the Soviet Union initiated on the basis of article 3 of the Threshold Treaty are protracted indicates that this is not an easy question to solve. However, we would like to emphasize that the fact that this question remains unsettled should not be made to serve as an excuse for delaying the bringing into force of the underground weapons test ban. From this viewpoint, and in order to bring the negotiations between the two super-Powers to an early and successful end, studies should be made among others on the possibility of authorizing all PNEs exclusively under international observation or banning PNEs over the threshold, tentatively for a given period, for instance five years.

The remaining problem is that of lowering the threshold over which weapon tests are banned. Taking into account the concern that the threshold of 150 kilotons set out in the Threshold Treaty may be too high, it would be necessary to conduct a serious technical examination and comparison of the yields down to which explosions can be detected and identified. In all events, the treaty text should be flexible enough to allow the gradual reduction of the threshold. If we continue negotiations and reduce gradually the threshold in accordance with the provision on the continuation of negotiations upon the conclusion of the multilateral treaty limiting underground weapon tests on the one hand, and if we negotiate simultaneously the conclusion of the agreement on peaceful nuclear explosions on the other hand, it will become eventually possible to realize a treaty banning all underground nuclear weapon tests."

Japan has been actively engaged with the technical aspects of verification and did in 1972 initiate a tripartite cooperation on this subject between Canada, Japan, and Sweden. Formal agreements between research institutes in the three countries were made to promote a close cooperation on scientific and technical issues. This cooperation has resulted in reviewing the ongoing activities in relation to a CTB. Joint working papers have also been tabled at the CCD as a result of this cooperation (CCD/376, 1972; CCD/457, 1975). These have in particular stressed the significance of the utilization of data obtained through international exchange. Japan in 1973 also proposed the holding of an informal meeting with participation of experts in connection with the question of a CTB (CCD/PV. 559, 1973).

Sweden

Sweden has given high priority to the achievement of a CTB without any further partial measures:

> "Thus all of us carry a heavy responsibility to do our utmost to achieve this aim. It seems obvious that a Comprehensive Test Ban Agreement (CTB), generally adhered to, would represent a decisive contribution towards this goal. At the same time a CTB would constitute an important step towards nuclear disarmament.
>
> The test ban issue has been on the agenda of the CCD as a matter of high priority for years. The CCD was again urged by the General Assembly last autumn to give the highest priority to the conclusion of a CTB agreement. Action is long overdue.
>
> My delegation regrets that this matter, as well as other disarmament problems, is nowadays being dispersed into sub-issues, with the clear danger of leading nowhere. I said already in the summer of 1974 that the bilateral Threshold Ban Treaty under negotiation between the United States and the Soviet Union will be of little practical value in preventing nuclear weapon developments. Most of the test explosions carried out during the last year have yields below that threshold. The value of a threshold treaty may also be seriously questioned, as it might slow down efforts to put an end to all test explosions of nuclear weapons." (CCD/PV.689, 1976).

In the same statement the Swedish opinion on the verification issue is stated in the following words:

> "My delegation fails to see any insurmountable technical obstacles with regard to the verification of a CTB, a problem which purportedly so far has held up progress towards an agreement. Scientific progress in the field of seismology has been such that a global monitoring system for a CTB can be established to provide adequate deterrence to States parties to a Test Ban Treaty from

carrying out clandestine testing. In the opinion of the Swedish delegation, it is possible to establish a monitoring system by which most earthquakes and explosions corresponding to a hard rock yield of about 1 kt can be detected, located and identified with a high degree of accuracy. This figure, incidentally, stands in drastic contrast to the 150 kt figure established by the Threshold Test Ban Treaty. In our view, the possibility of involving the United Nations in the operation of such a system should be explored."

In the Swedish opinion, the US and the USSR have a particular responsibility to achieve a CTB.

"It is customary to refer to various technical obstacles when the political will to undertake a specific measure towards disarmament is lacking. Another method consists in placing such political conditions for the realization of a particular project that it is doomed from the outset. In the context of a CTB, my delegation thus cannot agree with the concept that a CTB must be signed by all present nuclear-weapon States from the beginning. It follows that we also are extremely doubtful as to any suggestion to negotiate a CTB outside the CCD. Although a universal adherence to a CTB obviously is a most desirable goal, it is up to the two nations which possess a vast superiority in nuclear arsenals, namely the United States and the Soviet Union, to be the first to start this process. Otherwise, we can be certain that no progress will be achieved. Considering their vast superiority in nuclear arsenals, they would not by taking such a step expose themselves to any risks as far as their military security is concerned. At the same time the security of the world at large would be considerably improved." (CCD/PV.689, 1976).

Sweden has also over the years tabled draft treaties for a CTB. The most recent one, presented in 1971, updated an earlier version from 1969 (CCD/348, 1971). This proposal, which still reflects the Swedish position, states that after entry into force the nuclear powers should phase out their testing programs according to a time schedule laid down separately. This is just to avoid the practical problems involved in a sudden cessation of nuclear-weapon testing. According to the draft treaty, peaceful nuclear explosions are not prohibited and shall be carried out according to special rules. For verification of such a treaty the parties should undertake international exchange of data and co-operate in clarifying the nature of ambiguous seismic events. The signatories have a right to inquire and receive special information. The parties should also invite inspection in difficult cases. This concept is usually called *verification by challenge*. Failure to cooperate for clarification of difficult cases could be reported to the UN Security Council and to other parties of the treaty.

Like Canada and Japan, Sweden has emphasized the significance of international cooperation in the field of detection seismology, and Sweden has proposed the setting up of a network of seismological stations and an international data center, which should collect, organize, and disseminate verification data to interested parties (CCD/PV.610, 1973; CCD/482, 1976).

Union of Soviet Socialist Republics

In 1975, the USSR proposed to include in the Agenda of the XXXth Session of the UN General Assembly the question of concluding a treaty on the complete and general prohibition of nuclear weapon tests. With the proposal, USSR also attached a draft treaty, according to which any nuclear explosion at any place and in all environments should be prohibited. The contents of this draft treaty do not imply any changes in the USSR position on the test-ban issue.

Control of the compliance with this treaty should, for instance

". . . be conducted by the Parties through their own national technical means of control in accordance with generally recognized norms of international law. In order to promote the objectives of and ensure compliance with the provisions of the Treaty the Parties shall cooperate in an international exchange of seismic data . . ."

and

". . . undertake, when necessary, to consult one another, to make inquiries and receive appropriate information in connection with such inquiries. Each Party of the Treaty which ascertains that any other Party acts in violation of the obligations under the Treaty may lodge a complaint with the Security Council of the United Nations."

The concept of national technical means of control has not been explained more exactly, and there has been no description of how this control should be accomplished.

The USSR has not presented any working paper describing verification in technical terms.

Whereas the US requires on-site inspection for adequate verification, the USSR is opposed to this idea. On-site inspection is interpreted by the USSR as a way of intrusion, and the USSR maintains that national means in combination with international exchange of seismic data are fully adequate for verifying a CTB, as was expressed by the USSR representative (CCD/PV.638, 1974):

"The position of the United States and the Western countries supporting it, which insist on international inspection to verify compliance with an agreement on the cessation of underground nuclear tests, continues to block progress. The representatives of some Western States in the committee try to present the situation as one in which the non-acceptance by the USSR and other socialist countries of on-site inspections to verify the cessation of underground tests allegedly impedes the reaching of agreement on this problem. We have repeatedly explained that existing national means of detecting nuclear explosions, in combination with international cooperation in the exchange of seismographic data, create appropriate safeguards for compliance by the parties to the agreement with their obligations regarding the cessation of underground nuclear tests."

Nuclear explosions for peaceful purposes in relation to a CTB is another point with conflicting positions of the two superpowers. In one of the articles of the USSR draft it is stated that the treaty should not be applicable to underground nuclear explosions for peaceful purposes, the procedures of which have to be established by a special agreement to be concluded "as speedily as possible". The USSR delegate at the CCD commented on these procedures as follows (CCD/PV.688, 1976):

"The complete and general prohibition of nuclear-weapon tests must not, of course, create obstacles to benefiting from the peaceful uses of nuclear explosions. Nuclear Powers must carry out peaceful nuclear explosions in conformity with the provisions of article V of the Treaty on the Non-Proliferation of Nuclear Weapons. Of course, the carrying out of peaceful nuclear explosions must be subordinated to the task of preventing the spread of nuclear weapons. In the opinion of the Soviet Union, the procedure for carrying out peaceful nuclear explosions must be in keeping with the task of ensuring the non-proliferation of nuclear weapons. In establishing the procedure, it will be necessary to have due regard for the recommendations of IAEA, which is the most competent and qualified international body to work out such recommendations."

A third point of a CTB on which the superpowers have conflicting opinions is the universality of a treaty. The USSR draft treaty states that it shall not enter into force until it is ratified by all nuclear-weapon states. Whether India should be considered to belong to this category or not is unclear. The USSR ambassador at the CCD elaborated on this point (CCD/PV.688, 1976):

"It was proposed, for example, that not all, but only some nuclear States should unilaterally declare a moratorium on nuclear tests or suspend nuclear-weapon tests altogether, while other nuclear States would be in a position to continue such tests. Proposals of this kind could not, of course, solve the problem of ending nuclear-weapon tests by all and everywhere. They would only result in creating unilateral advantages for some States to the detriment of others. Such steps were proposed in violation of the principle of ensuring security for all parties to the corresponding agreement for the achievement of a particular measure of disarmament. The carrying out of such partial measures would not, of course, contribute to the strengthening of international peace and the security of States. On the contrary, such an approach to solving the problem of nuclear-weapon tests would be fraught with an aggravation of the international situation and a greater threat of nuclear war. Disarmament measures must be carried out without detriment to anyone's security. The benefit to be derived from such measures must accrue to the States which participate in carrying them out. This principle is an indispensable condition for making successful headway in the limitation of arms and in disarmament. Any efforts to solve this problem in a manner contrary to this principle, and all attempts to push through agreements which disregard the said principle, are clearly doomed to failure. The task is to find ways of solving disarmament problems without detriment to anyone's security and to the benefit of all."

United Kingdom

Being a close ally to the US, the UK has taken a position on the CTB issue similar to that of the US. The UK opposes the USSR draft treaty mentioned above on three points: verification, PNE, and universality. This was phrased as follows by the UK representative to the UN (UN 1975b):

"The draft treaty seems to us to be defective in two important respects. First it does not include verification provisions which would meet the real needs for confidence that all parties are respecting all the provisions of the treaty.

Secondly, despite the inclusion of the fourth preambular paragraph, for which we voted, the draft treaty does not deal adequately with the question of peaceful nuclear explosions. And finally, whilst my Government is always willing to proceed towards a universal nuclear test ban, we are not convinced that the arrangements which are proposed in the draft resolution for securing the support of all the nuclear-weapon States represent the best way to proceed."

Adequate verification means in the interpretation of the UK delegation at the CCD a procedure which satisfies all parties concerned.

The UK delegation at Geneva has received comprehensive support from experts in seismology and has produced a series of working papers on seismological detection (CCD/296, 1970; CCD/363, 1972; CCD/386, 1972; CCD/401, 1973; CCD/402, 1973; CCD/440, 1974). These working papers cover a broad spectrum of issues on seismological detection, including basic problems as well as a proposal for a seismological monitoring system.

United States

Ever since the test-ban negotiations started, the US has required provisions for on-site inspection to supplement seismic methods, in order to achieve what has been called "adequate verification", a term which has never been specified in a more quantitative way. This is reflected by a statement by the US delegate at the CCD in 1976 (CCD/PV.704):

> "In view of the existing limitations of national technical means of verification, we believe that adequate verification of a CTB continues to require some on-site inspection. In many instances, on-site inspection would be the only means of providing conclusive evidence—for example, through sampling for radioactivity —that a detected seismic event was a nuclear explosion rather than an earthquake or a conventional explosion. Thus, a verification system that included on-site inspection would provide not only a substantial deterrent to clandestine testing by increasing the risks that any significant violation would be discovered, but also a means of assuring confidence in the treaty regime in those cases where seismic methods may have misidentified earthquakes as explosions or presented ambiguous evidence concerning the nature of a seismic event.
>
> Unmanned seismic observatories, sometimes called "black boxes", have also been suggested as a means of verifying a CTB. USO's could lower the threshold magnitude for detection and identification, improve the capability to locate events, and thereby provide additional deterrence to a violation. However, they could not provide conclusive evidence that a seismic event was a nuclear explosion. Thus, USO's could make an important contribution to seismic verification of a CTB, but they are not the equivalent of, and should not be regarded as a substitute for, on-site inspection."

The US delegation at the CCD has stressed that from a technological point of view it is not possible to develop nuclear explosives for peaceful purposes without acquiring a nuclear-weapon capability, and has also pointed out the problem involved if PNE's were to be accommodated under a CTB. This was expressed as follows by a US delegate in the UN commenting on the USSR draft proposal for a CTB (UN, 1975d):

> "... the draft does not specify verification measures for peaceful nuclear explosions, but merely states that such explosions would be governed by a separate agreement. This approach leaves unresolved the critical question of whether, under a comprehensive test ban, an adequately verifiable accommodation for peaceful nuclear explosions can be worked out. At the CCD this summer my delegation pointed out that if peaceful nuclear explosions were to be accommodated under a comprehensive test ban, a verification system would have to be devised that, at a minimum, could provide adequate assurance that peaceful nuclear explosions did not involve (1) the testing of a new weapon concept; (2) the use of a stockpiled weapon to verify its performance; or (3) the carrying out of studies of the effects of nuclear weapons. Quite frankly, no solution to this problem has yet been found."

The US has not taken up a definite position on the question of whether a CTB should be adhered to or not by all nuclear powers (CCD/PV.704, 1976):

> "A question that has recently surfaced as a significant issue in CTB discussions is whether the adherence of all nuclear weapon states, or all nuclear testing powers, would be required before a CTB could enter into force. In light of the serious security implications of nuclear weapons testing, the question of participation would obviously have to be addressed in considering any CTB proposal. Among the factors that would presumably be taken into account in arriving at a position on this matter would be a testing state's perception of its own nuclear capabilities and testing experience relative to the nuclear capabilities and testing experience of other testing powers. However, we question the desirability and timeliness of taking a position on the participation issue in the abstract, before resolving the principal problems holding up a CTB—namely, verification difficulties, including PNEs. Once they are resolved, several options would be available. These include: an agreement that would enter into force upon the adherence of all nuclear powers, a limited duration agreement not requiring adherence by all nuclear powers that would provide for review and extension, and an unlimited duration against not requiring adherence by all nuclear powers but containing a provision for withdrawal in the event that treaty parties considered their supreme interests to be jeopardized. For its part, the United States has not made a determination whether a CTB should require the participation of all nuclear powers before it could enter into force."

The US has declared that it is prepared to agree on a test ban, as is shown by the following statement by the US delegate at the CCD (CCD/PV.704, 1976):

> "In the absence of a reliable, mutual prohibition, we believe that our nuclear testing program serves as an important means of maintaining the effectiveness of our nuclear deterrent. However, as representatives of the United States have stated on several previous occasions, we would be prepared to give up whatever benefits exist in continued testing if this were done pursuant to an adequately verified agreement that provided reasonable confidence that other parties to the agreement were also giving up those benefits. Inability to reach a common understanding on verification measures capable of providing such confidence has, in our view, been the principal reason why a CTB has remained beyond our grasp."

The US has given close attention to the technical verification issues at CCD. The great US efforts devoted to the improvement of the seismological monitoring capabilities include basic research as well as development and installation of instrumentation. The results of the comprehensive US research program on seismic detection and identification have been summarized in working papers CCD/330 (1971), CCD/388 (1972) and CCD/404 (1973).

3. NUCLEAR EXPLOSIONS

Since July 16, 1945, six countries have exploded over 1000 nuclear charges. The countries are: China, France, India, UK, US, and USSR. Most of these explosions were nuclear-weapon tests, but some had also peaceful purposes.

In this chapter we give a short introduction to the basic principles of a nuclear explosion, without going into any detailed discussion of the fundamental physics or the effects of nuclear weapons. The literature on nuclear physics is immense, whereas publications on the effects of nuclear weapons are rather scarce (the most extensive compilation of data on this subject being Glasstone, 1964). We also briefly review the nuclear testing activity and discuss the technical, military, and political significance of testing. The use of nuclear explosions for peaceful purposes is discussed in Chapter 12.

3.1 A AND H BOMBS

There is a fundamental difference between a nuclear and a chemical reaction inasmuch as a nuclear reaction involves the nuclei, but a chemical reaction only the electrons of the atoms. This means that in a chemical process the elements are the same after the reaction as before, although they may appear in new combinations, whereas in a nuclear reaction new elements are created.

The mass of a nucleus is less than the sum of the masses of the nucleons (protons and neutrons) that constitute the nucleus, and this mass difference represents the binding energy of the nucleus. According to the famous formula of Einstein, the mass difference, m, corresponds to an equivalent amount of energy, $E = mc^2$, where c is the velocity of light. The mass difference, and thus the binding energy, is different for different nuclei. The binding energy per nucleon is given in Fig.

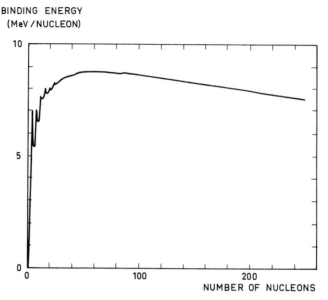

FIGURE 3.1. *Binding energy per nucleon as a function of the total number of nucleons in the elements. The energy is given in mega electron volts (MeV) per nucleon. (1 MeV = 1.6 × 10⁻¹³ joule.)*

3.1 as a function of the total number of nucleons in the elements. This figure shows that the binding energy reaches a maximum around 60 nucleons, and these elements, among which we have the most common metals, are the most stable elements. There are two principally different ways to release energy by a nuclear reaction, either to split an element with many nucleons, more than, say, 200, into two elements having less nucleons but higher binding energy per nucleon, or to combine elements with few, say, 2–6, nucleons into one or more elements having more nucleons and thus higher binding energy per nucleon. The process by which heavy elements are split into lighter ones is called *fission*, and the process by which light elements are combined into more heavy ones is called *fusion*.

We will first briefly discuss fission of heavy elements, and concentrate on the processes involved in nuclear explosions. A fission explosion is often, somewhat inadequately, called an atomic explosion, and the exploded charge is called an "A-bomb". The elements used as explosive material in nuclear fission devices are uranium (U) and plutonium (Pu). Two isotopes of uranium, $^{235}_{92}$U and $^{233}_{92}$U, with 235 and 233 nucleons, respectively, and the plutonium isotope $^{239}_{94}$Pu,

NUCLEAR EXPLOSIONS

with 239 nucleons, can be used. Different isotopes of an element have the same number of protons in the nuclei, but different number of neutrons, which may give them quite different properties as far as nuclear reactions are concerned, although they are chemically identical. The nuclear reactions of these three isotopes are in principle the same and can typically be written:

neutron + $^{235}_{92}$U → 2 fission fragments + about 3 neutrons +
+ γ-radiation + energy

(and similarly for $^{233}_{92}$U and $^{239}_{94}$Pu). This means that a uranium or plutonium nucleus captures a neutron, splits up into two fission fragments of different kinds, and produces on the average three neutrons as well as γ-radiation and energy. The energy is to a large extent kinetic energy of the fission fragments. The released energy per fission is of the order of 200 mega electron volts or MeV (1 MeV = 1.6×10^{-13} joule). A most important point is that the fission process generates more neutrons than it consumes, and due to this effect it is possible to create and sustain a *chain reaction*. A minimum amount of fissile material is needed to sustain the reaction, so that the produced neutrons do not escape from the process. This so-called *critical mass* depends on the composition and geometry of the fissile material, on the presence of impurities, and on the presence or absence of a neutron reflector, a *tamper*, surrounding the fissile material and thus reducing the neutron loss. The following critical masses for homogeneous, untamped spheres of the three isotopes have been published: $^{235}_{92}$U 47 kg, $^{233}_{92}$U 15 kg, and $^{239}_{94}$Pu 16 kg. These masses can be reduced by more than 50 % by the use of a suitable tamper.

In the fusion process the energy is derived from forcing together nuclei of light elements. Two isotopes of hydrogen, deuterium, $^{2}_{1}$D, having one proton and one neutron, and tritium, $^{3}_{1}$T, having one proton and two neutrons, are most commonly used in the fusion process. As isotopes of hydrogen, with the chemical symbol H, are used in the fusion process, these nuclear devices are often called "H-bombs". An example of a fusion reaction is:

$^{2}_{1}$D + $^{3}_{1}$T → $^{4}_{2}$He + neutron + energy,

where one deuterium and one tritium nucleus are combined to produce one helium nucleus, one neutron, and about 17 MeV of kinetic energy, the latter being shared between the helium nucleus and the neutron. To start the fusion process, the nuclei have to be very close to each other, which means that they must have energies high enough to make them overcome the Coulomb repulsion that occurs between equally charged nuclei. Such energies can be achieved by increasing the temperature to tens of millions of degrees, where the material will consist of free nuclei and electrons, a plasma, and the nuclei have enough thermal energy to start a fusion reaction. A reaction based on the thermal energy of the nuclei is often called a *thermonuclear reaction.* To sustain a thermonuclear reaction, the energy produced must be larger than that radiated from the process: otherwise the temperature will decrease, and the reaction will stop.

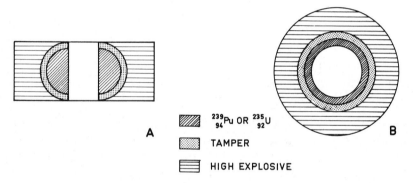

FIGURE 3.2. *Principal outline of a fission device (Larsson, 1974). (A) Gun method. (B) Implosion method.*

3.2 BOMB DESIGNS

A fission device consists in principle of (a) a subcritical system arranged in such a way that it can almost instantaneously be made supercritical, and (b) a strong neutron source to initiate the supercritical system. There are two quite different ways to construct such a device. In the so-called gun method, illustrated in Fig. 3.2.A, two or more pieces of fissile material, each of subcritical size, are brought together and compressed by conventional explosives to form a supercritical system, whereupon the chain reaction is initiated by a neutron source. In the so-called implosion method, the nuclear material is in the form of a subcritical spherical shell, which, at the explosion moment, is compressed by conventional high explosives to a compact, supercritical mass and initiated, Fig. 3.2.B.

The yield of a fission explosion depends on the construction and on the amount of nuclear material available. Yields of nuclear explosions are usually given in kilotons (kt) or megatons (Mt), where 1 kt corresponds to an energy release of 4.2×10^{12} joules, equal to the energy released by the explosion of 1000 tons of TNT. 1 Mt equals 1000 kt. There seems to be no definite lower limit to the yield of a fission explosion, and explosion yields down to and below 0.1 kt have been reported. The maximum explosion yield is limited by the amount of nuclear material in the device. It is not generally known whether there is a technical upper limit to the yield of a fission explosion; however, fission devices having yields in the megaton range would probably be considerably more expensive than fusion devices of the same yield.

The principal construction of nuclear-fusion devices is not generally known in as much detail as is that of fission devices. At the present stand of technology it appears, however, that only a fission explosion is capable of producing the temperature necessary for a thermonuclear reaction to start and thus initiate fusion. The ratio of fission to fusion energy determines the amount of long-term radioactive contamination. The less fission, the less long-lived radioactive fallout, which in turn is important in considerations of the tactical use of nuclear weapons.

A fusion bomb with a fission part contributing only a few percent of the total yield seems to be possible today. The possibility of initiating a fusion explosion by other means than a fission reaction, for example high-energy laser radiation, and thus to obtain a pure fusion explosion, has been discussed, but no substantial progress has so far been reported. It is generally assumed that present technology would make it difficult to construct fusion bombs with yields below a few kilotons. The largest nuclear charge detonated so far had an estimated yield of 58 Mt. There is no reason to believe that 58 Mt represents the upper limit to the yield of fusion explosions.

The fissile material needed for a nuclear explosion can be acquired in various ways. Less than 1 % of natural uranium consists of $^{235}_{92}$U. Two different ways are presently being used to separate $^{235}_{92}$U from natural uranium. The gas-diffusion method, by which gaseous uranium hexafluoride is filtered through several thousand filters, requires very large plants. Such gas-diffusion plants exist in China, France, the UK, the US, and the USSR. Another method, which can be utilized on smaller scales, is the gas-centrifuge method, where uranium hexafluoride is enriched in a centrifuge rotating with very high speed. Also here the enrichment has to be carried out in several steps. This technique is quite new, and small pilot plants for this process have been built in Japan, the Netherlands, the UK, and the US. Plutonium has been found only in very small quantities in nature. The plutonium isotope $^{239}_{94}$Pu, of interest for nuclear charges, is obtained from nuclear reactors working with $^{238}_{92}$U, which is the most common uranium isotope in nature. In reactors working with $^{238}_{92}$U, not only $^{239}_{94}$Pu but also other plutonium isotopes are formed. To obtain plutonium suitable for weapon production the reactors must be operated in a way different from that which is optimal for power production. Plutonium is chemically separated from uranium and other fission products in special reprocessing plants. Such plants exist in the six countries that so far have carried out nuclear explosions, and also in Belgium. There are also a number of small plants in several other countries. Uranium 233 does not exist in nature, but is obtained from thorium in nuclear reactors. The available resources of thorium seem to be larger than those of uranium, but $^{233}_{92}$U has so far been little used as fissile material.

Deuterium is found in natural hydrogen in a concentration of 0.015 %; it can be enriched by several conventional methods. Tritium exists only in very small quantities in nature, and the amounts required for fusion reactions must therefore be produced; this is done by neutron irradiation of lithium in nuclear reactors. Lithium exists in low concentrations in many minerals and can be mined in many countries.

The fundamental principles on which a fission device is built are generally known. In this context, the possibility that a single person or a group of persons would be able to construct a working nuclear device, for use in, e.g., extortion, has been discussed by several experts, who, however, have arrived at somewhat different conclusions (Gyldén & Holm, 1974; Willrich & Taylor, 1974). This discussion has led to considerable concern as to how to protect the fissile material not only by establishing various agreements and control procedures within the framework of the Non-Proliferation Treaty, but also by more rigorous physical means.

To build a nuclear bomb having the necessary fissile material requires a high degree of competence in various disciplines, such as nuclear chemistry and physics, explosion physics, etc., and we therefore think that it is unrealistic to imagine that one single person can build such a bomb. On the other hand, we believe it to be technically possible for any nation having competent people in relevant fields and possessing modern laboratory facilities, to construct at least a primitive fission device within a reasonable time. For probably most countries, the construction of a fusion bomb today would present great technical difficulties. Whether a terrorist organization having ample funds could buy the necessary competence for acquiring a nuclear device is an open question.

The easiest way for a terrorist group to get a nuclear device may well be to steal it from one of the many stores of nuclear weapons that are located throughout the world. It has been argued whether the safety devices implemented to prevent unauthorized use of nuclear warheads would make it impossible for a terrorist group to explode such a device. It may be that such technical precautions are highly effective, blackmailing with a stolen weapon would nevertheless present a real threat to responsible politicians and the general public.

3.3 NUCLEAR EXPLOSION PHENOMENA

In the preceding paragraphs we have seen that there are fundamental principal differences between chemical and nuclear explosions. The explosion phenomena themselves and the effects of the two kinds of explosion also exhibit considerable differences. For a chemical TNT explosion the detonation velocity is about 6000 m/s, which means that a 1-kt sphere, having a radius of 5.5 m, will detonate within a millisecond. The produced explosion gases will have an initial pressure of about 200 000 atmospheres and a maximal temperature of some 3000 K. Almost all energy released in a chemical explosion will be converted to shock-wave energy. In a nuclear explosion the energy is produced within a very short time, less than a microsecond. Immediately after the explosion the energy is deposited within a small volume of hot and compressed gas. In a fission explosion the initial gas temperature amounts to tens of millions Kelvin, and the pressure is of the order of a million atmospheres (Glasstone, 1964). The hot gas emits electromagnetic radiation in the form of X-rays. This radiation is absorbed very rapidly and heats the surrounding media, so that at this stage the energy propagates outward as thermal radiation.

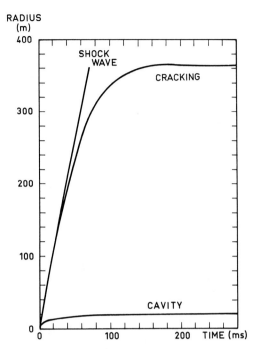

FIGURE 3.3. *Mechanical effects of an underground nuclear explosion. Calculated shock-wave, cracking, and cavity radii versus time for a 5-kt nuclear explosion in granite. (After Rodean, 1971 a.)*

The heated material also compresses the surrounding media, and a very hot shock wave is created. For an explosion in the atmosphere, 50 % of the explosion energy is converted to blast and shock-wave energy, 35 % is converted to thermal radiation, and the remaining 15 % is initial and residual nuclear radiation. Fairly detailed discussions of the various effects of atmospheric nuclear explosions on humans and structures are given by Glasstone (1964) and in a UN expert report (UN, 1970).

In an underground nuclear explosion, the thermal radiation will be absorbed by the surrounding dense material. Therefore, the sphere of action will in this case be much smaller than in the case of an atmospheric explosion. As a consequence, the shock-wave front will overtake the radiation front at a much earlier time in an explosion conducted underground then in an atmospheric one. The close-in explosion phenomena for a fully contained underground explosion are illustrated in Fig. 3.3. The intense shock wave of high temperature melts or vaporizes the material in the immediate vicinity of the explosion and forms

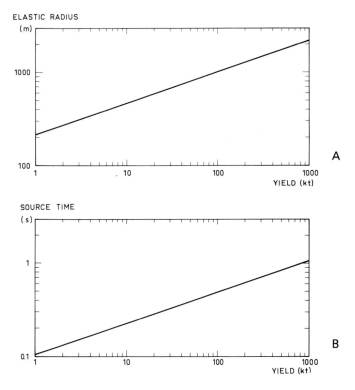

FIGURE 3.4. *Calculated elastic radius (A) and source time (B) versus yield for underground nuclear explosions in granite.*

an approximately spherical cavity. Outside the cavity, inelastic deformation and cracking of the rock will take place up to a certain distance beyond which the medium behaves elastically. Analysis of post-shot rock temperatures in eight nuclear explosions in the US has indicated that 90 to 95 % of the released explosion energy is deposited as thermal energy in the vicinity of the explosion (Rodean, 1971a). Considerable efforts have been made to describe and predict the inelastic behavior of ground material close to an underground explosion, and large computer programs for such predictions have been developed (Cherry & Petersen, 1970).

In studies of the generation of seismic waves by underground nuclear explosions, the detailed behavior of the inelastic zone close to the explosion is generally not considered. The seismic source of an underground explosion is usually considered to be an elastic sphere having a radius equal to the distance from the explosion point to the point where the material begins to behave elastically. For a 5-kt explosion in granite this radius is somewhat less than 400 meters, as can be seen from Fig. 3.3 (Rodean, 1971a); it varies with the yield, Y, as $Y^{1/3}$. As is seen from Fig. 3.4A, this means that also for an explosion of 1 Mt the apparent radius of the source is of the order of a few kilometers, which is small compared to the source dimensions generally attributed to earthquakes of similar strength (see Chapter 5). Figure 3.3 also shows that the phenomena associated with an underground explosion occur within a short time. For a 5-kt explosion in granite, the shock wave passes the inelastic zone in less than 0.1 s. This time the *source time,* also varies with the yield as $Y^{1/3}$. Figure 3.4B shows that the source time for megaton explosions is of the order of 1 s. Also the source time for nuclear explosions is considerably less than what is generally assumed for an earthquake of similar strength. A more detailed description of a nuclear explosion as a seismic source is given in Chapter 5.

In the following discussion of nuclear explosions as seismic sources, we do not distinguish between fission explosions and fusion explosions. There is also no way of telling from seismological data whether an underground explosion is a chemical, a fission, or a fusion explosion, unless the estimated yield does give any clue to its nature.

3.4 NUCLEAR TEST ACTIVITIES

The first nuclear charge, "Trinity", was detonated on the 16th of July 1945 in the Alamagordo desert in New Mexico, USA. The only two combat explosions performed up to now almost destroyed the two Japanese cities Hiroshima and Nagasaki on the 5th and 9th of August 1945, respectively. Since then, over 1000 nuclear test explosions have been carried out by the six countries which so far have tested nuclear devices. The exact number of tests conducted by these countries has not been officially released. The figures on test activities presented below are estimates based on data at present available and should be regarded as minimum estimates.

The years when the countries tested their first fission and fusion devices are given in Table 3.1. It can be noted that the time interval between the test of the first fission and the first fusion explosion is rather different for the different countries, being only three years for China, but eight years for France. The time intervals for the other nuclear countries are between four and seven years.

TABLE 3.1. *Year of first test of fission and fusion explosions by the nuclear countries.*

Country	First fission explosion	First fusion explosion
USA	1945	1952
USSR	1949	1953
UK	1952	1957
France	1960	1968
China	1964	1967
India	1974	–

The comprehensive list of nuclear explosions carried out during the period 1945–1972 published by Zander och Araskog (1973) was based on data available in 1973. Later data has made it appropriate to revise this list as of the date of the Partial Test Ban Treaty, and a revised list of nuclear explosions August 5, 1963–December 31, 1975 is given in Appendix 2. This list contains both officially announced explosions (Springer & Kinnaman, 1971, 1975) and explosions which we and others have inferred, from internationally available seismological data, to have taken place. One main difference between this list and earlier published lists is that we have added some, not officially announced, US explosions at the Nevada Test Site during these twelve years. Some of these explosions were identified as explosions by means of the seismological identification methods described in Chapter 10, whereas, in the case of the other events, the circumstantial evidence is such that they can be interpreted, with little doubt, as unannounced underground explosions. Some of these US explosions as well as some explosions in Western USSR were so weak that they might have been chemical explosions. It is, of course, possible that additional low-yield tests have been conducted without having been officially announced or without having been detected by seismological methods.

The atmospheric and underground test activities of the various countries are illustrated in Fig. 3.5. China and France, which are not parties to the Partial Test Ban Treaty, have up to 1975 carried out atmospheric tests, whereas the other countries have since August 5, 1963 conducted only underground tests. However, France has officially announced that French tests as from 1975 will be conducted underground. It is seen that on the whole the test activity of the two main nuclear powers has remained unchanged in frequency throughout

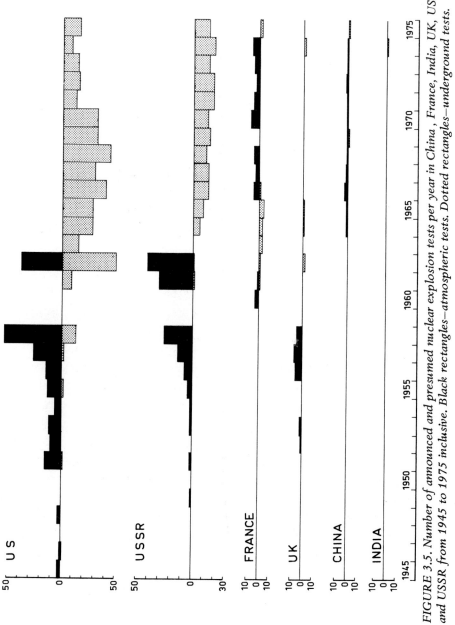

FIGURE 3.5. Number of announced and presumed nuclear explosion tests per year in China, France, India, UK, USA, and USSR from 1945 to 1975 inclusive. Black rectangles—atmospheric tests. Dotted rectangles—underground tests.

the years. The Partial Test Ban Treaty in 1963 did not cause a reduction in test activity, but led only to a switch from atmospheric to underground testing. It is, however, worth noting that charges of very high yield, up to tens of megatons, that were tested in the late fifties and the early sixties are not tested today. This decrease in interest in high-yield explosions depends largely on the development of accurate delivery systems, which make it possible to obtain the desired effect, even against well-protected targets, with explosions with lower yields.

TABLE 3.2. *Known and presumed nuclear explosions up to December 31, 1976, including the two of World War II and the PNE.*

Country	July 16, 1945– Aug. 4, 1963	Aug. 5, 1963– Dec. 31, 1976	Total
China	–	21	21
France	8	56	64
India	–	1	1
UK	23	4	27
USA	293	321	614
USSR	164	190	354
Total	488	593	1081

The total number of nuclear explosions carried out by the different countries before and after the signing of the Partial Test Ban Treaty is shown in Table 3.2. The US and the USSR have carried out a considerable number of explosions. According to this table, the US has conducted almost twice as many nuclear explosions as the USSR. Even if it is difficult to assess the accuracy of these figures, especially for the period before 1963, they probably reflect a significant difference in test activity. It can be noted that the UK appears to have conducted very few explosions during the last twelve years.

Nuclear explosions have been carried out at many sites all over the world, as is shown in Fig. 3.6. Most of them have, however, been conducted at the main test sites of the two superpowers. These test sites are at present the Nevada Test Site in Western US, and the Eastern Kazakh and Novaya Zemlya test areas in the USSR. From August 1963 up to December 31, 1976, 313 of the (321 in all) reported US explosions have been conducted at the Nevada Test Site, and of report-

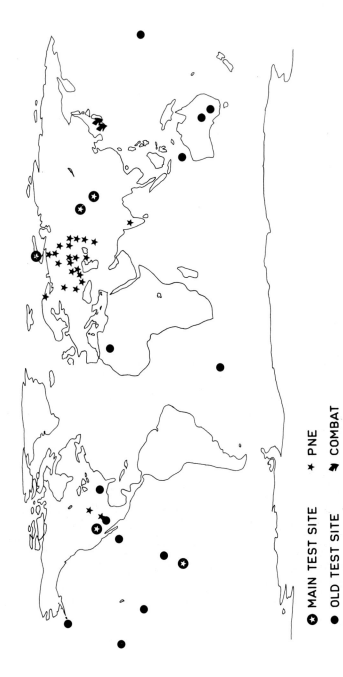

FIGURE 3.6. Map showing sites where nuclear explosions have been carried out. PNE indicates explosions for peaceful purposes.

ed explosions in the USSR (190 in all) 150 have been conducted on Novaya Zemlya or in the Eastern Kazakh test area. The various places in Western USSR where explosions have been carried out, assumedly for peaceful purposes, are further discussed in Chapter 12. The French tests have since 1966 been carried out on two Polynesian islands, Mururoa and Fangataufa, in the Pacific, where tests up to 1974 were conducted in the atmosphere. After the French decision in 1974 to test underground, the first underground nuclear explosion in the southern hemisphere was conducted on the Fangataufa Island on June 5, 1975. Up to February 1966, French tests were conducted in the Sahara desert in Algeria.

The Chinese nuclear explosions have all been carried out close to the Lop Nor lake in the Sinkiang province in Western China. Three of these explosions, one in 1969, one in 1975, and one in 1976, were carried out underground.

The four British tests announced to have taken place since 1963 were conducted at the Nevada Test Site in the US.

3.5 TECHNICAL, MILITARY, AND POLITICAL SIGNIFICANCE OF NUCLEAR-WEAPON TESTS

Throughout the many years of CTB negotiations, the verification problem has been the subject of much discussion, whereas discussions of the significance of test explosions from military and political points of view have been much less frequent. There may be both technical, military, and political reasons for carrying out nuclear test explosions. For the established nuclear powers, the most important benefit from nuclear tests is of technical and military nature. Nuclear test explosions probably are also of greater importance technically the smaller the number of explosions a country has conducted. France and China would thus probably learn more from an additional test than would the US and the USSR. The continued underground nuclear testing by the superpowers has little political significance, whereas a halt in the testing or a resumption of atmospheric tests might have a considerable political impact. From a political point of view, the first nuclear explosion test of a country is most significant, as it demonstrates that country's capability in this particular field.

Various reasons for a continuation of nuclear-weapon testing have been discussed (SIPRI, 1972; Panofsky, 1973; Scoville, 1973; CCD/

492, 1976), and we will here summarize and comment on some of these. Stockpile tests, to check the reliability of existing weapons, have been suggested as one argument for continued nuclear testing. Scoville (1973) refers to a statement by Dr. Walske, Assistant to the Secretary of Defense, indicating that up to 1971 no nuclear test had been conducted in the US with stockpiled weapons solely for the purpose of checking their reliability. The status of the nuclear material can be checked by chemical methods, and the reliability of the conventional explosives can be controlled separately. Stockpile testing therefore does not seem to be a valid argument for continued testing.

Tests in connection with research and development of new weapons seem to be of great importance. Although the two main nuclear powers have a great experience in constructing nuclear weapons, performance tests of new weapons based on known technology, but using, e.g., new geometries to suit new weapon systems, must be of considerable value to them. Full-scale testing seems to be necessary for the development of completely new weapons, such as direct-initiated fusion bombs having no fission part, or of weapons with rather extreme effects, e.g. a strongly enhanced neutron radiation.

Further tests to study the effects of ordinary nuclear weapons on different targets should be of limited value, at least for the two superpowers, as their experience from a large number of tests both in the atmosphere and underground is considerable. Effect tests of entirely new weapons or of weapons promoting a special effect, e.g. neutron radiation, must however be of great interest also for these countries. For less experienced nuclear powers, additional tests will probably have greater significance not only for weapon development but also for studies of the effects of the weapon on different targets. For the countries that are going nuclear, the first test explosion must be of very great importance to confirm that the device is working properly. A series of tests is probably required to develop a militarily significant nuclear-weapon system.

In our opinion, a halt in nuclear testing would not significantly affect the reliability of existing nuclear weapons, nor would it affect the credibility of the present nuclear forces of the two main nuclear powers. New nuclear weapons based on earlier-tested technology might be developed by the nuclear powers without any testing, but to a less extent and with less confidence than if full-scale testing were carried out. The development of fundamentally new nuclear weapons will be severely limited by a comprehensive test ban, and it seems unlikely that such new weapons can be transferred to operative units without any full-scale test. A halt in nuclear testing will probably very strongly limit any further nuclear-weapon development in the less experienced nuclear-weapon states. A test ban could also prevent states who intend to acquire a nuclear potential from testing their devices and thereby demonstrate their nuclear capabilities.

4. SEISMOLOGICAL BACKGROUND

Over the last decade significant advances have been made in our knowledge about and understanding of seismic waves and the mechanism of earthquakes. This is to a large extent a result of research programs initiated for the purpose of explosion verification. International projects covering a broader spectrum of the earth sciences have also had a significant impact in this respect. Here it would lead too far to outline the more general geophysical aspects of seismicity, and the present chapter is intended only as a short and elementary introduction to the properties of seismic waves and the mechanism of earthquakes. Emphasis will be put on characteristics of seismicity relevant to explosion verification, to serve as support of the subsequent chapters to readers unfamiliar with the subject. For a more comprehensive review, the reader is referred to treatises on seismic waves (Ewing et al., 1957; Bullen, 1963; Bolt, 1972a, b) and on physics of the earth (Jeffreys, 1970; Le Pichon et al., 1973; Press & Siever, 1974).

4.1 EARTH STRUCTURE

Seismicity in general and the properties of seismic waves are related to the structure of the interior of the earth, and this section is therefore devoted to giving a brief outline of this structure. A common model of the earth describes it as consisting of three main parts, the *crust*, the *mantle*, and the *core*, as indicated in Fig. 4.1. The crust and the solid outer parts of the mantle, down to a depth of about 100 km, form the *lithosphere*, which rides on a weak and partially molten zone of the mantle. The lithospheric layer is an important element in the theory of plate tectonics, which in a simple way explains many characteristics of earthquakes.

In short, the theory of plate tectonics is based on the hypothesis

SEISMOLOGICAL BACKGROUND

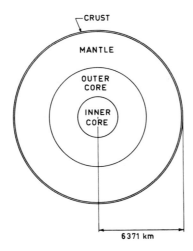

FIGURE 4.1. *A simple model of the interior of the earth.*

that the lithospheric layer consists of a small number of rigid plates, as outlined in Fig. 4.2. These plates are in motion relative to one another with an average speed of a few cm per year. The directions of the current motions relative to the African plate are indicated in Fig. 4.2. There are three main types of boundary between the plates: converging, diverging, and transcurrent boundaries, depending on whether the plates are colliding, drifting apart, or sliding along each other. These boundary types are schematically illustrated in Fig. 4.3.

Converging plate boundaries are found, for example, in the Kuril––Kamchatka–Japan region, where an oceanic plate moves under a continental plate. Divergence zones are mostly found under water along the ocean ridges. Iceland sits on such a ridge. One of the most intensively studied transcurrence zones is the one following the rim of the North American west coast, where the oceanic Pacific plate is moving northwards relative to the continental American plate.

In seismology the earth is modelled as an elastic body. The capability of the medium to transmit waves is described by various elastic para-

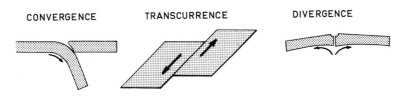

FIGURE 4.3. *The three main types of lithospheric plate boundary.*

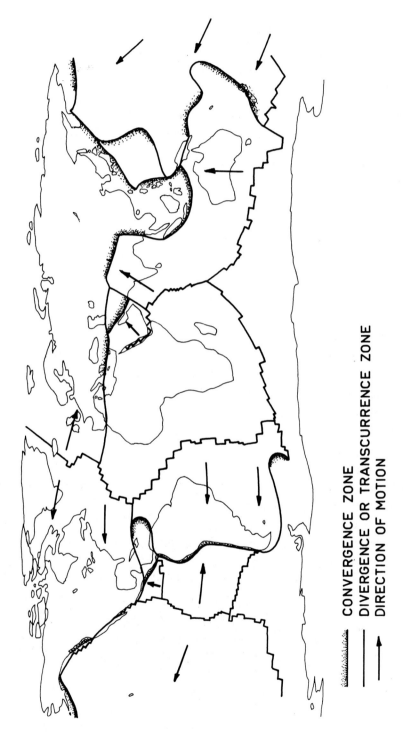

FIGURE 4.2. Boundaries of major lithospheric plates. The relative motions, assuming the African plate to be stationary, are indicated. After Le Pichon et al. (1973) and Press & Siever (1974).

SEISMOLOGICAL BACKGROUND

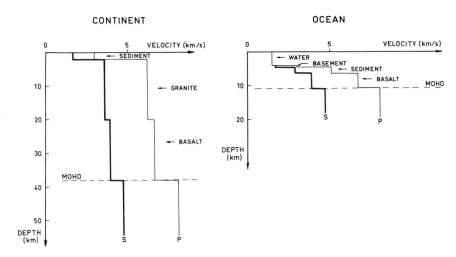

FIGURE 4.4. Models of oceanic and continental crusts giving the velocities of compressional, P, and shear, S, waves. After Shor & Raitt (1969) and Kanamori (1967).

meters. Here we discuss the elastic properties of the earth in terms of the velocities of *compressional waves,* for which the motion of the individual particles of the medium coincides with the direction of wave propagation, and *shear waves,* for which the particle motion is at right angles to the direction of wave propagation.

We start by outlining the wave velocities of the crust, the bottom of which is marked by a sharp velocity discontinuity called *Moho* after its discoverer Mohorovičić (1909). The crust is much thicker under the continents than under the oceans, as can be seen from Fig. 4.4. Moho lies at a depth of about 35 km under the continents and about 10 km under the level of the oceans (Lee & Taylor, 1966). The continental crust below the superficial sediments usually consists of two layers. The top layer, which reaches a depth of 15–25 km, is granitic, and the bottom layer is basaltic. The oceanic crust consists only of a basaltic layer with superimposed sediments and a thin basement layer, as can be seen in Fig. 4.4. Wave velocities are assumed to be constant within each layer. The thickness of the crustal layers varies regionally, and more complicated structures than those shown in Fig. 4.4 are frequently found. In some continental parts, for example, it is found that the crust contains an additional layer characterized by a rather low wave velocity (Bamford, 1972; Mueller, 1973b). Moreover, a significant anisotropy of the wave velocities associated with Moho has been observed, particularly under the oceans (Kosminskaya et al., 1972).

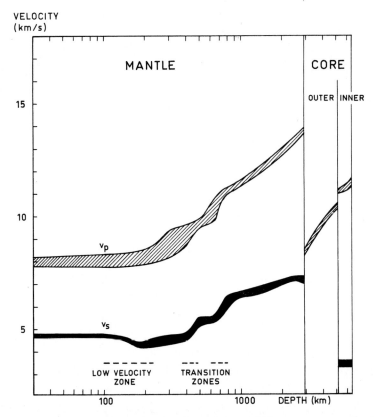

FIGURE 4.5. *The average compressional-wave, v_p, and shear-wave, v_s, velocities as functions of the depth in the mantle and the core. Due to the uncertainty of our knowledge, which is within a few percent, the velocities are shown as bands rather than lines. After Press & Siever (1974).*

As would be expected from the approximately spherical shape of the earth, wave velocities in its deep interior are spherically symmetrical, or very nearly so. Figure 4.5 demonstrates a model of typical average variations in the compressional-wave velocity, v_p, and shear-wave velocity, v_s, as functions of the depth. There are discontinuities in the velocities at the core–mantle and outer-core–inner-core boundaries.

Figure 4.5 shows that there is a minimum in the shear-wave velocity in a certain zone. This low-velocity zone covers the depth interval from about 100 km down to about 250 km and is believed to be characterized by a weak resistance to shearing stress. A decrease also in the compressional-wave velocity, much less pronounced, however, than for shear waves, sometimes occurs in the low-velocity zone. The part of the earth above the roof of the low-velocity zone constitutes the lithospheric layer of rigid plates mentioned earlier. Figure 4.5 also shows that there are two transition zones at about 400 km and 700 km, respectively. The increase in velocity at 400 km is assumed to be associated with a phase transition in the material (Press & Siever, 1974).

The earth's core, the outer boundary of which represents a principal discontinuity, consists of an outer, fluid, and an inner, solid, zone. According to Fig. 4.5, v_p and v_s are steadily increasing with depth in the mantle, except in the low-velocity zone. However, at the core–mantle boundary, v_p undergoes a sharp drop. In the outer core, v_p again increases with depth, and at the inner-core boundary there is a small but sharp increase. Through the inner core, v_p stays almost constant down to the center of the earth. No shear waves are transmitted through the fluid outer core, and the shear-wave velocity in the inner core is quite low. The dramatic change in wave velocities in the core causes complex propagation conditions for waves penetrating deep into the earth's interior. Waves being affected by the core are therefore of limited interest in detection seismology.

Besides the lateral variation of the crust shown in Fig. 4.4, there are also significant lateral heterogeneities in the uppermost part of the mantle. Lateral variations in the shear-wave and compressional-wave velocities by circa 10 % have been observed. These variations are often associated with gross differences between the oceanic and the continental lithospheric structures as well as with the complex structures at the plate boundaries, as delinated in Figs. 4.2 and 4.3. The lateral heterogeneities in the mantle are not restricted to the upper 100–200 km. Recent studies suggest more deep-seated lateral variations, and it has been argued that differences between continental and oceanic structures extend to depths exceeding 400 km (Jordan, 1975). Variations in compressional velocity by at least 1 % have been assumed to occur at depths below 2000 km, corresponding to inhomogeneities over large distances (Davies & Sheppard, 1972; Julian & Sengupta, 1973; Jordan & Lynn, 1974).

Seismic waves are absorbed to some extent in the earth's interior. A quality factor or anelastic parameter, Q, has been introduced to describe the degree of absorption. The larger the absorption the smaller becomes the value of Q. The physical mechanism behind the observed absorption does not appear to have been fully explained (Jackson & Anderson, 1970). Estimates of Q in the mantle show large variations both vertically and laterally. The results from about twenty studies,

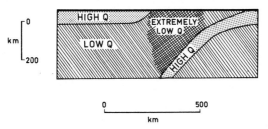

FIGURE 4.6. *Vertical section through the down-going lithospheric plate at the Tonga Islands in the Pacific, showing layers of varying anelastic parameter, Q. Modified after Barazangi & Isacks (1971).*

as compiled by Smith (1972), show the following ranges of variation in Q:

	Compressional waves	Shear waves
Upper mantle	100–2 000	150– 350
Lower mantle	1 600–6 000	1 400–2 200

Figure 4.6 illustrates the complexity of the variations in Q at a converging plate boundary. The plate going down carries a zone with high Q into the low-velocity zone, where Q is small.

4.2 SEISMIC WAVES

Part of the energy released by an explosion or an earthquake is converted into elastic waves. Different types of such waves are observed, which can be broadly characterized by their kinematic and dynamic properties. Kinematic characteristics refer to travel-time and travel-path properties of the waves, and dynamic characteristics to features such as amplitude, period, particle motion, and wave form. The kinematic characteristics, which are determined by the structure of the earth, are equal for explosion-generated and earthquake-generated waves. The dynamic characteristics, which are determined both by the earth's structure and by the seismic source, differ for explosions and earthquakes. It is demonstrated in this section that the present knowledge about the earth structure and the theory of elastic waves allow very accurate predictions of the kinematic characteristics, whereas predictions of dynamic characteristics are much more uncertain. Since the location and identification of seismic events are based on kinematic and dynamic characteristics, respectively, this means that it is easier to locate seismic events than to identify them.

SEISMOLOGICAL BACKGROUND

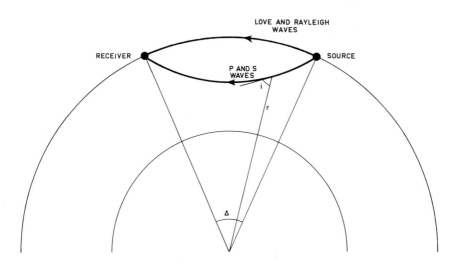

FIGURE 4.7. Travel paths of P and S waves through the earth's interior, and of Love and Rayleigh waves along the earth's surface. The radius r is the distance from the center of the earth to a point on the ray path, i is the angle between the direction of the ray and the radius r, and Δ denotes the angular distance between the source and the receiver.

In this section we discuss three types of seismic waves according to their travel paths. Two of these wave types travel over large distances, i. e., between 1 000 and 10 000 km, and are called *body waves* and *surface waves*, propagating through the interior and along the surface, respectively, as is indicated in Fig. 4.7. The third type, *crustal waves*, consists of waves propagating along the crust over distances less than about 1 000 km.

4.2.1 Body waves

The principal types of body waves are the earlier mentioned compressional waves and shear waves. It was also mentioned above that these two kinds of waves differ with respect to the particle motion involved. For the compressional waves, or *P waves* (primary waves), the particle motion coincides with the direction of wave propagation, as is the case with ordinary sound waves. For the shear waves, or *S waves* (secondary waves), the particle motion is at right angles to the direction of wave propagation, similar to electromagnetic waves. The P-wave velocity, v_p, is about 1.7 times the S-wave velocity, v_s. This means that the P waves will appear before the S waves in seismic records, as can be seen from Fig. 4.8. Since the usefulness of S waves in detection seismology is limited, we will in the following deal mainly with the P waves.

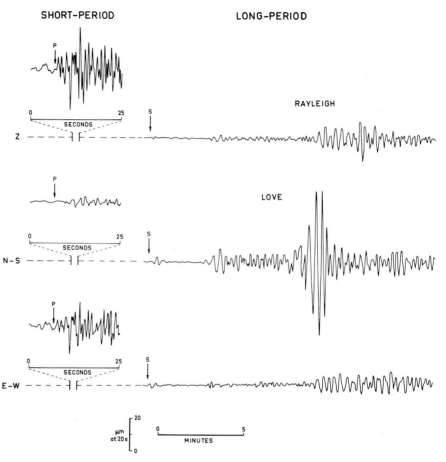

FIGURE 4.8. Records of the vertical and horizontal components of P, S, Rayleigh, and Love waves obtained at Hagfors Observatory from an earthquake in the Kashmir –Tibet Border Region on 750428 at 110644, magnitude m_b(USGS)=6.3. Note the expanded time scale used for the P waves.

The frequency band of the P waves most commonly used in detection seismology lies at 0.5–5 Hz. The corresponding wavelength is of the order of 10 km, which is quite small compared to the length of the travel path. For many purposes, therefore, P waves can be approximately described as rays, a concept used in optics. A fundamental principle of wave physics says that a ray propagating between two points follows the path minimizing the travel time. If the wave velocity within the earth were uniform, the travel path would be a straight line. As can be seen from Fig. 4.5, the velocity is essentially increasing with depth. To minimize their travel time, body waves therefore follow a deeper

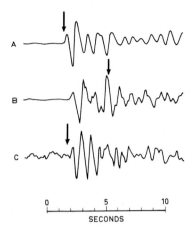

FIGURE 4.9. Comparison of observed and predicted (indicated by an arrow) arrival times of short-period P waves recorded at Hagfors for three underground explosions with known origin times and locations.

Signal	Date	Origin time	Epicenter	Magn. m_b	Δ (deg.)	Travel time (min)	(sec)
A	700326	190000.2	37.3N 116.5W	6.5	74.7	11	40.9
B	711106	220000.1	51.5N 179.1E	6.8	68.1	11	0.9
C	730517	160000.0	39.8N 108.4W	5.4	69.7	11	10.7

path, making the whole travel path curved, as is indicated in Fig. 4.7. The travel time, T, of seismic body waves is a function of the angular distance, Δ, between the source and the receiver. The definition of Δ is shown in Fig. 4.7. Standard tables of the travel time $T(\Delta)$ have been compiled for different body waves (Jeffreys & Bullen, 1967; Herrin et al., 1968). The travel-time tables predict arrival times for body waves quite accurately. In Fig. 4.9, seismic records of P waves from nuclear explosions with known origin (time and location) are shown together with indication of the P-arrival time predicted from the travel-time tables. The differences between the observed and calculated P-arrival times in Fig. 4.9 are within 3 s, which is about 0.5 % of the total travel time from the source.

At a boundary with a sharp velocity discontinuity, such as the core–mantle boundary, body waves can undergo reflection and refraction. Figure 4.10 shows the ray paths of P waves being reflected and refracted at the core–mantle boundary. Such reflected and refracted P waves are usually called PcP and PKP, respectively. Due to the sharp velocity decrease at the core–mantle boundary, refracted rays will follow a shallow path in the core. This has the effect that at angular

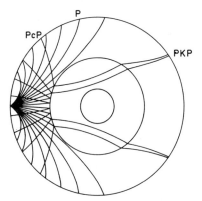

FIGURE 4.10. Calculated ray paths for P, PcP, and PKP phases of the compressional-wave-velocity models for the mantle (Herrin et al., 1968) and for the core (Jeffreys & Bullen, 1967). After Jackson (1970).

distances from about 100 to 140 degrees, no rays will reach the earth's surface. This sector is therefore usually called the core-shadow zone. The low-velocity zone of the mantle has a shadow effect on body waves which is similar in nature to, but much weaker than, that of the core.

A sharp increase in velocity with depth has a focusing effect on the rays. This leads to the simultaneous arrival of several rays propagated along different paths, and to complex records. The effect is associated with the large velocity gradient around 400 km depth and has been observed at distances of about 20 degrees. The velocity gradient around 700 km depth gives rise to a similar effect at distances of about 25 degrees.

Besides reflection and refraction occurring at the core–mantle boundary, there are a few other similar effects of particular importance in detection seismology. For example, a P wave which propagates directly upwards from the source to the earth's surface, is reflected back into the earth's interior, and then follows a path similar to the direct P wave, is of great importance in the estimation of the depth of an earthquake. This reflected wave is called pP. In connection with reflection and refraction, P waves can also be converted to S waves, and vice versa. An S wave propagating directly upwards from an earthquake to the earth's surface and on reflection there becomes converted to a P wave is denoted sP. The sP wave, which follows a travel path similar to that of the direct P wave to the receiver, is likewise of great importance in estimating the depth of an earthquake.

Using the minimum-travel-time principle, it can be shown that for each path the ratio $r \cdot \sin i(r)/v(r)$ is a constant and in fact equal to the

SEISMOLOGICAL BACKGROUND

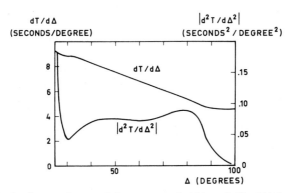

FIGURE 4.11. *The first and second derivatives, $dT/d\Delta$ and $d^2T/d\Delta^2$, of the travel time, T, as functions of the epicentral distance, Δ, for P waves according to Herrin's travel-time tables (Herrin et al., 1968). After Shlien & Toksöz (1973).*

first derivative of the travel time, $dT/d\Delta$, which usually is called the slowness parameter of the ray. (The radius, r, and the angle, $i(r)$, are defined in Fig. 4.7; $v(r)$ denotes the velocity as a function of r.) Since each ray path has a unique value of $dT/d\Delta$, it is possible to estimate the epicentral distance if the value of the slowness can be measured. Figure 4.11 shows $dT/d\Delta$ and $d^2T/d\Delta^2$ as functions of Δ. From the formula above one can also show that the incident angle, i, at the earth's surface is quite steep, ranging from 38 degrees to 15 degrees for paths with angular distances between 20 degrees and 100 degrees, respectively. This means that the particle displacement at the surface due to P waves is largest in the vertical direction, since the direction of the incident particle motion coincides with the travel path of the compressional waves. This effect can be seen in the records reproduced in Fig. 4.8.

As the body waves propagate, they expand over a surface, and the amplitudes will therefore be reduced with increasing travel distance. This is called *geometrical spreading*. The curvature of the travel-time curve, $T(\Delta)$, is described by the second derivative, $d^2T/d\Delta^2$, as shown in Fig. 4.11. A large value of $d^2T/d\Delta^2$ means that the rays will be focused, and the amplitudes of the recorded waves become large. For a low value of $d^2T/d\Delta^2$, the rays will spread out, resulting in lower amplitudes. For the detailed account of geometrical spreading, the reader is referred to Bullen (1963).

The reduction in amplitude of seismic waves due to anelastic absorption is usually described by an exponential factor, $\exp(-\pi f T(\Delta)/\bar{Q})$, where f is the frequency. $\bar{Q}$ represents the average Q value of the

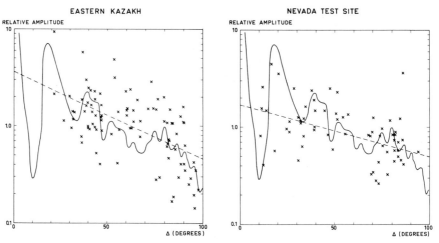

FIGURE 4.12. *P-wave amplitudes as a function of the angular distance,* Δ. *The data have been obtained from bulletins of about 150 seismic stations on about 400 underground nuclear explosions in the Eastern Kazakh test area (left) and at the Nevada Test Site (right). The dashed lines were estimated from the data and represent least-squares fits to linear relations between the logarithm of the amplitude and the distance. The solid curves represent the amplitude decay,* $-B(\Delta,h)$, *according to the Gutenberg–Richter magnitude formula discussed in Section 4.2.1.*

quality factor (anelastic parameter) along the path. Q is usually assumed to be independent of the frequency. The formula tells that in a case where the waves traverse two or more zones of different Q values, the zone with the lowest Q will have a dominant influence on the absorption. Because of the strong lateral variations in Q, estimates of average Q values along P-wave ray paths will vary quite strongly. Average Q values for travel paths from the North American continent and from the Eurasian continent, respectively, to Scandinavia differ by a factor of five, the larger Q values being found for the Eurasian continent (Filson & Frasier, 1972; Frasier & Filson, 1972).

Geometrical spreading in combination with anelastic absorption largely determine the reduction in body-wave amplitude as a function of distance. Several empirical curves of the attenuation for P-wave amplitudes have been estimated, some of which are based on observations from nuclear explosions (Evernden, 1967; Carpenter et al., 1967; Kaila, 1970; Veith & Clawson, 1972; Kaila & Sarkar, 1975). Such curves based on explosions only, are less sensitive to uncertainties in the radiation from the source than curves based on earthquakes. Figure 4.12 shows observed P-wave amplitudes for nuclear explosions as functions of the epicentral distance, together with amplitude–distance

SEISMOLOGICAL BACKGROUND

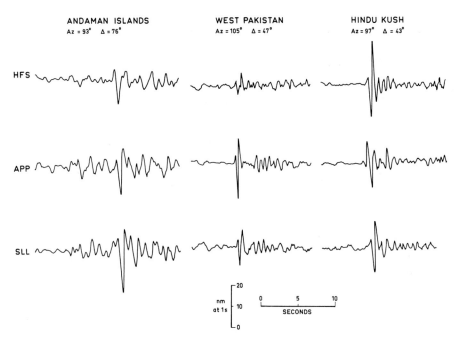

FIGURE 4.13. *Short-period P-wave records taken at the three Hagfors substations HFS, APP, and SLL from three different events, illustrating amplitude variation over short distances. According to USGS the events had the following source data:*

Location	Date	Origin time	Epicenter	Depth (km)	m_b
Andaman Islands	730726	200633	10.5N 93.9E	33	5.1
West Pakistan	730815	150909	36.4N 70.8E	200	4.8
Hindu Kush	730830	210430	30.2N 68.1E	33	4.7

The distance, Δ, and direction of approach of the P-wave front at Hagfors, Az, are indicated in the figure.

curves. The amplitudes in Fig. 4.12 were corrected for explosion strength, and reflect only the effect of propagation. Details of the computational procedure are given in Chapter 11. Even if there is a general amplitude decay with distance, as indicated by the dashed lines, the most striking feature of the diagrams is the large scatter of the data. This spread clearly illustrates the strong sensitivity to lateral heterogeneities of P-wave amplitudes. For stations at the same distance, the observed amplitudes can differ by a factor of 10.

P-wave amplitudes recorded also at closely spaced stations are subject to significant variation. Figure 4.13 illustrates the variation between three substations at the Hagfors Observatory, separated by about 50 km. Here the amplitudes vary by a factor of about 3, but variations

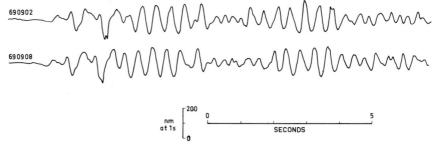

FIGURE 4.14. *Short-period P-wave records at Hagfors from two closely spaced explosions in Western Russia, with the following source data according to USGS:*

Date	Origin time	Epicenter	m_b
690902	045957	57.4N 54.9E	4.9
690908	045956	57.4N 55.1E	4.9

by up to a factor of 10 have been observed in Norway over distances of about 100 km (Husebye et al., 1974). Figure 4.13 also illustrates that the wave forms of the records at the three Hagfors stations are rather different. The variation in amplitude and wave form is often believed to be an effect of multi-pathing and scattering by fine-scale inhomogeneities in the earth (Mack, 1969; Aki, 1973). Even if P-wave signals can be subject to significant variations over only 50–100 km, the wave forms from seismic events that are closely spaced can be quite consistent. This is shown by Fig. 4.14, comparing P-wave signals recorded at the same station from two closely spaced explosions. It can be seen that although the signals are quite complex, they are almost identical. The high degree of similarity of the signals shown in Fig. 4.14 is, however, an exception rather than a rule. These two explosions are further discussed in Chapter 12.

Although the P-wave amplitudes are subject to significant lateral variations, various standardized amplitude–distance curves are used to normalize amplitudes observed at different epicentral distances. Originally, this distance correction was introduced for the calculation of an indirect measure of the strength of an earthquake, based on observations in the far field. The most widely used definition of such a measure, usually called body-wave magnitude, m_b, can be written:

$$m_b = {}^{10}\log\frac{A}{T} + B(\Delta,h),$$

where A refers to the zero-level to peak-level amplitude of the vertical displacement (in μm) and T to the period of the P wave (in seconds).

The correction term, B, developed by Gutenberg (1945), was obtained empirically from a large sample of earthquake observations, and is a function of the epicentral distance, Δ, as well as the depth, h, of the earthquake. The values of B were chosen to agree with some other definitions of seismic magnitudes presented in Sections 4.2.2 and 4.2.3. In Fig. 4.12, the function $-B(\Delta, h)$ for $h = 0$ is compared with actual $^{10}\log(A/T)$ data. From the comparison in Fig. 4.12 it is clear that the parameter m_b is only a crude, quantitative measure of the strength of a seismic source. The magnitude m_b should be regarded only as a parameter defined by convention, having no direct, physical relation to the seismic source. It is, however, frequently used as a simple tool for rapid but rough calculations.

4.2.2 Surface waves

There are two main kinds of seismic waves propagating along the surface of the earth, as is shown in Fig. 4.8. One of the wave types is characterized by a retrograde, elliptical particle motion in a vertical plane coinciding with the direction of wave propagation; these waves are called *Rayleigh waves*. Thus the Rayleigh waves have both vertically and horizontally oriented displacements, as can be seen from Fig. 4.8. The Rayleigh waves are transmitted because of the discontinuity formed by the surface of the earth.

The other kind of surface waves is called *Love waves*. Here, the particle motion is perpendicular to the direction of wave propagation and is entirely horizontally polarized. The Love waves are the result of the horizontal stratification of the crust. If the structure of the upper 100 or 200 km of the earth were uniform, Love waves would not be generated.

The frequency band of Love and Rayleigh waves of relevance in detection seismology covers periods from 10 to 100 s, but mostly a narrow band around the 20-s period is used. The surface waves travel with a velocity of about 3–4 km/s, corresponding to a wavelength of 60–80 km. The surface waves are therefore less sensitive than P waves to fine-scale variations in the surface structure.

The longer the wavelength of surface waves the deeper do they penetrate into the interior of the earth. The compressional-wave and shear-wave velocities increase with increasing depth, which means that a velocity differentiation of the spectral components of the surface waves will be obtained. Waves with longer periods travel faster along

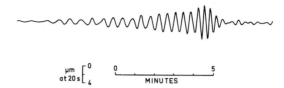

FIGURE 4.15. *Record obtained at Hagfors of the long-period vertical-component Rayleigh waves from an earthquake in the Taiwan region, with the following source data according to USGS: Date 731002; Origin time 025942; Epicenter 23.9N 121.6E; Depth 49 km; m_b 5.1.*

the earth's surface. The period-dependent velocity will result in a dispersed wave train, which means that the waves are sorted out into groups of waves, as is clearly seen in Fig. 4.15. The degree of dispersion is controlled mainly by the structure along the travel path. Figure 4.16 shows some examples of group velocity as a function of period, where "group velocity" denotes the velocity with which a group of radiated wave trains travels. Due to the relatively low velocity of the surface waves, they appear later in seismic records than do the body waves, as is shown in Fig. 4.8.

In contrast to the case of the body waves, a standard travel-time table for surface waves cannot be compiled; the lateral variations in the structure of the crust and the upper mantle make it impossible to apply a single dispersion curve on a global basis. The largest lateral

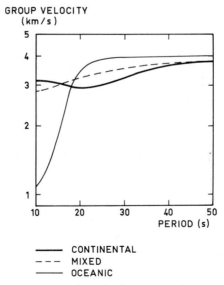

FIGURE 4.16. *Group velocities of Rayleigh waves as functions of the period for three different types of travel path. Modified after Filson (1974).*

variations appear between oceanic and continental structures. For periods above 20 s, the velocities over oceanic structures are usually larger than the velocities over continental structures, as can be seen in Fig. 4.16. Many wave paths are mixtures of oceanic and continental structures, and to be able to use the arrival times for surface waves, one has to map the regional velocities for the entire globe and integrate the effect of the structures along each individual path (Julian, 1973a). Because of their comparatively long periods, surface waves are difficult to use for precise location of seismic sources.

The surface-wave front is an expanding circle on the surface of the earth. The reduction in surface-wave amplitudes will therefore be less pronounced than the reduction in body-wave amplitudes. The effect of anelastic absorption of the surface waves is similar to that of the body waves discussed above. The amplitude factor due to anelastic absorption for the surface-wave amplitudes at frequency f can be expressed by $\exp(-\pi f R \Delta / Q v_g)$, where R is the radius of the earth, and v_g is the group velocity, which latter depends on the period, as shown in Fig. 4.16. In the literature only a few estimates of the quality factor Q have been published for surface waves. Average values between 200 and 500 at periods around 20 s have been obtained by Marshall & Carpenter (1966) and by Burton (1975a). Estimated Q values seem to be slightly regionally dependent. It has also been found that Q is frequency-dependent for surface waves. This appears to be an effect of the low-velocity zone discussed earlier, since periods corresponding to penetration depths around 70–150 km show smaller Q values than do longer or shorter periods (Smith, 1972; Burton & Kennett, 1972).

A variety of functions have been proposed for describing the amplitude reduction of surface waves with distance (Basham, 1969). These are used to normalize the observed displacement and then to estimate the magnitude of the seismic events. One such formula for surface-wave magnitudes is:

$$M_s = {}^{10}\log\frac{A}{T} + 1.66\,{}^{10}\log\Delta + 3.3.$$

Again, A denotes the maximum zero-to-peak displacement (in μm), and T is the period (in seconds). The distance correction was estimated from empirical data. The formula above was originally proposed to be based on the horizontal component of Rayleigh-wave ground amplitudes at periods of about 20 s, but is now mostly applied to the vertical component of Rayleigh-wave amplitudes. The formula is used for periods

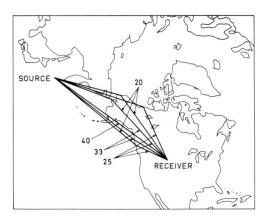

Fig. 4.17. Propagation paths for Rayleigh waves with periods around 20, 25, 33, and 40 s to the Large Aperture Seismic Array in Montana, US, from an earthquake in the Kuril Islands, with the following data according to USGS: Date 661121; Origin time 121927; Epicenter 46.7N, 152.5E; Depth 40 km; m_b=5.6. The distance between the earthquake epicenter and the recording point is 64.3 degrees. (After Capon, 1970.)

around 20 s and distances larger than 25 degrees. Other formulas have been proposed for use at shorter distances (Basham, 1971; Evernden, 1971b). The M_s formula above does not account for any lateral variations in the transmission, which are known to be quite pronounced between oceanic and continental structures. Formulas which do take the travel path into account are of the following form (Marshall & Basham, 1972):

$$M_s = {}^{10}\log A + B'(\Delta) + P(T),$$

where B' is a revised correction function for the distance, which for $\Delta < 25$ degrees is proportional to $0.81\,{}^{10}\log\Delta$, and for $\Delta \geqslant 25$ degrees is quite close to $1.66\,{}^{10}\log\Delta$. P is a correction which depends on the period T and has been tabulated for T values between 10 and 40 s for four different travel paths: continental Eurasia, continental North America, mixed oceanic-continental, and oceanic. This formula has been found to reduce the spread of data obtained at many different places.

The amplitude reduction of surface waves is a function not only of the distance but also of the depth of the earthquake. Corrections due to depth are, however, usually not added to the M_s formulas above. One reason may be that the depth effect can be difficult to separate from other effects, for example the effect of crustal structure. This point will be further discussed in Chapter 5.

At the boundaries between continental and oceanic structures, the

SEISMOLOGICAL BACKGROUND

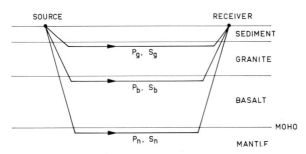

Fig. 4.18. Schematic representation of crustal-wave paths. (The thickness of the crust is exaggerated.)

surface waves are subject to refraction and reflection. This means that surface-wave records not only are affected by the direct surface waves, but for complex structures there is also a mixture with significant components of laterally refracted and reflected waves. Figure 4.17 demonstrates this effect. It can be seen that the waves do not arrive from a single direction, but from an azimuthal sector of about 20 degrees. It should be stressed that the interpretation made in Fig. 4.17 is not unique (Capon, 1970).

4.2.3 Crustal waves

Wave propagation over short distances, of up to about 1000 km, from the source primarily involves P and S waves, and to some extent also the far-travelling surface waves previously discussed. We concentrate here on crustal P and S waves, the dominant frequencies of which lie in the 1–10 Hz band. The wave lengths involved are shorter than 5 km.

The P and S waves can be approximated as rays travelling through or near the crust. The short distances involved make it also practical to approximate the crust by a structure of plane layers. The wave velocities are supposed to be constant within each layer, but there are differences between layers, usually with progressively increasing velocity with depth, as shown in Fig. 4.4. The main travel paths are shown schematically in Fig. 4.18.

Refraction occurs at the layer interfaces. Depending on whether the waves travel mainly through the granite layer, through the basaltic layer, or below the Moho discontinuity, they are indexed g, b, or n, respectively. The travel time for the different paths are proportional to the inverse velocities of the respective layers, with a correction applied for the time needed by the waves to reach the layer through which

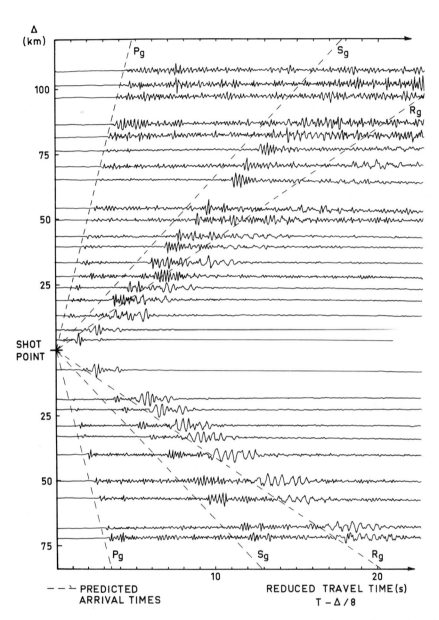

Fig. 4.19. Records of the vertical component of crustal waves from a chemical explosion. The records are arranged according to the distance, Δ, from the source to the points where they were taken (vertical axis). The measure on the horizontal axis is the travel time reduced according to the formula $T - \Delta/8$, where T is the true time elapsed after the shot, and $\Delta/8$ is the travel time of a wave with a velocity of 8 km/s. The records were obtained during the "Blue Road" profile experiment to investigate the crustal structure in Scandinavia. (After Lund, 1975.)

they travel. Typical records of crustal waves obtained at different distances from the source are shown in Fig. 4.19. Both the P_g and the S_g waves can be clearly seen in the figure. The R_g wave, which is a crustal Rayleigh wave, is particularly clear in the lower group of records. The broken lines in Fig. 4.19 represent arrival times predicted according to a layered crustal model. Figure 4.19 indicates that crustal P and S waves may be quite complex. It can therefore be quite difficult to identify the different phases and to accurately measure the arrival times.

The discussion above has been based on a simplified model, useful for many purposes. As was mentioned earlier, strong lateral variations in the crustal structure may, however, occur, which may make the picture more complex.

The *magnitude* concept was originally applied to records of seismic events at short distances from the source (Richter, 1935). Estimates of magnitudes based on data of crustal waves are usually written M_L, and are defined by the relation:

$$M_L = {}^{10}\log A - {}^{10}\log A_o.$$

Originally, A denoted the maximum amplitude (in mm) of the trace obtained with a certain seismometer (Richter, 1958). The correction factor, A_o, is a function of Δ and is based on the amplitude-decay pattern of an arbitrarily defined magnitude-zero earthquake. The correction term, ${}^{10}\log A_o$, as given by Richter, is roughly proportional to $2.5 {}^{10}\log \Delta$ for values of Δ between 100 and 500 km. More recent studies of P_n amplitudes suggest a decay inversely proportional to Δ^3 for distances between 100 and 1000 km (Evernden, 1967).

4.3 SEISMICITY

Not until the beginning of this century, when a worldwide network of seismological stations had been built up, was the global pattern of earthquake-foci distribution recognized (Thirlaway, 1975). Since that time, the capabilities of seismological instrumentation and methods have steadily developed and thus helped to increase our knowledge about earthquakes. Figure 4.20 shows the distribution of the current seismicity. In a geological time perspective the period for which earthquake observations have been recorded with seismological instruments is negligible. There is as yet no phenomenon known that could be used to study paleoseismicity. Such studies may, however, become possible

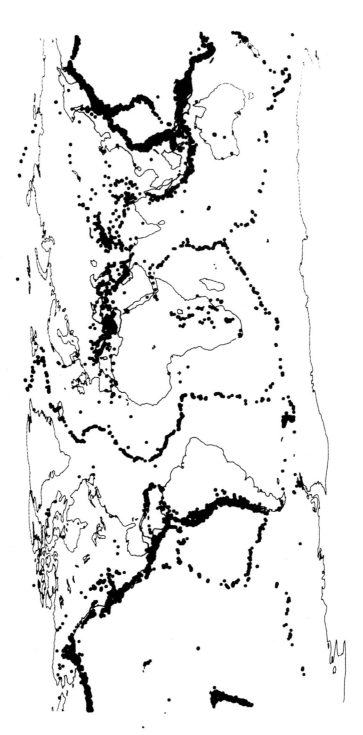

Fig. 4.20. World seismicity for the period 1969–1974, as reported by the USGS. Only events with body-wave magnitudes, m_b, larger than 4.0 and depths shallower than 100 km or restricted to normal depth are plotted.

SEISMOLOGICAL BACKGROUND

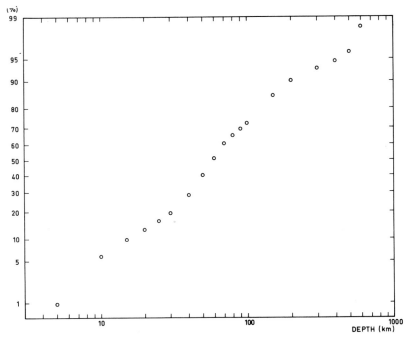

Fig. 4.21. Depth distribution of world seismicity for the period 1969–1974, as reported by the USGS. Events with a body-wave magnitude $m_b < 4.0$ or with depths restricted to 33 km have been omitted.

to some extent, as indicated by Engelder (1974), who studied frictional wear of surfaces which have slid against each other. In order to investigate seismicity before the days of instrumental seismology, one has to resort to historical records of earthquakes damaging man-made structures. Only few regions have, however, a well recorded history (Ambrasyes, 1971). Moreover, there is little correlation between seismicity causing damage to structures on the earth's surface and the total seismic activity.

In the following we will outline some basic characteristics of the worldwide seismic activity and deal in particular with geographical distribution, pattern of time occurrence, and strength distribution. We will also discuss the seismicity that occurs within the territories of some of the nuclear countries.

4.3.1 Geographical distribution

Seismic activity is mainly associated with the relative movements of the lithospheric plates. Most earthquakes are concentrated to zones

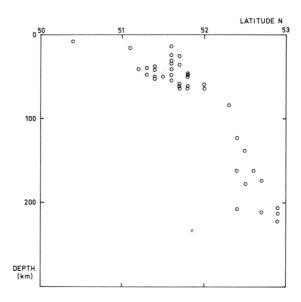

Fig. 4.22. Depth distribution of earthquakes along a cross section between longitudes 174 W and 175 W of the Aleutian Islands. The earthquakes represent data reported by the USGS for the period 1971–1973, and events with body-wave magnitudes, $m_b <$ 4.0 or focal depths restricted to 33 km are omitted. It can be seen that the earthquake foci delineate a downgoing trend.

along the plate boundaries, as can be seen from Figs. 4.2 and 4.20. Only a small percentage of the earthquakes originate within the plates. The seismic zones at the plate boundaries are quite narrow, only of the order of 100 km at divergence boundaries along the oceanic ridges. In convergence zones, however, in particular in those consisting solely of continental structures, seismic zones can be much broader, as is shown by the earthquake epicenters in Central Asia, which are scattered over distances of thousands of kilometers. (An *epicenter* is the point on the earth's surface located directly above the focus of an earthquake.)

Several thousands of earthquakes with a magnitude larger than 4 are reported each year. The seismic activity differs remarkably in frequency and strength between the various plate boundaries. It has been estimated that 75 % of the global release of seismic energy is concentrated to the circumpacific belt (Gutenberg & Richter, 1954). The circumpacific belt follows the coastlines of the Pacific, and is often called the "ring of fire". Another 20 % of the energy release is concentrated to the Alpine belt extending from the Mediterranean area across the Middle East and into Central Asia.

4.3.2 Depth distribution

Earthquakes have been found to occur down to a depth of about 700 km, which represents the limit of the subduction of downgoing plates. The depth distribution for a worldwide sample of earthquakes is shown in Fig. 4.21. It can be seen that about 30 percent of them originate below a depth of 100 km. Traditionally, earthquakes are categorized as shallow, intermediate, or deep, if the depth is less than 70 km, 70–300 km, or more than 300 km, respectively. These intervals have been chosen arbitrarily, and quite often the limit for shallow earthquakes is put at 100 km, being the approximate thickness of the lithosphere. Intermediate and deep earthquakes are usually associated with downgoing plates, and their distribution pattern marks in fact the plate-subduction zone, as indicated by Fig. 4.22. In some areas the earthquakes are less uniformly distributed with depth, and are rather concentrated to nests along a downgoing trend, as for example in the Hindu Kush region. Deep earthquakes, i. e. those below 300 km, are found in a few regions only, such as the Sea of Japan, the Fiji Islands, the Banda Sea, and Southern Argentina. Intermediate-depth earthquakes are not only concentrated to downgoing plates, but are also found where plates collide in regions, like in Central Asia, as shown by the map in Fig. 4.23. The deep earthquake activity seems to be weaker than that of shallow earthquakes.

4.3.3 Magnitude distribution

Since very few earthquakes break the surface of the earth, very little data are available for direct measurement of the strength and extent of the earthquake source. Estimates of the strength have therefore to be based on seismic recordings, often at large distances from the source. The most widely used measure of the strength is the seismic magnitude presented in Section 4.2. It should be emphasized that the seismic magnitude is no more than an empirical *ad hoc* measure, which only crudely and by convention describes the relative strength of earthquakes. As mentioned earlier, the magnitude convention cannot be directly related to physical parameters, like deformation or strain release. It can, however, be used to outline the broad features of the relationship between strength and distribution of earthquakes.

The largest seismic magnitudes so far measured are about 8.5. At the most sensitive seismological stations it is possible to detect earthquakes

Fig. 4.23. Geographical distribution of earthquakes deeper than 100 km. The earthquakes represent data recorded by the USGS for the period 1969–1974, for events with body-wave magnitudes $m_b > 4.0$.

SEISMOLOGICAL BACKGROUND 79

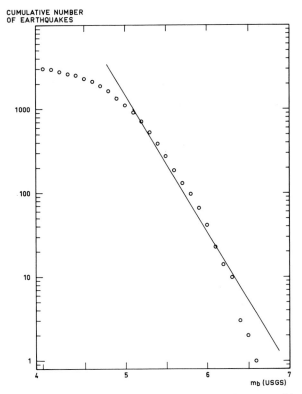

Fig. 4.24. Distribution of body-wave magnitudes m_b for a worldwide sample of earthquakes during 1972, as reported by the USGS. Events with a focus deeper than 100 km have been omitted. The straight line is an eye fit.

at teleseismic epicentral distances, i. e., between 20 and 100 degrees, with magnitudes down to about 4. At near distances it is possible to detect earthquakes with magnitudes below 0.

The relative occurrence of small and large earthquakes can roughly be described by the statistical distribution of their magnitudes. A variety of studies have shown that an exponential distribution is a fair approximation over a limited magnitude range. Mostly, this is expressed by relating the number of earthquakes, $N(m)$, equal to or greater than magnitude m by the formula:

$$^{10}\log N(m) = a - bm,$$

where a and b are constants related to a given region and to a given period of time. Estimates of a and b have been made both for worldwide and for regionalized data sets (Evernden, 1970). Similar relations have also been used for microearthquakes down to magnitudes around

−2 (Scholz, 1968). Figure 4.24 shows a fit of the exponential $N(m)$ formula to a worldwide data sample.

The exponential distribution is the most widely used approximation of empirical data, in spite of the fact that the parameters a and b are not directly related to any physical properties. Formally, the constant a is equal to the logarithm of the number of earthquakes with magnitudes greater than 0 and is a measure of the total seismicity for a given region. The value of b is a measure of the relative number of small-magnitude and large-magnitude earthquakes. Attempts have been made to interpret the physical meaning of the constant b, and it has been suggested that a low value of b indicates a high stress in the source region and vice versa (Wyss, 1973). Variations in the value of b between 0.5 and 1.5 have been observed. It has been found also that a larger value of b is related to a larger seismic activity. Estimated values of b vary not only from region to region, but also with depth. Small changes in b have been observed in the depth interval 0–15 km (Eaton et al., 1970). At greater depths, the value of b appears to decrease with depth (Curtis, 1973).

It is clear from Fig. 4.24 that at smaller magnitudes the factual number of earthquakes observed falls below the linear approximation of the $N(m)$ formula. This is due to the incomplete detection of weak earthquakes. It is also seen that at large magnitudes, the linear fit is poor. Similar deviations from the $^{10}\log N$ relation have been observed by several investigators. It has been suggested that the deviation from linearity at higher magnitudes is due to underestimation of the magnitudes of large earthquakes (Brazee, 1969), or that the log-frequency–magnitude relationship is in fact not linear in that magnitude range (Chinnery & North, 1974). Because of the nonlinearity at low and high magnitudes, attempts have been made to replace the exponential distribution by a normal distribution, centered around a mean magnitude, m, and having some standard deviation σ (Lomnitz, 1974). Such normal-distribution models have a drastic cut off at lower magnitudes.

4.3.4 Space–time patterns

Considerable efforts have been spent in a search for models in time and space of the occurrence of earthquakes. In some regions, and for strong earthquakes ($m_b > 7$), some regular features have been observed both in space and in time. Along the Aleutian Islands a regular, westerly

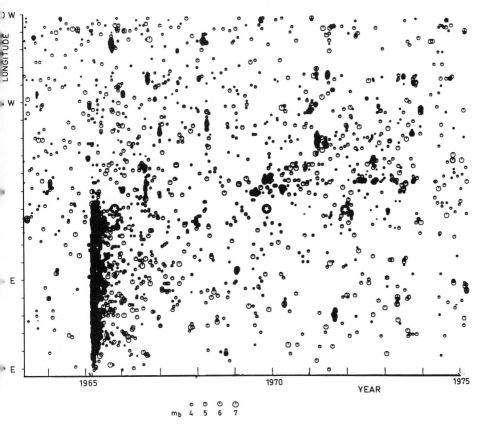

Fig. 4.25. Space-time pattern of earthquakes along the Aleutian Islands for the period 1963–1974, as reported by the USGS. Earthquakes with body-wave magnitudes m_b <4.0 have been omitted. The earthquakes represent data within latitudes 48 N–52 N, and are plotted as a function of longitude and time of occurrence. The explosions "Longshot" in 1965, "Milrow" in 1969, and "Cannikin" in 1971 are also indicated in the figure.

progression of the locations has been observed for the five major earthquakes in 1938–1965 (Kelleher, 1970). Each of these major earthquakes was accompanied by a large number of smaller aftershocks, spread out over a few years in time, as well as over a zone of rupture of the order of hundreds of kilometers, as shown in Fig. 4.25. The edges of these aftershock zones correlate with the structural boundaries. Moreover, the zones usually abut without any significant overlapping. Similar effects have been observed in other active seismic regions (Sykes, 1971).

It is also well-known that strong earthquakes often are preceded by weaker shocks, usually occurring from a few hours to weeks prior to the major earthquake. In some cases the time difference amounts to only a few seconds; for such cases it has been suggested that the main earthquake was in fact triggered by a foreshock (Strelitz, 1975).

Studies of the seismicity in the vicinity of the epicenter of a large earthquake during the preceding months and years indicate that the prior seismic activity often occurs near the epicenter of the main event and near the edges of the rupture zone, but not at the epicenter of the main event (Kelleher & Savino, 1975).

Several attempts have been made to find periodicities in seismicity. No striking regularities have been found in the time interval of from a few days to a few years. It can be noted that clear periodicities have been observed in the lunar seismic activity, mainly attributed to tidal triggering (Toksöz, 1975).

4.3.5 *Earthquake prediction*

By earthquake prediction is usually meant the forecast of location, time, and magnitude of a future earthquake. Notwithstanding the fact that a few successful predictions have recently been reported, the prediction problem is still far from its solution (Aggarwal et al., 1975). The geographic distribution of earthquakes is mainly limited to narrow intraplate zones, which limits the possible location of most earthquakes. It has been demonstrated in several cases that a gap in a seismic zone is a likely location for a future earthquake (Kelleher et al., 1973). It is still more difficult to predict the time of occurrence and the strength of a future event. Much of the efforts in earthquake prediction during the last decade has been devoted to studies of premonitory effects of various geophysical phenomena. Rikitake (1975) has compiled a summary of almost 300 papers dealing with premonitory effects. It has, for example, been reported that the value of b, i. e., the coefficient of the $^{10}\log N$ relation mentioned above, did decrease prior to main earthquakes. Changes in the ratio v_p/v_s of the compressional-wave and shear-wave velocities have also been detected prior to earthquakes.

During the 1970's increasing efforts have been put into the field of earthquake prediction. Large national programs in earthquake prediction have been launched in, for example, China, Japan, the US, and the USSR. Moreover, the US has entered into a formal cooperation and data exchange with Japan and the USSR in this field (Kisslinger & Wyss, 1975).

4.3.6 Man-made earthquakes

Some of the strong underground nuclear explosions, with yields in the megaton range, that have been made were followed by weak aftershocks, the largest of which was about two magnitude units below the explosion magnitude. This type of activity has been observed after explosions at the Nevada Test Site (Hamilton et al., 1972), the Aleutian Islands (Engdahl, 1972) and Novaya Zemlya (Israelson et al., 1974). Two kinds of afterevent have been discerned: one is associated with a deterioration of the cavity and with the development of the chimney, the other is caused by release of tectonic strain energy. The largest afterevents have been associated with cavity deterioration and have occurred within a day or two of the main shock. Tectonic afterevents were usually unrelated to collapse phenomena and occurred within 15 km of the explosion cavity. The tectonic afteractivity is highly dependent on regional conditions. The largest level has so far been observed for the Nevada Test Site, where the activity can go on for several weeks after a megaton explosion (Hamilton et al., 1972).

The question whether underground nuclear explosions trigger unreleased natural earthquakes or change the natural pattern of seismicity has received close attention. Particular attention has been given to the Aleutian Islands, where large explosions in the megaton range have been carried out on top of one of the most active seismic zones in the world. In Fig. 4.25 the two megaton explosions in 1969 and 1972 on the Aleutian island Amchitka have been plotted in a space-time diagram together with the earthquakes for the period 1965–1972. No correlation can be seen between the explosions and the occurrence of earthquakes. In more detailed analyses, no indications were found of any influence of the explosions on the seismicity other than some shallow aftershocks in the low-magnitude range occurring close to the explosion (Willis et al., 1972; Engdahl, 1972).

Finally it should be mentioned that earthquakes have been triggered by human activities, such as excavation of mines, dam building to form large reservoirs, and injection of fluid into crustal rocks. A summary of earthquakes caused by these types of activity has been prepared by Gough (1975). Seismic events, or rock bursts, in mines are assumed to be caused by changes in the distribution of stress in the rock. Gough estimates that at a depth of 3 km the lithostatic stress field would be sufficient to generate a seismic event of a magnitude of about $m_b = 5$. The largest earthquake so far induced by the filling of large reservoirs was estimated to have had a magnitude of m_b = 6.5. This earthquake occurred at the Koyna dam in India in 1967, about three years after the initial impoundment. Fluids have been injected at high pressures into rocks at depths of 1–3 km. A remarkable correlation was observed between injection of waste fluid and seismic activity at a deep disposal well in Colorado, US (Evans, 1966). A similar correlation was observed at a controlled water-injection experiment, also in Colorado (Gibbs et al., 1973). The largest magnitude of an earthquake associated with injection was close to m_b = 5.

4.3.7 Seismicity within the territories of some nuclear countries

Geophysical and tectonic boundaries usually do not coincide with the national borders of countries. Some countries, for example, are crossed by convergence zones with a great, both shallow and deep, seismic activity, whereas others are exposed to a seismicity associated mainly with transcurrence zones with primarily shallow earthquakes. Other countries, again, are virtually aseismic. In this section we outline briefly the seismic activity within the territories of some of the nuclear countries. We are omitting the UK and France, since both are virtually aseismic. It should be noted that since 1963 the UK has conducted its tests at the Nevada Test Site in the US, and that the French tests have been carried out outside France. We are here mainly dealing with the distribution of earthquakes regionally, in depth, and in strength. The locations of test sites and seismic zones are also compared.

China

Historical records of earthquakes are probably more comprehensive in China than in any other region in the world (Bolt, 1974; Allen et al., 1975). Many of the Chinese earthquakes have caused great destruction. The tectonics of China is quite complex, and the geological structure varies considerably over the country. From the map, Fig. 4.26, it can be seen that most of the earthquake epicenters are concentrated to the western parts, in the provinces Tsien-Shan, Sinkiang, and Kanshu. The epicenters are, however, scattered over quite large areas. The earthquakes are usually quite shallow, and depths below 100 km have been reported only for a small number of events in the Hindu Kush area close to the USSR border. The Lop Nor test site in Southern Sinkiang lies within a seismic region. The average number of earthquakes per year from 1970 to 1973 reported by the USGS to have a body-wave

SEISMOLOGICAL BACKGROUND

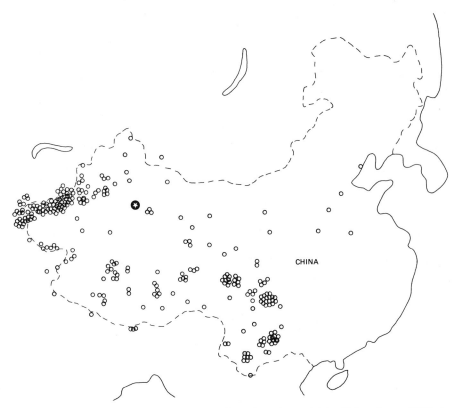

Fig. 4.26. Geographical distribution of earthquakes within China from 1970 to 1973 according to data extracted from the USGS monthly bulletins. Only events with a body-wave magnitude $m_b > 4.0$ are included. The explosion test site at Lop Nor in southern Sinkiang is indicated by a star.

magnitude larger than 4.0 is around 70. The magnitude distribution of the earthquakes for this period, normalized to a period of one year, is shown in Fig. 4.27.

France (Pacific test area)

The early French underground explosion tests in the sixties were conducted in Algeria, but since 1975 France conducts her underground explosions in the Tuamoto Archipelago in the Pacific. Here we discuss only the seismicity of the Pacific area. The Pacific test site is located in an area of the central part of the Pacific plate, which is virtually aseismic. A search in the USGS listings from 1969 to 1974 of events within 20 degrees in both latitude and longitude from the test site has yielded only eight events with a magnitude m_b greater than 4.0. All these events, which apparently were quite shallow, occurred near the

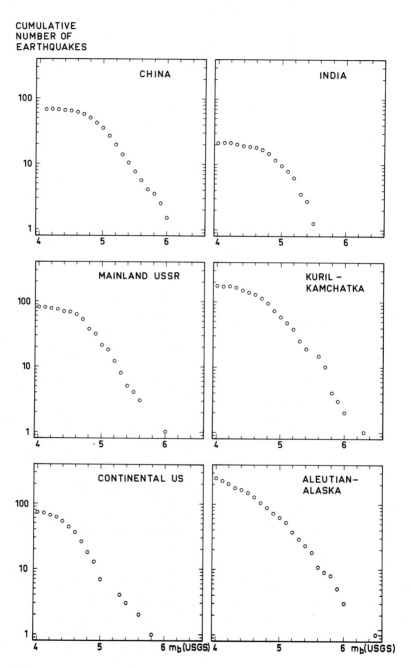

Fig. 4.27. Distribution of body-wave magnitudes m_b for earthquakes in China, India, the US, the USSR, the Kuril-Kamchatka region, and the Aleutian-Alaska region. The data were extracted and compiled as described in the legends to Figs. 4.26 and 4.28–4.31, and have been normalized to one year.

SEISMOLOGICAL BACKGROUND

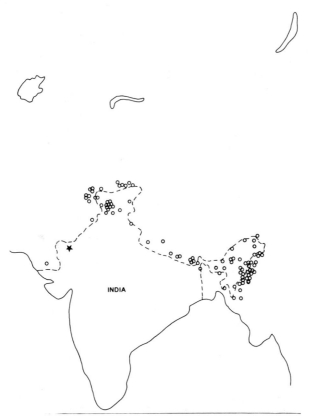

Fig. 4.28. Geographical distribution of earthquakes within the Indian continent from 1970 to 1973, as extracted from the USGS monthly bulletins. The site of the explosion on May 18, 1974, is also indicated (star).

Line Islands, at a distance of about 1800 km from the Mururoa and Fangataufa test islands. Weak, previously unnoticed seismicity in the Central Pacific at latitudes 5S–30S and longitudes 125W–160W was recently detected by a local, seismic network in French Polynesia (Talandier & Kuster, 1976). This activity occurs as swarms of events and is associated with active submarine volcanoes.

India

Most of the seismic activity in India (Fig. 4.28) is concentrated to the northern part of the country along the Himalayan mountain range, where the Indian and Eurasian plates are colliding, according to current studies of plate tectonics (Fig. 4.2) (Le Pichon et al., 1973). The activity is also quite strong along the border to Burma. The activity is quite shallow, except for the very northern part, where earthquakes occur

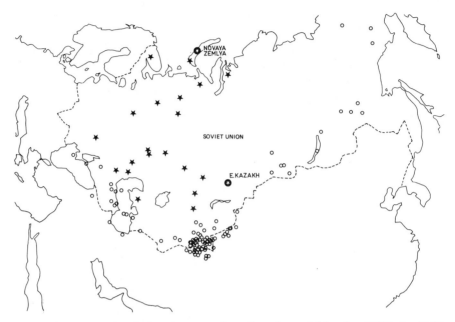

Fig. 4.29. Geographical distribution of earthquakes within the USSR in 1972, according to a compilation by Austegard (1974). The main test sites in Eastern Kazakh and Novaya Zemlya together with sites of one or a few explosions are also indicated (stars).

down to depths of between 100 and 200 km (Kaila et al., 1972). The activity in the area around the site of the first Indian explosion in the Rajasthan desert and in the central and southern parts is quite low. It can be seen from Fig. 4.27 that the average number of earthquakes per year from 1970 to 1973 reported by USGS with a body-wave magnitude larger than 4 is around 20. It can be noted that the Laccadive and Minicoy Islands, west of Southern India, are virtually aseismic, whereas the seismicity at the Andaman and Nicobar Islands to the east of the Indian continent is quite high.

Union of Soviet Socialist Republics

The locations of earthquake epicenters observed within the USSR in 1972 are shown on the map, Fig. 4.29 (Austegard, 1974). For comparison, the main USSR test sites in Eastern Kazakh and Novaya Zemlya as well as some other sites where one or a few explosions have been conducted are also shown on the map. Earthquakes along the Kuriles and Kamchatka reported by the USGS in 1972 are shown in Fig. 4.30. It can be seen that most of these events occurred at sea. The seismic

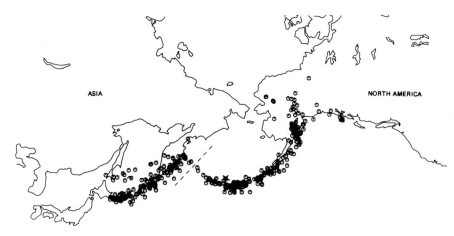

Fig. 4.30. Geographical distribution of earthquakes in 1972 in the Kuril-Kamchatka and Aleutian-Alaska regions, as reported by the USGS. The dashed line marks the border between the two regions. The site of the explosions "Longshot", "Milrow", and "Cannikin" on the Amchitka island in the Aleutians is also indicated (star).

activity in the Kuril–Kamchatka region is higher than in any part of the mainland of USSR. The geographical distribution of epicenters, as shown by Figs. 4.29 and 4.30, is quite similar to that found in other studies (Savarensky et al., 1962). It should be noted that Figs. 4.29 and 4.30 represent data for only one year, whereas the distribution data for China and India in Figs. 4.26 and 4.28 cover four years. A comparison of the relative number of events in China, India, and the USSR can be made from Fig. 4.27, where the data have been normalized to a period of one year. Earthquakes on the mainland of USSR are concentrated to three major regions: Pamir–Balkash, Caucasus–Caspian Sea, and Lake Baykal. There are also a few epicenters scattered over Northeastern Siberia and the Black Sea region. It can be noted that there is also a weak seismic zone in the Ural Mountains, although no events were defined for this region in 1972. It is clear from the map in Fig. 4.29 that the largest parts of the USSR are aseismic. Most of the seismicity is shallow, except in a few areas in the Pamir–Balkash and Kuril–Kamchatka regions, where earthquakes occur at intermediate depths (100–300 km). According to the magnitude-frequency data for 1972 (Fig. 4.27), about 80 events in the mainland of USSR and about 180 events in the Kuril–Kamchatka region had a body-wave magnitude m_b (USGS) > 4.0.

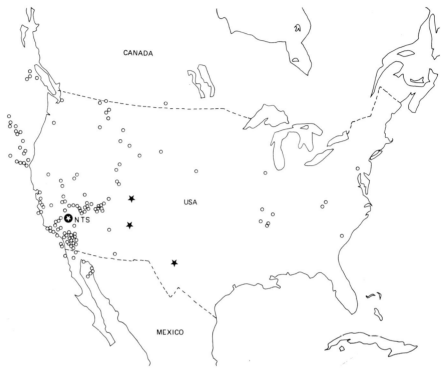

Fig. 4.31. Geographical distribution of earthquakes in continental US in 1972 according to a compilation by Israelson et al. (1976). The Nevada test site (NTS) and three other places where nuclear explosions have been conducted are indicated on the map (stars).

United States of America

The epicenters of events in continental US in 1972 are shown in Fig. 4.31 (Israelson et al., 1976). The locations of events in the Aleutian–Alaska region for the same year as reported by USGS, are shown in Fig. 4.30. Earthquakes in the US are concentrated to the western part of the country. Their epicenters are located in a rather broad zone, extending nearly northwards through Nevada and in a straight, narrow zone extending along the west coast. This latter zone is associated with the San Andreas transcurrence fault. There are also minor concentrations of events further to the north along the coast of Oregon and near Vancouver Island. In the central and eastern parts of the US, there are only a few, scattered events in the Mississippi valley and the Northeastern states. The seismicity is quite shallow all over the US; depths below 30 km are seldom reported, and most of the depths are less than 10 km. Fig. 4.27 shows that in 1972 about seventy earthquakes on the

SEISMOLOGICAL BACKGROUND

mainland of US and about 250 in the Aleutian–Alaska region had a body-wave magnitude m_b (USGS) > 4.0. Besides continental US, Alaska and the Aleutians are highly seismic, as shown in Figs. 4.27 and 4.30. Earthquakes in the Aleutian Islands and Alaska occur at depths both in the shallow and the intermediate depth ranges.

5. EXPLOSIONS AND EARTHQUAKES AS SEISMIC SOURCES

The term "seismic source" is a comprehensive name for earthquakes and explosions, or, more generally, for any radiator of seismic waves. A seismic source can be described by its strength and its spatial and temporal characteristics, i. e., by parameters such as source dimension, geometry, and time function of the radiation. An explosion is a much simpler phenomenon than an earthquake. The spatial as well as the temporal dimensions of an explosion are smaller than those of an earthquake of comparable strength. Moreover, an explosion is almost truly symmetric, whereas a great many earthquake sources are highly unsymmetric.

A variety of models have been proposed to describe the properties of seismic sources. Such models are either empirical, being evolved from near and distant measurements, or obtained from a theoretical description of elastic dislocations. Seismic-source models provide the theoretical basis for the discrimination between explosions and earthquakes. In seismological research, source models are used to interpret far-field observations in terms of physical processes at the source.

In this chapter we present and discuss a few examples of source models for explosions and earthquakes. A brief outline is also given of theoretical calculations of seismic waves. For a more comprehensive account, the reader is referred to the work on seismic-source theory by Archambeau (1976) and the volumes on theoretical calculations of seismic waves by Bolt (1972 a, b).

5.1 EXPLOSION-SOURCE MODELS

In Chapter 3 it is pointed out that most of the energy released by an underground nuclear explosion is deposited as thermal energy in the

SEISMIC SOURCES

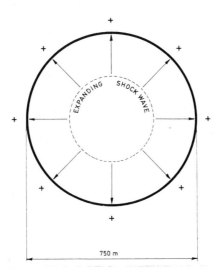

Fig. 5.1. Symmetry of an explosion source. The plus signs indicate that the polarity of the initial motion of the P waves is compressional. The dimension corresponds to the elastic sphere of an explosion with a yield of 5 kilotons.

immediate vicinity of the explosion, and that only a small fraction is radiated as seismic waves. Several attempts have been made to estimate this fraction of the seismic energy, and have resulted in estimates ranging from about 0.01 % to about 5 % of the explosion energy (Berg et al., 1964; Trembly & Berg, 1966; Haskell, 1967; Pasechnik, 1968).

The geometry of an explosion source is usually described by the surface of the elastic sphere, i. e., the boundary between the inelastic and elastic regions of the explosion. As the radial pressure of the explosion shock-wave acts on the surface of the elastic sphere, seismic waves are generated. If the sphere were truly symmetric, only compressional waves with constant amplitude in all directions would be generated, as is indicated in Fig. 5.1. For such a model, the initial motion of the compressional waves recorded at the surface of the earth should be compressional. The pressure, $p(t)$, as a function of time, t, has been represented by the following analytic approximation on the basis of free-field observations (Mueller & Murphy, 1971):

$$p(t) = (P_0 \exp(-at) + P_{0c}) H(t),$$

where $H(t)$ is the unit step function, $P_{0c} = p(\infty)$ is the steady-state pressure, $P_0 + P_{0c} = p(0)$ is the peak-shock pressure, and a is a decay constant. The parameters of this pressure function depend on the yield

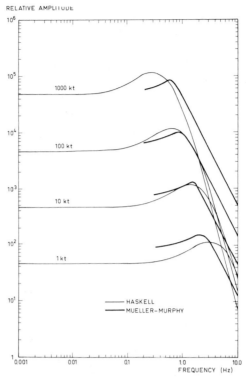

Fig. 5.2. Relative displacement-amplitude spectra of the radiation according to explosion-source models by Haskell (1967) and Mueller & Murphy (1971) for 1, 10, 100, and 1000 kt explosions in tuff and rhyolite.

of the explosion and on the properties of the medium in which the explosion is conducted (Mueller & Murphy, 1971). The elastic far-field displacement as a result of the applied pressure $p(t)$ has been calculated for explosions with yields from 1 kt to 1 Mt. The resulting amplitude spectra of the far-field displacements are displayed in Fig. 5.2. The spectra were normalized to 100 kt and 1 Hz, and the parameters of the pressure function were chosen to describe explosions in tuff and rhyolite. There are also other ways of representing an explosion source than by the pressure function. Another, widely used, concept for describing an explosion source is the so-called "reduced displacement potential" of the source, introduced for mathematical convenience. The reduced displacement potential, F, is related to the displacement, U, at distance r from the shot point by the expression

$$U(r, t) = \frac{\partial}{\partial r} \left\{ F(t - \frac{r}{v_p}) \cdot \frac{1}{r} \right\},$$

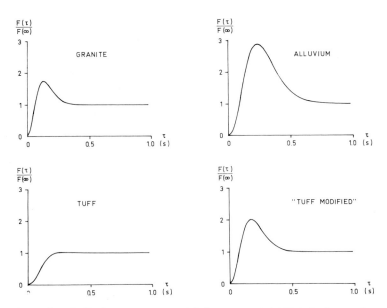

Fig. 5.3. Reduced displacement-potential functions, $F(\tau)/F(\infty)$, for nuclear explosions with a yield of 5 kt in alluvium, granite, and tuff (from Haskell, 1967), and a modified potential function for tuff (from Dahlman, 1974).

where v_p is the compressional-wave velocity (Werth & Herbst, 1963). The term "potential" refers to a mathematical representation of the seismic waves, and the term "reduced" to the fact that F is independent of the distance from the shot point.

Some such displacement functions based on experimental data have been given by Werth & Herbst (1963). Haskell (1967) has approximated these functions by analytic expressions. Examples of such source functions are shown in Fig. 5.3 for 5-kt explosions in granite, tuff, and alluvium. The curve for tuff differs from those for alluvium and granite, and also from that for salt, in that it does not show any overshoot. A modified curve for tuff having an overshoot similar to those for other media has been suggested, and is also included in the figures as "tuff modified" (Dahlman, 1974). Relative displacement-amplitude spectra of the source models for granite, tuff modified, and alluvium are shown in Fig. 5.4 for a 5-kt explosion. The spectrum of the displacement potential of an explosion in a hard-rock material such as granite is predicted to contain more high frequencies than would the spectrum of an explosion in a less competent material such as alluvium. The model also predicts larger amplitudes from an explosion in granite than from one in tuff or alluvium. This difference, which is due to

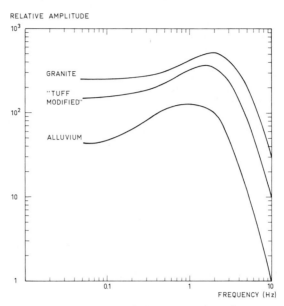

Fig. 5.4. Relative amplitude spectra of the reduced displacement-potential functions shown in Fig. 5.3 for explosions of 5 kt in alluvium, granite, and "tuff modified".

so-called seismic coupling between the ground materials, is further discussed in Chapter 11.

The amplitude spectra of the displacement described by Haskell are in Fig. 5.2 compared with those obtained with the Mueller–Murphy model. For both models the peak of the spectrum is shifted towards lower frequencies with increasing yield; this shift is stronger for the Haskell model than for the Mueller–Murphy model. For frequencies above the spectral peak the amplitude decreases with frequency, f, as f^{-2} for the Mueller–Murphy model and as f^{-4} for the Haskell model. An amplitude roll-off proportional to f^{-2} has been proposed for some earthquake models discussed in Section 5.2. A roll-off proportional to f^{-4} is extremely strong, and predicts, as can be seen in Fig. 5.2, larger amplitudes for a 10-kt explosion than for a 1000-kt explosion at frequencies above 4 Hz.

In descriptions of the radiation of long-period waves, the explosion is often approximated by a point source. An explosive point source can be described mathematically by three mutually perpendicular dipoles of forces, as indicated in Fig. 5.5 (Love, 1944). The two forces in each dipole act along the same line. Each dipole of such a source model can be described by a moment function, $M_0(t)$. This moment

SEISMIC SOURCES

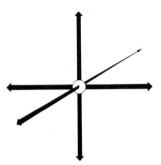

Fig. 5.5. Geometry of three dipole forces being equivalent to an explosive point source.

function can be thought of as the product of the force and the distance between the two equal and opposite forces when this distance diminishes indefinitely. It can be shown that the moment function $M_0(t)$ is directly proportional to the reduced displacement potential earlier discussed (Mueller, 1973a).

$$M_0(t) = 4\pi\rho v_p^2 F(t),$$

where ρ denotes the density of the medium.

The final value of M_0, $M_0(\infty)$, is often called the seismic moment of the explosion (Mueller, 1973a). This seismic moment is analogous to a parameter defined for earthquake-source models discussed in Section 5.2.

5.2 EARTHQUAKE-SOURCE MODELS

The majority of earthquakes occur near the lithospheric plate boundaries and are due to the relative motion of plates. One can relate the earthquake source to a process of strain accumulation. As the elastic strength of the material in the source region is exceeded, the stored strain energy is released and transformed into frictional heat and various elastic waves. The energy release is usually believed to occur in a narrow zone along a plane, usually called the *fault plane*. Since the fault rarely breaks the surface of the earth, the geometry and dynammic parameters of a fault plane are seldom known. Often, however, the earthquakes occur along preexisting faults representing zones of weakness.

The focus of an earthquake is defined as the point where the energy release originates. The focus is often called the *hypocenter* of the earthquake; the *epicenter* is the vertical projection of the hypocenter onto

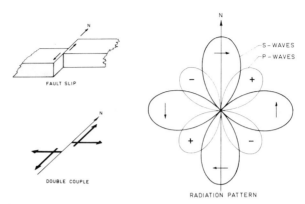

Fig. 5.6. A double-couple point-source model and the resulting pattern of P and S waves. The lobes of the radiation pattern show the variation in radiated amplitude as functions of the azimuth. Compressional and dilatational motions of the P waves are indicated by + and −, respectively, and the arrows indicate the polarity of the S waves.

the earth's surface. The spatial coordinates of an earthquake focus are fully determined by the epicenter and the depth below the earth's surface.

Most of the present knowledge about the details of earthquake-source phenomena is based on the radiated seismic waves. Close-in measurements of earthquake sources exist only for a few cases. A variety of models for the earthquake source have, however, been proposed.

The *point-source earthquake* is supposed to be a fair approximation of the radiation of long-period waves from large earthquakes. This model has also turned out to be a quite useful tool for establishing the orientation of the fault and the direction of slip along the fault. The model consists of a system of forces acting at a point along the fault, as is shown in Fig. 5.6. Four equal forces are symmetrically arranged into two couples. Each couple can be described by a moment function, $M_0(t)$, defined as the product of the force and the distance between the two equal and opposite forces. This moment function is analogous to the moment function for the explosion point-source model discussed in Section 5.1. The compressional and shear waves are radiated in a pattern according to Fig. 5.6. The initial motion of the P waves is compressional and dilatational in alternating quadrants, and the polarity of the shear waves alternates in a similar way. It should be noted that this radiation pattern is quite different from that of the explosion model in Fig. 5.1.

SEISMIC SOURCES

An earthquake source can be theoretically described also by a function, $D(\bar{r},t)$, of the slip displacement along the fault. Here, $\bar{r}$ is the position along the fault relative to the focus, and t is the time during the development of the slip. The slip-displacement function is often modelled as a step or ramp function of time. The slip is usually supposed to propagate along the fault with a constant rupture velocity of around 2–3 km/s. In fault models, the geometry of a fault has been modelled as rectangles, ellipsoids, and circles, and the movements have been assumed to be unidirectional or multidirectional, which means that the slip is moving in one direction or in several directions, respectively.

An attempt to describe an earthquake source by a displacement function has been made by Trifunac (1974), using close-in seismic records from the San Fernando earthquake in 1971. According to Trifunac, the amplitude of the dislocation, which propagated southward and upward with a speed of about 2 km/s, was about 10 m in the hypocentral region at 10-km depth in one end of the fault, decayed to about 1 m around the center of the fault, and built up again to about 6 m when the fault reached the surface.

For most earthquakes it is impossible to estimate both the slip function, $D(\bar{r},t)$ and the fault geometry, and an even more simplified procedure has to be used. The approximate area of the fault plane, A, and the final average displacement, $D_\infty = D(\bar{r},\infty)$, across the fault are used to estimate the amount of elastic energy released. Table 5.1 summarizes estimates of D_∞ and A, based on near-field and far-field observations giving the order of magnitude of areas and displacements involved. The seismic moment, defined as $M_0 = \mu D_\infty A$, where μ is an elastic parameter describing the rigidity of the source region, is equivalent to the final value $M_0(\infty)$ of the moment function for the double-couple point source discussed above (Burridge & Knopoff, 1964), see Fig. 5.6. The stress condition in the source region is described by the stress drop, $\Delta\sigma$, defined as the difference in stress before and after the earthquake. The stress drop is related to the other source characteristics by the expression $\Delta\sigma = c\mu D_\infty A^{-\frac{1}{2}}$. The constant, c, depends on the exact geometry of the fault; its value is usually around 1 or 2. Numerical values of $\Delta\sigma$ and M_0 are also given in Table 5.1.

The displacement-amplitude spectrum of a seismic source is a valuable tool for displaying some of the physical source parameters. The variety of models proposed for the earthquake source all have the

TABLE 5.1. Estimated source parameters for some earthquakes.

Date	Origin time	Epicenter		Depth (km)	m_b (USGS)	M_s	A (km^2)	M_0 (10^{19} Nm)	D (cm)	$\Delta\sigma$ (10^5 N/m^2)	Reference
630803	102137	7.7N	35.8W	33	6.1		352	1.22	105	21	Udias (1971)
631117	004803	7.6N	37.4W	33	5.9		243	0.38	48	12	Udias (1971)
660704	121528	37.5N	24.8W	33	5.4		108	0.02	5	2	Udias (1971)
660704	183336	51.7N	179.9E	13	6.0		420	2.26	163	30	Udias (1971)
660922	164103	39.4N	120.2W	10			100	0.83	30		Tsai & Aki (1970b)
671210	225124	17.7N	73.9E	33	6.0	6.3	252	0.82	108	6–19	Singh et al. (1975)
690428	232043	33.3N	116.3W	20	5.7	5.2	30	0.05		77	Thatcher & Hamilton (1973)
710209	140041	34.4N	118.4W	13	6.2	6.5	440	1.3		30	Trifunac (1974)

A = Area of fault plane. M_0 = Seismic moment. D = Average displacement. $\Delta\sigma$ = Stress drop.

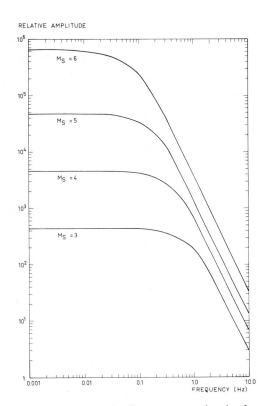

Fig. 5.7. Spectra of the relative displacement amplitude for earthquakes with different surface-wave magnitudes, M_s, according to a source model given by Aki (1972).

gross spectral shape outlined in Fig. 5.7, regardless of differences in details of the models (Randall, 1973). This general spectral shape agrees with observations from a large number of earthquakes. Even if a spectrum is almost flat at lower frequencies, its maximum occurs at zero frequency according to theory. This maximum at zero frequency is due to the assumption that the dislocation along the fault is unidirectional, maintaining its direction with no oscillations (Savage, 1966). The amplitude level at zero frequency is proportional to the final value, $M_o(\infty)$, of the moment function, M_o, defined earlier. This final value can be used as a measure of the size or energy of the earthquake.

At frequencies above the so-called corner frequency, f_c, the amplitude drops quite rapidly. To get a physically valid model, where the total radiated energy remains finite, the amplitude should decrease with increasing frequency as a power of f of at least -1.5. Most source models suggest a slope of -2 at frequencies above f_c, but full agreement on this point has not been reached so far (Aki, 1972). The slope is related to the degree of discontinuity of the dislocation. A quadratic slope means a step in the velocity, and a cubic slope a step in the acceleration of the dislocation (Savage, 1972).

The duration of the source may be construed to be the time the rupture takes to travel across the fault, which is usually supposed to occur at some constant velocity. The corner frequency, f_c, is inversely proportional to the dimension of the source. The spectral bandwidth is roughly equal to f_c and inversely proportional to the duration of the displacement. The effect of the dimension of an earthquake source on the radiated energy is reflected in the displacement-amplitude spectrum. An increase in the dimension of an earthquake leads to an increased amplitude level at zero frequency and to a lower corner frequency, which means that the spectrum is shifted towards lower frequencies. Large earthquakes, therefore, radiate relatively little energy at high frequencies. A similar corner-frequency shift is noted for the explosion spectra in Fig. 5.4.

5.3 COMPARISON BETWEEN EXPLOSION-SOURCE AND EARTHQUAKE-SOURCE MODELS

In this section we summarize the differences in mechanism, symmetry, and spatial and temporal dimensions between models of explosion sources and models of earthquake sources. In particular, we point to the implications of these differences on seismic-wave radiation.

The compressional and shearing nature of explosion and earthquake models, respectively, implies that explosions should radiate only P waves, whereas earthquakes should radiate both P and S waves. The difference in mechanism also implies that the polarity of the initial motion of P waves from an explosion should be entirely compressional, whereas there should be both compressional and dilatational initial P-wave motions for earthquake shear models.

Explosion-source models, which are truly symmetric, predict equal P-wave radiation in all directions. Earthquake-fault models, on the other hand, give a highly unsymmetric radiation of both P and S waves.

Both the spatial and the temporal dimensions of earthquake models are larger than those of explosion models at comparable strengths. This implies that the waves from earthquakes should have a longer duration than those from explosion waves. The explosion P waves are in fact predicted to be pulse-like. The difference in dimension also im-

SEISMIC SOURCES

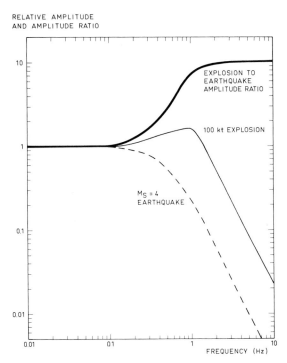

Fig. 5.8. Comparison of the relative displacement-amplitude spectrum predicted by Aki (1972) for an earthquake with $M_s = 4$, and by Mueller & Murphy (1971) for an explosion of 100 kt in tuff, also having an M_s value of about 4.

plies spectral differences in the radiated waves, as can be seen in Figs. 5.4 and 5.7.

The relative displacement-amplitude spectrum predicted by a model of Aki (1972) for an earthquake with $M_s = 4$ is in Fig. 5.8 compared with the spectrum predicted by the method of Mueller & Murphy (1971), for a 100-kt explosion in tuff. An explosion with a yield of 100 kt will give an M_s value of about 4. (For a discussion of the relationship between explosion yield and magnitude, see Chapter 11.) The two spectra have been normalized at a period of 20 s (which means that the amplitude at 0.05 Hz has been set equal to unity). The figure shows that the predicted amplitudes at frequencies above 1 Hz are ten times as large for the explosion model as for the earthquake model. This pronounced difference between short-period and long-period waves from explosions and earthquakes predicted by some models has also been observed, and is utilized for discrimination by the so-called $m_b\,(M_s)$ method discussed in Chapter 10.

5.4 THEORETICAL CALCULATION OF SEISMIC WAVES

The travel times of seismic waves through the interior of the earth are well known, and the arrival times of seismic signals can be predicted with an uncertainty of the order of 0.1 percent. The variations in amplitude and period of the seismic signals, on the other hand, are less well-known, and only the large-scale dynamic characteristics of seismic signals can be predicted today. During the last decade, significant efforts have also been devoted to theoretical calculations of the dynamic characteristics of seismic waves. These calculations have become an increasingly useful tool in the interpretation of seismic records. It is generally quite difficult to get the calculated signals to agree with those observed. The seismic waves are influenced along their travel paths by radial and lateral inhomogeneities in the earth. Another complicating factor is the inherent uncertainty about the detailed physical process at an earthquake source. Nuclear explosions have, however, comparatively well-known source characteristics, and records from explosions can therefore be used to study propagation-path effects. Below, we give a few examples of results that have been obtained from theoretical calculations of P and Rayleigh waves. For detailed accounts of the underlying theory and of advances in computational methods, the reader is referred to Bullen (1963) and Bolt (1972 a, b), respectively.

Theoretical signal computation can be regarded as a synthesis of three elements: the source, the propagation path, and the recording system. Of these, only the recording system is usually known with any accuracy. The effect of the propagation path can in turn be considered as the result of several factors. One is the layers of the source crust, another the anelastic attenuation of the mantle, and a third the receiver crust. The source models used in such computations are of the kind described for explosions in Section 5.1 and for earthquakes in Section 5.2.

Several of the models of short-period P waves are based on the assumption of linearity between the various factors shaping the P-wave forms. This means that these factors can conveniently be treated by Fourier transforms, and the transform of the resulting P signal is just the product of the individual factors. Figure 5.9 gives an example of how the P waves, as they propagate from the source to the receiver, are developed by the different factors. Calculation of the effect of the crust, which is usually represented by plane layers, involves complicated computations, but effective algorithms for the purpose have been

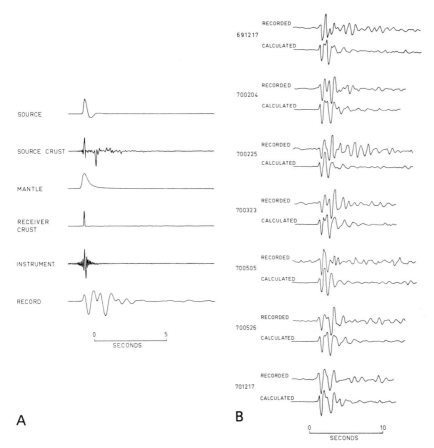

Fig. 5.9. A—An example of a calculated short-period P-wave record (bottom of figure) and the theoretical waveforms of the shaping factors. B—Comparison of recorded and calculated short-period P-wave signals at the Hagfors Observatory from seven explosions at the Nevada Test Site. (After Slunga, 1976.)

developed (Kogeus, 1968). The effect of the anelasticity of the mantle is mostly described by an exponential, frequency-dependent term, $\exp(-\pi f T(\Delta)/\overline{Q})$, where $\overline{Q}$ is the average Q along the path, f is the frequency, and $T(\Delta)$ is the travel time at distance Δ. Although the broad characteristics of an observed signal can sometimes be reproduced, as illustrated by Fig. 5.9, there are still details which are difficult to describe (Hasegawa, 1971).

It has been observed that P signals from earthquakes quite often look more complex than the explosion signals shown in Fig. 5.9. Theoretical calculations of P signals from earthquake sources have shown that the dimensions and the asymmetry of the source model influence the P-signal complexity. The theoretical results have, however, also

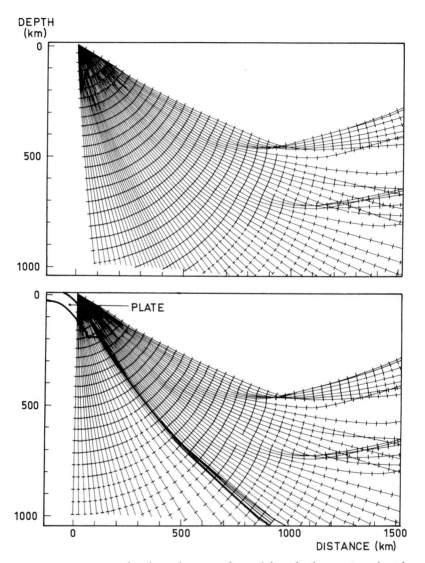

Fig. 5.10. P-wave ray paths through an earth model with (bottom) and without (top) a downgoing plate. (After Julian, 1970b.)

suggested that the signal complexity is even more sensitive to the particular crust model used, especially when the model includes a low-velocity layer (Douglas et al., 1974 a). Some earthquakes have, however, been found to generate quite simple P signals. The results of theoretical calculations suggest that this can be due partly to high-Q travel paths in the mantle and partly to the orientation of the source relative to the receiving station (Douglas et al., 1973, 1974 b).

The models discussed above presume radial heterogeneity. Numerical experiments with travel-time and amplitude effects of lateral heterogeneities have been carried out by Davies & Julian (1972) and by Jacobs (1972). In Fig. 5.10 the effect on the radiated P waves of a high-Q plate dipping down into a low-Q zone is shown. It can be seen that the rays show considerable asymmetry in the geometrical spreading. More recently, various techniques have been developed to model not only the ray paths, but also the shapes of body-wave signals traversing a laterally heterogeneous medium (Kennett 1973; Smith 1975).

Theoretical calculations on Rayleigh waves have to a great extent been focused on the amplitude spectrum of the waves, rather than on the wave forms. Computed and observed Rayleigh waves do show many common features (Douglas et al., 1972 b).

Results of numerical experiments with linear models similar to the ones for P waves mentioned above suggest that Rayleigh-wave spectra from explosions are quite flat in the period range of 10–50 s and do not vary much with the source medium and the yield of the explosion (Tsai & Aki, 1971). Theoretical Rayleigh-wave spectra for earthquakes have been found to be particularly sensitive to the depth of the earthquake focus (Tsai & Aki, 1970 a), and significant notches in the spectrum occur in the period range of 10–50 s (Douglas et al., 1971). Other factors, such as crustal thickness at the source, spatial and temporal behavior of the source, and attenuation along the travel path, appear to have much less influence on the shape of the Rayleigh-wave spectra from earthquakes than has the focal depth.

6. SEISMOLOGICAL STATIONS

During the sixties considerable efforts, money and manpower were devoted to the development and implementation of new, highly sensitive seismological stations in several countries. This was done almost entirely for the purpose of investigating the possibilities of using seismological data to monitor a Comprehensive Test Ban Treaty. Most of the efforts formed part of the US *Vela Uniform Project* (Foster, 1971; Lukasik, 1971). In the new stations and their equipment, the most recent achievements in electronics and computer technology were applied. Thus, seismology suddenly changed from being a technologically rather underdeveloped science to a science utilizing the most modern facilities available for collection and analysis of data.

This technological breakthrough also brought to seismology a new generation of scientists, coming from disciplines such as physics, communication-engineering, numerical analysis, and computer science. The inflow to seismology of new instruments and of scientists from other disciplines has vitalized seismology not only by improving the technical facilities, but also by introducing new ways and methods to approach scientific problems. Although most new efforts were more or less directly devoted to the problem of detecting and identifying nuclear explosions by seismological means, data from highly sensitive stations and modern methods of analysis are now frequently used also in other geophysical research projects.

In this chapter, we present seismic instruments available today as well as existing stations. As any international seismic monitoring system must be based on a global network of stations from which data are generally available through an open, international data exchange, we discuss present seismic stations in terms of networks rather than single stations.

6.1 INSTRUMENTATION

When discussing the recording of seismic signals, it is important to consider both the frequency and amplitude characteristics of the recording system. The frequency characteristic is often discussed in terms of *bandwidth*, that is, the frequency interval within which the system is designed to work. The amplitude characteristic is often defined as the ratio of the largest to the smallest signal a system can record. This ratio is usually called the *dynamic range* and is expressed in decibels (dB), where the dynamic range in dB is 20 times the base-10 logarithm of the amplitude ratio.

The bandwidth of seismic signals is generally small, and the dynamic range wide. The bandwidth of teleseismic signals usually lies in the range of from 0.01 Hz to 5 Hz, the actual width depending to some extent on the size of the event and on the transmission properties of the earth between the source and the recording station. The dynamic range of teleseismic signals goes from around 1 nm for P-wave signals from weak events to more than 10 mm for long-period surface waves from strong events. The total dynamic range of the signals is thus more than 140 dB. Seismic records are usually split into separate frequency bands obtained at different sensitivities. These frequency bands are chosen so as to reduce the influence of seismic background noise and thus to optimize the conditions for signal recording. (This is further discussed in Chapter 7.) Three different types of seismological recording are presently used, known as short-period, long-period, and broadband recording. By a short-period record, we mean the record of seismic signals obtained by instruments having their maximum sensitivity at frequencies above 1 Hz. The signals most frequently recorded in this frequency band are the P waves. Short-period recording is of fundamental importance in the detection, location, and definition of seismic events. Long-period records contain signals obtained at periods around 20 s, in some instruments at still longer periods. These records, which show the surface waves most clearly, are valuable for the identification of earthquakes and explosions. The dynamic range of both the short-period and long-period signals is of the order of 100 dB, extending from 1 nm to 100 μm for short-period signals, and from 100 nm to 10 mm for long-period signals. Broadband records show seismic signals in the frequency band from 0.1 to 10 Hz and give the most complete presentation of the incoming seismic signals (Marshall et al., 1972; CCD/401, 1973). Due to strong disturbances from seismic noise, broadband re-

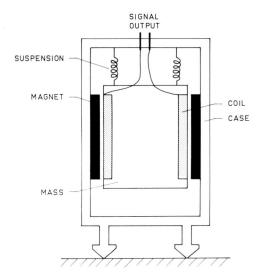

Fig. 6.1. Outline of a modern moving-coil seismometer.

cording is less capable of perceiving weak events than are the narrow-banded short-period and long-period recordings. Broadband instruments have so far been used almost entirely within the USSR, but the interest in such instruments is currently growing in other countries, such as the UK.

6.1.1 Seismometers

The weak vibrations of a seismic signal are detected by an electro-mechanical instrument called *seismometer*, the general principle of which is illustrated in Fig. 6.1. The main components of a seismometer are a magnet, which is fixed to the ground, and a spring-suspended mass with an electric coil. The seismic waves move the ground and the magnet attached to it, whereas they leave the mass with the coil virtually unaffected. The relative motion of the magnet and the coil generates a current in the coil which is proportional to their relative velocity. Most seismometers operating today are based on this principle and are thus sensitive to the particle velocity of the seismic signals. The size of the seismometer mass and the spring constant determine in principle the period of the seismometer.

Without going into a detailed technical discussion of the various seismometer designs that have been developed, we will give a few fundamental data on them. Readers interested in technical descriptions

Fig. 6.2. Portable short-period seismometer. (Courtesy Teledyne, Geotech.)

and discussions of seismometer designs are referred to special reports (Melton & Kirkpatrick, 1970; Ward, 1970; Willmore, 1975). Two main types of seismometer for recording of teleseismic signals can be distinguished: the short-period seismometer and the long-period seismometer. A modern short-period seismometer is shown in Fig. 6.2. The short-period instrument has a free period of around 1 s and the long-period instrument a period usually adjustable in the range of from 20 s to 30 s. Seismometers having a constant response over a broad frequency range have recently been designed (Block, 1970; Burke et al., 1970). Response curves for some presently used instruments are shown in Fig. 6.3. The curves show the response of the entire systems, including the recording equipment.

Seismometers generally have a rather limited bandwidth and a

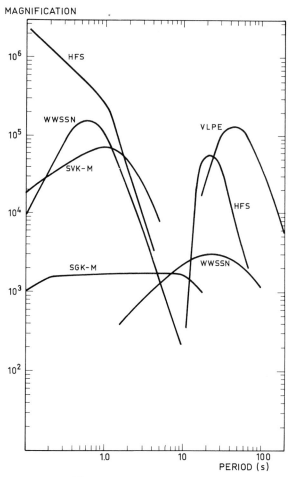

Fig. 6.3. Response curves of long-period, short-period, and broadband instruments. The curves show the overall response of the seismometer–amplifier–recorder system. The absolute magnifications shown are typical of those used. The WWSSN long-period and short-period curves are typical of conventional seismological stations. The HFS curves show the response of the instruments of the Hagfors Observatory in Sweden, and are typical also of other array stations. The SVK-M short-period and SGK-M broadband instruments are used in the USSR. The VLPE curve shows the response of the so-called High Gain Very Long Period Experiment stations.

very high dynamic range. Modern short-period seismometers connected to high-quality amplifiers are capable of recording signals as weak as 0.1 nm (1 Ångström), which is less than 0.1 % of the wavelength of visible light, a remarkable sensitivity for an electromechanical instrument. The maximum ground motion that can be recorded by a short-period instrument is of the order of a few mm, corresponding to a dynamic range of the order of 150 dB. Long-period instruments have a dynamic range of the order of 120 dB, corresponding to a displacement range of from 10 nm to 10 mm.

The maximum dynamic range of presently available amplifiers is of the order of 80 dB, that is, considerably less than the dynamic range of the seismometers. To utilize the dynamic range of the latter, and thus match the dynamic range of the seismic signals, two or more amplifiers with different gains have to be connected to one seismometer. Another way of increasing the dynamic range of the amplifiers is to use a gain-ranging amplifier, that is, an amplifier which adjusts its gain according to the size of the incoming signal. Even the modern stations of today may, however, have difficulties to obtain amplifying and recording systems with an adequate dynamic range capable of handling signals from very large events.

The instrumental noise in a high-quality short-period seismometer-amplifier system is more than a factor 10 below the level of seismic background noise recorded at any place on the globe. This means that instrumental noise does not limit the short-period-detection capability. Long-period seismometers are, due to their long eigenperiods, affected by local environmental disturbances, such as variations in temperature and pressure and local atmospheric convection currents (Pomeroy, 1970). These local disturbances limit the sensitivity of the long-period instruments, especially at long periods. By installing such instruments in environmentally well controlled subsurface vaults or in boreholes, the sensitivity may be increased by a factor 10 or more.

Modern seismometers have been designed to withstand severe environmental conditions, e.g. at the bottom of boreholes kilometers deep, at the bottom of the ocean, or on the moon. Seismometers available today are therefore likely to meet any requirement of a monitoring network.

6.1.2 Recording equipment

As was mentioned above, seismic signals cover a very wide dynamic range, but have a rather limited bandwidth. This general characteristic of the signals in combination with the need for round-the-clock, year-round recording of seismic signals put rather special demands on the recording equipment. Up to now, three different techniques have been utilized for the recording of seismological data: visual recording on photographic or strip-chart paper, analog magnetic-tape recording, and digital recording, usually on magnetic tape.

The type of photographic recording system used at most of the ca. 1000 conventional seismological stations now operating all over the globe gives one-day records on a 30 cm by 92 cm photographic paper. A typical time resolution of short-period records is 60 mm/minute and of long-period records 15 or 30 mm/minute (USCGS, 1960). This type of record has the advantage of giving a convenient general view of the recorded signals; it also utilizes the recording paper in an efficient way. The most serious drawbacks are the limited dynamic range and the low time resolution of high-frequency short-period signals. It was mentioned above that the dynamic range of both the short-period and long-period signals is of the order of 100 dB, whereas the dynamic range of a paper record is only of the order of 30 dB. A photographic record can thus cover only a small part of the dynamic range of the seismic signals at highly sensitive sites. Many of the conventional stations are, however, placed at sites where the level of short-period noise is considerably higher than 1 nm, and the dynamic range of the seismic signals to be recorded is therefore correspondingly reduced. By recording the signals on parallel recorders having different gains, the dynamic range, but not the resolution, can be increased.

Analog recording on magnetic tape can be done using either frequency or amplitude modulation (Davies, 1961). For recording low-frequency seismic signals, frequency modulation is the most suitable technique and is therefore the most commonly used one. The bandwidth of a frequency-modulated tape recorder depends on the recording speed. However, the bandwidth of recorders used for teleseismic signals is large, 10 Hz or more, compared to the bandwidth of teleseismic signals. Analog-tape recording will thus give enough time and frequency resolution for teleseismic signals. The dynamic range of an analog magnetic-tape recorder is around 40 or 50 dB. The exact value of the range depends on the recording speed, the technical quality of the unit, and, to some extent, on how the dynamic range is defined. Although significantly larger than the dynamic range of a photographic recorder, the range of 40 to 50 dB typical of the analog magnetic-tape recorder is still small compared to the range of the seismic signals. Thus, at least two recording channels with different gains are needed to cover the dynamic range of the signals. In the analysis of signals recorded on analog tape, the tape can be played back through different filters and other analog analysis equipment or through analog-to-digital conversion equipment, to make the data accessible for computer analysis. To allow monitoring of the seismic signals and of the performance of the equipment, magnetic-tape recording should be complemented by some type of visual recording.

In a digital data-acquisition system the analog signals are read (sampled) by an analog-to-digital conversion unit and converted into binary form. The sampling rate, that is, the number of samples taken per time unit, is controlled by an electronic clock. Today, sampling rates up to several million samples per second are technically achievable. The bandwidth is determined by the sampling rate; at the highest signal frequency occurring, at least two samples per period are needed. In seismological applications, sampling rates of 10 to 20 samples/s are usually used for short-period signals. These rates correspond to band-widths of 5 Hz and 10 Hz, respectively. Typical figures for long-period signals are 1-Hz sampling rate and 0.5-Hz bandwidth.

The dynamic range of a digital system is usually discussed in terms of bits, where the number of bits, n, for an amplitude, A, is defined as $A = 2^n$. The dynamic range in dB is equal to $6n$.

Analog-to-digital conversion systems with a dynamic range of up to 16 bits, or 96 dB, are available today. For seismological purposes, systems with a dynamic range of 12 or 14 bits, corresponding to 72 or 84 dB, respectively, are mostly used. A digital data-acquisition system can thus give any desired bandwidth and a wide dynamic range, usually wider than that of the amplifiers. However, the total dynamic range of a digital recording system is smaller than that of the seismic signals. Thus, unless special low-gain equipment has been installed also stations with digital recording systems might be overloaded by large events. An illustration of the high dynamic range achieved by the digital recording system of the Hagfors Observatory utilizing high-gain and low-gain channels is given by Fig. 6.4. The amplitude ratio of the largest explosion signal to the weakest after-event signal in the figure is 20 000, corresponding to 86 dB. The total dynamic range of the short-period system at Hagfors is 120 dB.

The most convenient way to record digital data is in a computer-compatible format on magnetic tape, so that the data can be analyzed on a digital computer (Hoagland, 1963). Different computers write tape in different format and data packing density. Even if two com-

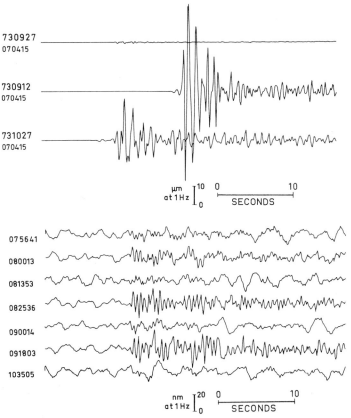

Fig. 6.4. Short-period seismic signals recorded at the Hagfors Observatory from three large underground nuclear explosions and seven afterevents on October 27, 1973, at Novaya Zemlya. The magnification used for the explosion records is one fivehundredth of that used for the afterevent signals. (After Israelson et al., 1974.)

puters have the same nominal tape standard, the problems of transferring tape from the one to the other should not be underestimated.

6.1.3 Array stations, principal design

The limit to the sensitivity of short-period recording is not set by instrumental noise, but rather by the omnipresent seismic background noise, the characteristics of which will be discussed more extensively in Chapter 7. This background noise limits the capability to detect weak seismic signals. To reduce the influence of ground noise and increase the detection capability, suggestions were put forward in the late 1950's to apply antenna theory to the design of stations for seismic monitoring.

In electronic communication systems involving transmission and/or reception of electromagnetic waves, antennas have long been used to increase the signal strength. Antennas are used to receive radio and television signals, and particularly extensive antennas are used in radar systems. The more advanced antennas consist of many elements put together in a geometric configuration which may differ according to the specific application and signal frequencies concerned. Most of the work on antenna theory that has been carried out has concerned the development of radar antennas.

The application of antenna theory to seismic stations implied the interconnection of several seismometers to form an antenna and thus to increase their capability to record weak seismic signals. Various numbers of sensors, geometries, and geographical dimensions of these so-called *seismic array stations* have been suggested and tested (Lacoss, 1965; Capon et al., 1967). The number of sensors has ranged from three to 525, and the aperture of the array from a few kilometers up to 200 km. The configuration of array stations operating today is discussed later in this chapter.

The fundamental principle of antenna theory is that each element of the array receives signals that are identical in form but may be shifted in time. By means of appropriate time delays, the signals are put on top of each other and added in a process usually called "beam-forming", which is expected to give a theoretical gain in signal-to-noise ratio by a factor equal to the square root of the number of sensors employed. Experience has shown that seismic signals are not identical at seismometers some kilometers apart, so the expected gain of array stations has in most cases not been reached. There seems therefore to be little impetus today for construction of further, very large, array stations.

Array stations did not bring only antenna theory into seismology, they also brought modern electronic equipment able to operate with high reliability under severe environmental conditions. Solid-state amplifiers with a dynamic range of about 80 dB and analog and digital data-transmission equipment, to bring the signals from remote seismometers to a data-recording center, either by cable or by radio link, have been developed and used in array stations. The development of analog and digital data-acquisition and data-analysis systems for seismological data has gone hand in hand with the development of the seismic array stations.

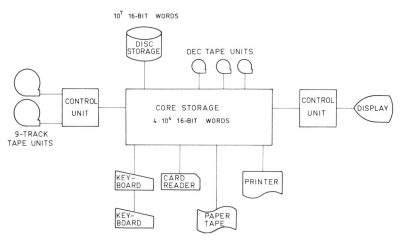

Fig. 6.5. Block diagram of the off-line data-analysis system used at the Hagfors Observatory.

A large seismic array station produces so much data that the handling and storage of it constitutes a severe problem. The Large Aperture Seismic Array, LASA, originally equipped with 525 short-period seismometers, produced a data stream corresponding to about 150 magnetic tapes per day. Even an array station with a rather limited number of seismometers, such as the Hagfors Observatory in Sweden, now having 15 short-period and 5 long-period seismometers, produces 200 million binary numbers or one and a half magnetic tape per day. It might be difficult for those who are not familiar with digital recording to see what these figures mean, but the data collected at Hagfors during one year would, if recorded on punched cards, require about 300 tons of such cards. Having to handle and for many years store such amounts of data is a heavy burden on institutes and organizations running array stations. Many array stations therefore retain their original data only for a limited time, and store permanently only those data that are of interest to them. Although this in some cases might be almost a technical necessity, it may nevertheless create problems in the international cooperation, as we do not have any international agreement on what data should be retained. Today, data of significant importance for one institute could well be erased at another station. As we will see later, it is of fundamental importance for event-identification studies, and also for geophysical studies in general, to be able to utilize data on events recorded by a certain station network. It may take a long time to create data bases of ad-

SEISMOLOGICAL STATIONS 119

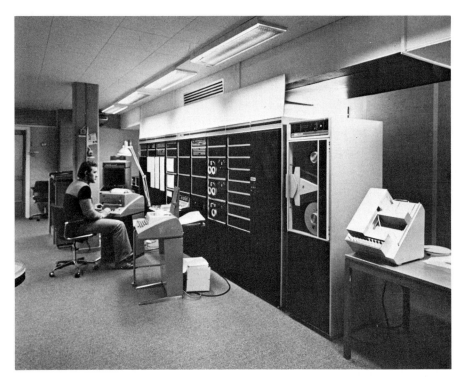

Fig. 6.6. The off-line data-analysis system of the Hagfors Observatory.

equate sizes, especially for source regions where the earthquake or explosion activity is low. It is therefore necessary to have stations running continuously for many years to create such fundamental data bases.

To utilize the considerable amount of data produced by the existing large and medium size array stations, special facilities for data handling and analysis have been developed. These systems are designed to handle, in an efficient way, large amounts of data. Such a system, a typical example of which is schematically outlined in Fig. 6.5, usually consists of a small computer with a limited core memory, a disk or drum mass memory, tape units, and one or more graphic displays. The basic idea behind these interactive systems is to allow a close contact between man and data and thereby facilitate the analysis of a large number of signals in a rather standardized, but for most seismological purposes adequate, way. Such off-line analysis systems are today used in Canada, Sweden, the UK, and the US. A photograph of the system used at the Hagfors Observatory is shown in Fig. 6.6. These systems and the computer programs developed to handle, primarily, time-series data not only are a considerable achievement for the analysis of seismological data, but are also useful for the analysis of other large data bases, e.g. those obtained in the monitoring of our environment.

6.1.4 Future instrumental development

Through the interest of a number of nations in the problem of monitoring a CTB by seismological methods, and especially through the US Vela Uniform program, an extensive development of seismic instruments and stations was carried out during the 1960's, a development that would certainly not have gone so fast and been so comprehensive without this political pressure. The speed of this instrumental development will probably slow down, mostly because the instruments and the data-recording and data-analysis equipments that are available today seem to be adequate to establish a monitoring system for a CTB. Some further developments do seem, however, possible and are discussed below.

The capability of detecting short-period waves is today limited by seismic and not by instrumental noise, and there is therefore no impetus to develop new and more sensitive short-period seismometers. The capability of detecting long-period signals, especially such of very long periods, might be further increased by the use of more stable instruments and more rigorous installations. An increased interest in broadband recording and a corresponding development of suitable seismometers can be expected. As far as the electronic equipment for signal amplification and transmission is concerned, it seems probable that the general progress in the field of electronic component design will influence also seismological instrumentation. Gain-ranging amplifiers might become more frequently used in the future. Such systems increase the dynamic range without increasing the resolution of the signal. The general development towards smaller and cheaper digital computers and data-collection systems will probably lead to a more extensive use of digital recording and analysis equipment, also at smaller stations. New systems capable of storing large amounts of digital data might also be developed for use particularly at large data centers.

6.2 STATION NETWORKS

One isolated seismological station, however advanced its equipment may be, can do little to solve the problem of adequately monitoring a CTB, and will also be of limited value in most seismological research projects. It is the joint interpretation of data from globally distributed seismological stations that has to be used in a CTB monitoring system, and also in most geophysical studies. It has long been realized among seismologists that arrival times from a number of stations have to be used in order to locate a seismic event with a reasonable accuracy. This has led to an extensive international exchange of preliminary seismic data, mostly arrival times, between the seismological stations and institutes of most countries on the globe. The need for accurate location of epicenters also led to the establishment of institutes specialized in computation and distribution of epicenter data. The main requirement on stations for location purposes is in practice limited to accurate reading and reporting of arrival-time data. As is discussed in Chapter 7, the global capability to detect and locate weak events is critically dependent on a limited number of highly sensitive and well-operated stations.

The requirements that must be satisfied by stations to be used for event identification and for monitoring purposes are much heavier. Stations used for event identification must have the ability to produce data that can be used to distinguish between earthquakes and explosions. Such data may be high-gain long-period data or spectral and other parameters determined from short-period records. Very little identification data of these kinds is included in today's international data exchange. The requirements for uninterrupted operation are also much stronger on stations in a monitoring network than on stations used only for research purposes.

In the following, we discuss the various types of seismic station networks that are in operation today and in what ways they can be used for Test-Ban monitoring.

6.2.1 *National networks of conventional stations*

By a conventional sesmic station we mean a station having at least short-period, but often also long-period, seismometers with photographic recording equipment. According to Covington (1974), 1116 seismological stations were in operation in 1974. The geographical distribu-

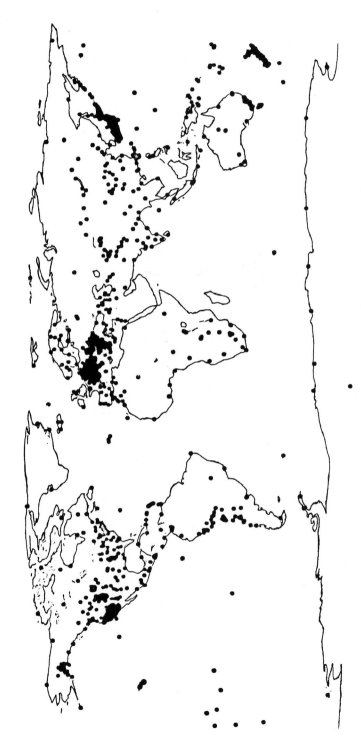

Fig. 6.7. Locations of the 1116 seismological stations reported to be in operation in 1974 (Covington, 1974.)

tion of these stations, Fig. 6.7, is rather uneven, with many stations located in Europe, Japan, and the US, and few in Africa, South America, and the northern part of the USSR. Most of these stations were established and are operated by national institutes and universities. They represent a rather inhomogeneous network, with large variation in instrumentation, sensitivity, and operating conditions. Many of the stations operate in or close to regions where earthquakes occur, and are thus of great importance for the monitoring of local earthquake activity and for research in connection with earthquake prediction. The local stations in Japan, New Zealand, the western US, and the southwestern USSR are examples of such networks. The networks in western US and southwestern USSR, consisting of 180 and 70 stations, respectively, might also be of great importance for a CTB-monitoring system. The last-mentioned networks are capable of detecting and accurately locating also small earthquakes in two earthquake regions of great interest in the CTB discussion.

The conventional stations were originally established to record seismic signals from fairly strong earthquakes. These stations, having photographic recording equipment with low dynamic range and often low sensitivity, will be of limited value in monitoring weak explosions. Only a few conventional stations, geographically well located and with high sensitivity, might be of importance in a CTB-monitoring network by increasing the location accuracy in certain areas.

6.2.2 The World Wide Standard Stations Network

To achieve a global network that would produce standardized seismological data, US authorities in 1961 decided to install a network, known as the *World Wide Standard Stations Network,* WWSSN, of seismological stations with identical short-period and long-period instrumentation recording on photographic paper. Most of the planned 125 stations were put in operation; Fig. 6.8 shows the 112 stations reported to be in operation in 1966 (Powell & Fries, 1966). This network of standardized stations, so far the most extensive of its kind to have been established, was a most important contribution to global seismology, as it gave reliable and comparable seismological data from globally distributed stations. Data from these stations, which are available on microfilm at low cost, have been extensively used by seismologists. The funds for supporting these stations have been withdrawn, however, and it is not known how long the individual stations will continue to

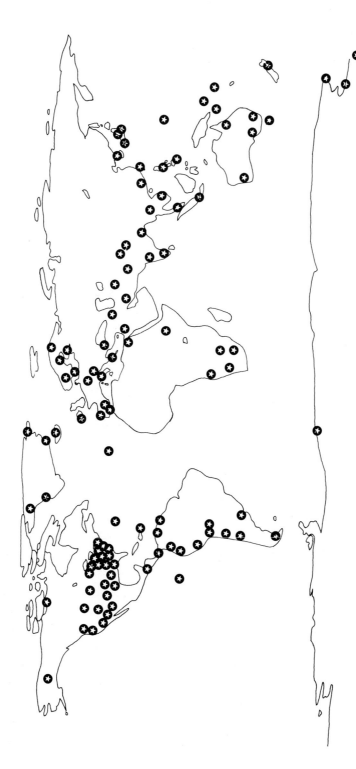

Fig. 6.8. Locations of the 112 WWSSN stations reported to be in operation in 1966.

125

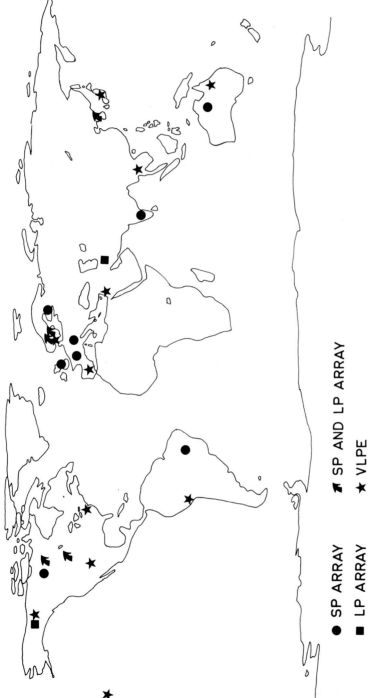

Fig. 6.9. Locations of array stations operating today or planned to be in operation in the near future, and of the VLPE stations. (SP=short period, LP=long period.)

TABLE 6.1. Summary of array-station data. (For explanation of some of the column headings see Notes below.)

Country	Station	Operating since	Coordinates	Total aperture (km)	No. of seismometers SP[a]	No. of seismometers LP[b]	Array configuration	Recording system[f]	Bull. prod.	Operating organization	References
Australia	Warramunga	1965	19°56'39"S 134°20'27"E	20	20	—	Cross	A tape	No	Australian National University, Canberra	UKAEA (1967)
Brazil	Brasilia	1970	15°50'28"S 47°49'12"W	20	17	—	Cross	A tape	No	Dept. of Geophysics, University of Brasilia	
Canada	Yellowknife	1963	62°29'34"N 114°36'16"W	20	19	3[d]	Cross (Fig. 6.10)	D & A tape	Yes	Earth Physics Branch, Energy, Mines and Resources, Ottawa	CCD/406 (1973) Weichert (1975a)
Finland	Jyväskylä	1976[c]	62°N 25°E	90	3	—	Triangle	D		University of Helsinki, Helsinki	Pirhonen (1976)
France	CEA	1974		500	20	—	Natl. netw. of 5 triangle stations	A tape	Yes	Commissariat à l'Énergie Atomique, Montrouge	Massinon & Plantet (1975)
F. R. of Germany	Grafenberg	1975	49°41'31"N 11°12'18"E	10	9	3[e]	Triangle	D & A tape	Yes	Central Seismological Observatory, Grafenberg, Erlangen	Grafenberg (1974)
India	Gauribidanur	1965	13°36'15"N 77°26'10"E	10	10	3	Cross	A tape	Yes	Bhabha Atomic Research Centre, Trombay, Bombay	UKAEA (1967)
Iran	ILPA	1976	(50 km SW of Teheran)	40	—	7[d]	Hexagon	D		Seismic Data Analysis Center, Alexandria, Va., USA[g]	CCD/491 (1976)
Norway	NORSAR (original)	1971	60°49'25"N 10°49'56"E	110	132	22[d]	(Fig. 6.10)	D	Yes	NORSAR, Kjeller	Bungum et al. (1971b)
	NORSAR (reduced)	1976	60°49'25"N 10°49'56"E	50	49	7[d]	(Fig. 6.10)	D	Yes	NORSAR, Kjeller	A. Nilsen, personal commun. (1976)
South Korea	KSRS	1976	(90 km E of Seoul)	40 (LP)	19	7[d]	Hexagon (LP)	D		Seismic Data Analysis Center, Alexandria, Va., USA[g]	CCD/491 (1976)
Sweden	Hagfors	1969	60°08'03"N 13°41'44"E	40	15	3	(Fig. 6.10)	D & A tape	Yes	Hagfors Observatory, National Defense Research Inst, Stockholm	Ericsson (1969)

SEISMOLOGICAL STATIONS 127

TABLE 6.1. (continued)

Country	Station	Operating since	Coordinates	Total aperture (km)	No. of seismometers SP[a]	No. of seismometers LP[b]	Array configuration	Recording system[f]	Bull. prod.	Operating organization	References
United Kingdom	Eskdalemuir	1962	55° 19'59"N 3° 09'33"W	20	22	—	Cross	A tape	Yes	Procurement Executive, Ministry of Defence, nr. Reading, Berks.	Whiteway (1965) UKAEA (1967)
US	LASA (original)	1965	46° 41'19"N 106° 13 20"W	200	525	21[d]	(Fig. 6.10)	D	Yes	Seismic Data Analysis Center, Alexandria, Va., USA[g]	Green et al. (1965) Forbes et al. (1965) Wood et al. (1965) Briscoe & Fleck (1965)
	LASA (reduced)	1974	46° 41'19"N 106° 13'20"W	80	180	13[d]	(Fig. 6.10)	D	Yes	Seismic Data Analysis Center, Alexandria, Va., USA[g]	CCD/491 (1976)
	ALPA (original)	1969	64° 53'58"N 148° 00'20"W	80	—	19[d]	Dodecagon	D	No	Seismic Data Analysis Center, Alexandria, Va., USA[g]	Teledyne (1970)
	ALPA (reduced)	1976	64° 53'58"N 148° 00'20"W	40	—	7[d]	Hexagon	D		Seismic Data Analysis Center, Alexandria, Va., USA[g]	CCD/491 (1976)

NOTES. *Country* refers to the host country—a station may have been established on the initiative of and with support from another country.
Coordinates refer to the generally used reference seismometer of the array station.
Operating organization is given to facilitate contact with responsible agency to obtain data and other information.
References refer to reports mainly containing technical details of the station.

[a] Short-period seismometer   [c] Planned array station   [e] Broadband instrument   [g] Organization from which
[b] Long-period seismometer   [d] Three-component seismometer   [f] A=analog, D=digital   data are available

contribute their data to the center. The WWSSN has a limited value for monitoring a CTB, mostly because the data are recorded on photographic paper and because the sensitivity, especially of the long-period seismometers, of most stations is low. From a scientific point of view, however, it is important that at least some of the more sensitive stations be maintained, in order to provide valuable, long and unbroken recording sequences. The Seismological Research Observatories, SRO, briefly described below might be considered as the successor to the WWSSN.

6.2.3 Array stations

In 1958 the first technical discussions were held in Geneva on the problem of seismological monitoring of underground nuclear explosions. The Geneva expert panel (see Chapter 2) recommended for this purpose the use of small, so-called Geneva-type array stations, with short-period and long-period, 3-component seismometers and a 10-element array of short-period seismometers having an aperture of 3 km. In October 1960, the first array of this type was opened at Wichita Mountains in Oklahoma, USA. Further four similar array stations were established in the US in the early sixties, as part of the US Vela Uniform Project. A Long Range Seismic Measurements (LRSM) program was also established in the US to record and analyze seismic signals from US underground explosions. Some 25 mobile stations in all were set up equipped with 3-component long-period and short-period instruments recording on film and magnetic tape. Most of the LRSM stations were placed in the US, but some were placed in other countries, e.g. Bolivia, Norway, and F. R. of Germany, and these were equipped also with a short-period array of seven vertical seismometers placed in a cross with an armlength of 3 km. Data obtained by the LRSM stations were published in standardized reports (e. g. UED Earth Sciences Division, 1966).

Array stations of various sizes and equipped with different numbers of seismometers have since then been installed in various countries, mostly as part of national efforts in detection seismology. The locations of the array stations operating or planned to be in operation in 1976 are shown in Fig. 6.9. Basic data on these stations are summarized in Table 6.1, and principal station geometries are shown in Fig. 6.10.

A few comments will supplement the information in Table 6.1. The so-called "British type" array stations in Australia, Canada, India, and

SEISMOLOGICAL STATIONS

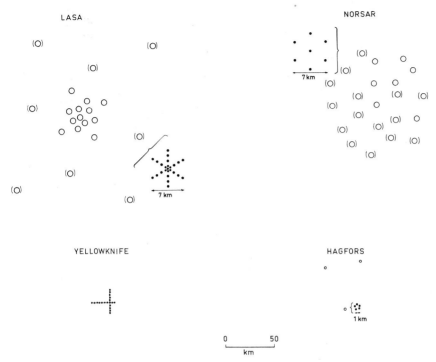

Fig. 6.10. Geometries of some array stations. The Yellowknife array is typical of the "British type" arrays. The inserts for LASA, NORSAR, and Hagfors show the geometry of the respective subarray. The LASA and NORSAR subarrays put within brackets are no longer in operation.

Scotland were originally established by the UK Atomic Energy Authority (CCD/401, 1973). Although the stations now are operated by institutes in the respective host countries, their data are still sent to the UK. This network has for many years been the only network of globally distributed, unclassified seismological array stations.

The Yellowknife array in Canada has recently been substantially modified and modernized, long-period seismometers have been added, and a new and advanced on-line digital data-collection and data-analysis system has been implemented. The array station in Brazil has been established by seismological institutes in the UK, who also provide support for its operation.

The LASA array originally consisted of 21 subarrays, each with 25 seismometers. Since then its size has been reduced by the discontinuance of recording at the two outer rings of subarrays. The present LASA consists of 13 subarrays made up of short-period instruments, each subarray also containing a three-component set of long-period instruments. The present aperture of the array is approximately 80 km.

Seven sites of the ALPA array are planned to be refitted with the newly developed long-period borehole seismometer. When this work has been completed, recording at the other sites will be discontinued, the aperture of the remaining 7-element array then amounting to 40 km.

The stations in Norway, Iran, and South Korea were built and are operated with American support. A substantial reduction of the Norwegian station, NORSAR, from 22 to seven subarrays is being planned for late 1976.

The US has established a Seismic Data Analysis Center in Alexandria, outside Washington, D. C., to carry out joint analysis of data obtained on-line from US supported array stations in various parts of the world.

France has five triangle stations with sides of 20—30 km and a few single stations spread over the country and connected by a radiocommunication system bringing all the information into the French atomic energy commission office at Bruyères-le-Châtel, in the vicinity of Paris.

The three subarrays of the Hagfors Observatory are equipped with unidirectional velocity-filtering detectors using an analog coincidence technique, see Chapter 7. A digital data-collection and data-analysis system that allows on-line computation, also of identification parameters, was installed in 1976.

The array stations briefly presented above are all of interest as fundamental parts of a future global monitoring network. Several stations will probably have to be somewhat upgraded if they are to meet the stronger operational requirements of a monitoring network. In some cases, also facilities and resources for a prompt preliminary analysis of the data have to be added.

6.2.4 Very-Long-Period Experiment Stations

In the beginning of the 1970's high-gain ultralong-period seismological stations, known as Very-Long-Period Experiment, or VLPE, stations were installed at ten places distributed around the globe (Fig. 6.9). The instruments used in VLPE stations have their peak response at a period of about 30 s, and operate with gains some 10 times as high as those of conventional long-period instruments, see Fig. 6.3. This increase in sensitivity is achieved by installing the instruments in well-sealed vaults below the surface of the earth and by applying appropriate frequency filtering. Data are recorded on both conventional photographic and digital recording equipment. Data from these stations were in 1975 collected at the National Geophysical Data Center in Boulder, Colorado, USA, from where microfilm copies of photographic records are available. These long-period stations represent a network well suited to record long-period seismic signals, especially signals from earthquakes. The detection capability of VLPE stations for long-period signals from explosions seems to be rather low, probably because explosion signals have a frequency content falling outside the range of response of the VLPE instruments. In spite of this, the VLPE stations may be able to play a role in a CTB monitoring network to produce data for the identification of earthquakes. Also VLPE data from explosions can be evaluated, using the negative-evidence identification method discussed in Chapter 10. To become useful for monitoring purposes, VLPE records must be routinely analyzed and reported, which is not done today.

Five VLPE stations will be upgraded by the installation of a vertical-component short-period seismometer and the SRO recording system described below. Such so-called Auxiliary Seismic Research Observatories (ASRO) will be installed at La Paz in Bolivia, Charters Towers in Australia, Kabul in Afghanistan, and at two further sites not yet selected (CCD/491, 1976).

6.2.5 Seismic Research Observatories

The Advanced Research Project Agency, ARPA, of the US has recently initiated a project to establish 13 so-called Seismological Research Observatories, SRO, in different parts of the world. Eleven of these stations have already been selected, viz. one in each of Australia, Colombia, India, Iran, Kenya, New Zealand, Taiwan, Thailand, Turkey, and at two sites in the US (CCD/491, 1976). Each one of these stations is equipped with a broadband borehole seismometer of a special design and filtered into short-period and long-period frequency bands. The system is designed to have a flat amplitude response from 1 s to 100 s, and a dynamic range of about 120 dB. Data are recorded both as conventional seismograms and in digital form. Short-period data are recorded in digital form first when an automatic detector at the station has detected an event. This new station network, which can be regarded as an updating of the WWSSN, is intended to have a common data center, from which data should be generally available.

These stations seem to be most valuable for data collection for seismic research and also of interest as part of a monitoring network. The weak points of the system, as far as monitoring is concerned, seem to be the automatic detector, working on only one seismometer, and the mode of intermittent short-period digital recording.

6.2.6 ARPANET

The ARPANET is a computer-communication network by which computers, primarily in different parts of the US, can be brought in contact with each other via high-speed data-transmission lines. In 1974, about 100 computers in the US were connected by this net, and there is also satellite connection to the NORSAR station in Norway and to some institutes in the UK (Kirstein, 1975; Rieber-Mohn, 1975). Although the ARPANET is a general network for intercomputer communication, one of its prime tasks is to transfer seismic data from selected seismic stations to a data-analysis center. The purpose of this Seismic Data Analysis Center is to check the quality of the data received and to produce an initial event summary, or seismic bulletin. The time series of long-period and short-period array and single-station data and the seismic bulletins will be stored on a large-capacity digital storage device, called "mass store" or "Datacomputer". Data from the SRO will be placed in the Datacomputer by the US Geological Survey Albuquerque Seismological Center with an expected delay of one or two months (CCD/491, 1976). If the ARPANET seismic data were made generally accessible, we believe that the network could become of practical value as an efficient link in a global network for monitoring a CTB.

6.2.7 Black and white boxes for test-ban monitoring

During the discussions of a seismic control system for monitoring a CTB, it has been suggested that so-called "Black Boxes" be lodged within the territories of the testing countries, primarily the US and the USSR. These "Black Boxes" are unmanned seismic stations equipped to transmit data to institutes outside the host country (Dumas, 1971). Data from "Black Boxes" in the USSR should be accessible to the US, and vice versa. Data could also be made generally available. The "Black Boxes" would have the advantage of operating close to the seismic events they are supposed to monitor, and could have a very high detection capability. The technical problem of collecting data from such stations by means of satellite communication systems is probably easily solved. The main problem is the political one: of a country establishing monitoring equipment in another country. Another problem is how to ensure that these stations will not be disturbed either by intrusion in the station equipment or by artificial seismic disturbance created outside the station. We cannot judge whether it is possible to make such installations intrusion-safe, but we are convinced that it will be possible to create ground vibrations outside the station that will prevent detection of the weak signals these stations are supposed to detect. The availability of adequate data from a "Black Box" is thus in practice dependent on the goodwill of the host country. We therefore think that data for use in test-ban monitoring could be obtained more conveniently by unclassified stations in certain countries, including the testing countries. We may call these stations, cooperating in an international data exchange, "White Boxes".

7. DETECTION

In the previous chapter it was mentioned that the seismic background noise generated by the general unrest of the earth puts an inherent limit to what can be detected by seismic instruments. The disturbing effect of seismic noise is illustrated in Fig. 7.1, where short-period records at the Hagfors Observatory from two Chinese atmospheric explosions in the Lop Nor area are compared. The two explosions were of approximately the same magnitude. The records in the upper part of the figure show clear P-wave signals, but even for an experienced seismologist it would be difficult, without prior knowledge of the event, to detect any signals at all in the lower records.

This chapter is devoted to the problem of signal detection. By way of introduction we outline the main characteristics of the seismic noise. Against this background, the most important signal-processing and detection techniques for single stations and arrays are then presented. Finally, we give estimates of the detection capabilities of seismic stations and networks of particular interest for monitoring a CTB.

7.1 SEISMIC NOISE

Over the years, seismic noise has received close attention among seismologists. Bibliographies covering the work up to 1964 contain more than one thousand papers (Gutenberg & Andrews, 1956; Hjortenberg, 1967). The nature of the seismic noise has been found to be quite complex, and even today there is not full agreement on some of the questions relating to the generating mechanisms. Here, however, the emphasis is placed on statistical properties of seismic noise pertinent to signal detection.

We start with an outline of the gross spectral features, illustrated in Fig. 7.2, which shows some representative displacement-amplitude

DETECTION 135

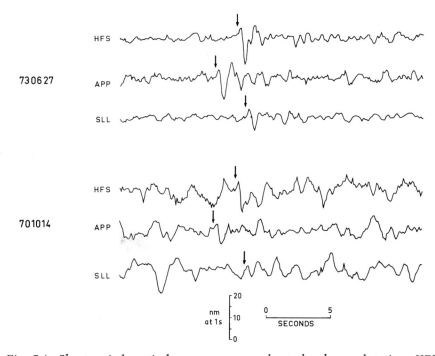

Fig. 7.1. Short-period vertical-component records at the three substations HFS, APP, and SLL of the Hagfors Observatory from two Chinese atmospheric explosions with the following source data according to USGS:

Date	Origin time	Epicenter	m_b
730627	035951	40.6N 89.5E	4.8
701014	072959	40.9N 89.4E	4.6

spectra of vertical-component seismic noise covering the frequency band 0.01–10 Hz. The general shape of the horizontal-component spectrum is essentially similar to that of the vertical-component, although the absolute level of the former usually is higher (Fix, 1972). All spectra in Fig. 7.2 have the same general shape, with a pronounced peak around 0.1–0.3 Hz, although the overall levels differ by 20–40 dB. On broad-band seismic records the background noise is usually dominated by amplitudes in this frequency band, commonly known as the microseismic band. Figure 7.2 also indicates that above 0.5 Hz the noise amplitude decreases strongly and approximately inversely with the third power of the frequency. At lower frequencies, down towards 0.01 Hz, the noise level gradually increases.

The origin and the generating mechanism of microseisms, which typically have Rayleigh-wave particle motion with periods between 3 and 10 s, are still subject to discussion (Korhonen, 1971). Micro-

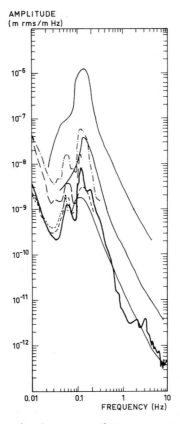

Fig. 7.2. Displacement-amplitude spectra of seismic, vertical-component earth noise, given as root-mean-square (rms) amplitudes in a 1-mHz passband. The spectra represent data from various investigations, as compiled by Fix (1972), and indicate the range of variation in shape and absolute level.

seisms are frequently associated with low-pressure areas at sea and surfing on the coastlines (Haubrich & McCamy, 1969). It has been observed that the dominating frequencies in the microseismic spectrum are very close to the frequency of the sea waves and twice that frequency (Haubrich et al., 1963). It has also been suggested that the characteristic frequencies of microseisms reflect the earth structure at the source as well as that along the travelpath and at the receiver (Savarensky et al., 1967). The amplitudes in the microseismic band are subject to large temporal variations; at high amplitude levels the microseismic spectra become broader and disturb higher and lower frequency bands.

Since the seismic noise is most dominating in the microseismic frequency band, it seems natural to use frequencies above or below

TABLE 7.1. *Summary of noise characteristics. Numerical values are only indicative, as large deviations may occur.*

Characteristic	Long-period noise		Short-period noise	
	Period of about 30–40 s	Period of about 20 s	Frequency of about 0.5–1 Hz	Frequency of about 2 Hz
Source	Surface loading by local pressure	Distant atmospheric sources	Distant and regional atmospheric sources	Cultural and local atmospheric sources
Max. distance of coherency	A few km	More than 100 km	2–10 km	<0.5 km
Propagating velocity		3–4 km/s	3–4 km/s	3–5 km/s
Temporal variation	Seasonal variation by 10 dB	Microseismic storm variation by 10–20 dB	Microseismic storm variation by 10–20 dB	Diurnal variation by 8 dB
Site variation	10 dB	20–40 dB	20–40 dB	20 dB
Depth reduction	A few dB per 100 m	Probably small	10 dB/1000 m in deep wells	A few dB for the first 10 m in consolidated bedrock

this band for signal detection. Traditionally, this is accomplished by long-period and short-period seismometers, both types having an instrumental filtering effect enhancing signals with periods above and below the microseismic band, respectively. The long-period band covers roughly the frequencies 0.01–0.1 Hz and the short-period band frequencies between 0.5 and 5 Hz. These two frequency bands are the ones most commonly used in seismic detection and discrimination; the properties of the seismic noise in these bands are therefore discussed in some detail below. As the structure and origin of short-period noise differ from those of long-period noise, they are discussed separately. The main characteristics of short-period and long-period noise are summarized in Table 7.1.

7.1.1 Long-period noise

Long-period noise covers approximately the period interval 10–100 s. The limits are not very rigorously defined, and quite often only periods around 20–40 s are considered. We will not be concerned with instrumental noise caused by buoyancy or other effects due to inadequate instrumental conditions.

It has been demonstrated that the long-period noise is composed of two major, rather different, components, one of which propagates like surface waves, whereas the other is completely disorganized (Capon, 1969 a, 1973). From studies at the LASA array it has been found that during half of the time, about 40 % of the power of long-period noise can be associated with disorganized noise (Capon, 1969 a). This means that the noise records taken by sensors separated by more than 7.5 km are virtually incoherent. The incoherent noise is well correlated with variations in the atmospheric pressure; the correlation seems to be closer with the horizontal component of the incoherent noise than with the vertical one (Ziolkowski, 1973). The incoherent noise is supposed to be of very local origin, it being caused by the elastic load of the atmosphere on the ground. This type of noise is also believed to be rather rapidly attenuated with depth, in particular the horizontal components. At a depth of 150 m a reduction by about 10 dB has been suggested (Ziolkowski, 1973). The main energy of the disorganized noise seems to lie at periods around and above 30 s (Savino et al., 1972a).

Although the disorganized noise is quite significant, the long-period noise is dominated by coherent noise propagating as Rayleigh and Love waves with velocities of 3–4 km/s over distances of more than 30 degrees (Lacoss et al., 1969). Love-wave and Rayleigh-wave noise components usually travel along the same path indicating the same generating source. The sources of the noise can often be associated with low-pressure areas at sea (Capon, 1973). The coherent, long-period noise possesses many of the properties of Rayleigh-wave signals from seismic sources. It has been found that the horizontal component of the noise generally has the largest amplitudes. The main part of the energy of the propagating noise is concentrated to periods around 20 s. The shape of the spectra in Fig. 7.2 indicates that there is a minimum at periods around 30 s. It has been suggested that this minimum of the long-period noise essentially represents a transition between the propagating and the disorganized components (Savino et al., 1972 a).

There are significant differences in temporal effect between various parts of the long-period band. Very few systematic and comprehensive studies of the time dependence have been conducted, and the temporal variation has to be estimated from case studies of occasions with extremely low noise and occasions with extremely high noise. For periods greater than 30 s, the annual fluctuation is quite small, with the amplitude level usually lying within 8 dB, maximum values occurring during

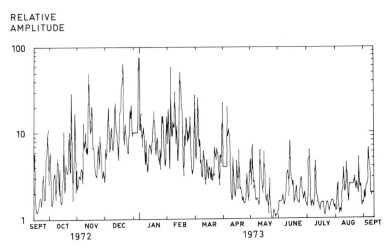

Fig. 7.3. Temporal variation during 12 months in the long-period vertical-component noise amplitude around the 20-s period at NORSAR, as given by Bungum & Tjöstheim (1976).

winter time and minimum values during summer time (Savino et al., 1972 b). At periods around 20 s, the amplitude level is also subject to a similar seasonal variation, but with a larger difference between maxima and minima, as shown by Fig. 7.3. More significant, however, is the observation that the amplitude level can change very quickly, by some 20 dB, over just a few hours. These peak values can often be associated with microseismic storms. In extreme cases, the total range of variation at periods around 20 s can be as large as 40 dB. At the Hagfors Observatory the logarithmic noise amplitudes around the 20-s period, taken over one year, are approximately normally distributed, with a mean value corresponding to an amplitude around 130 nm (Elvers, 1975). Bungum & Tjöstheim (1976) found that 20-s noise amplitudes at NORSAR fit a Rayleigh distribution better than a normal one.

The 20-s noise is partly related to regional weather disturbances, and large differences in amplitude can therefore be expected between different seismic stations. At long periods, above 30 s, the variation between different stations is less pronounced. The variation in the vertical-component noise level between stations in the VLPE network seems to stay within 10 dB (Murphy et al., 1972). At VLPE sites with gravel or similar material of about 200-m thickness above the bedrock, the horizontal components around 30 s have been found to be more stable than at sites with less overburden, where significant short-term variations have been observed (Savino et al., 1972 b).

Love-wave and Rayleigh-wave signals are usually accompanied by signal codas associated with dispersion, scattering, and multipathing of the surface waves. From a detection point of view, these coda signals, which for large earthquakes may persist for serveral hours, have to be considered as noise. This type of noise, which in fact is generated by earthquakes is further discussed in Section 7.3 in relation to the problem of interference between long-period signals from different seismic events.

7.1.2 Short-period noise

Figure 7.2 indicates that over the frequency range of 0.5–5 Hz, the noise amplitude decreases strongly and approximately inversely with the cube of the frequency. This uniform decay is one of the most typical characteristics of the spectrum of short-period noise, and is found in most spectra, regardless of their origin. The fine structure of this part of the spectra may, however, differ significantly. Some spectra may, for example, contain narrow peaks, usually above 2 Hz, which have been associated with characteristic frequencies at industrial power plants (Korhonen & Kukkonen, 1974). Because of their narrow bandwidth, such peaks usually have little influence on the detection capability.

A variety of generating sources have been proposed also for the short-period noise. Strong microseismic activity is supposed to be the major contributor in the low-frequency part around 0.5–1 Hz. Short-period noise around 0.5–1 Hz has also been associated with low-pressure areas (Dahlman, 1969; Pfluke & Murdock 1971; Bungum et al., 1971a). For the high-frequency noise various sources, like wind, local traffic, and industry, have been proposed.

The low-frequency part of the short-period noise is partly propagating with velocities corresponding to the velocity of Rayleigh-waves and local P waves (Vinnik & Pruchkina, 1964). The noise is thus to a large extent coherent at lower frequencies, but the degree of coherency decreases with increasing frequency, and so does the distance over which the waves are coherent. This effect is essentially due to a decrease in wavelength with increasing frequency, and signals with shorter wavelengths become more sensitive to variations in local structures. The coherency at frequencies around 0.5 Hz, for example, may be maintained out to distances of about 10 km; at 1 Hz the distance is reduced

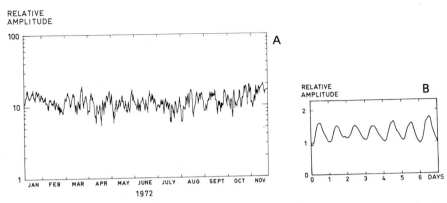

Fig. 7.4. Temporal variation in the short-period vertical-component noise amplitude in the frequency band 1.2–3.2 Hz at NORSAR, as given by Bungum & Ringdahl (1974). A—Long-term variation from January 7 to November 22, 1972. B—Variation as a function of time of the week, where the first day (0–1) is Friday. The curve represents an average over 80 weeks.

to about 1 km, and at 2 Hz it is less than 0.5 km (Capon et al., 1968; Dahlman, 1969).

The largest temporal variation in short-period noise is found at low frequencies. This variation, which is associated with microseismic storms and seasonal effects, can be as large as 20 dB (Burch, 1968). At higher frequencies the variation is less pronounced and also more regular (Bungum, 1972). This is indicated by Fig. 7.4 A, which shows the long-term variation in amplitudes in a frequency band between 1.2 and 3.2 Hz. At frequencies above about 1.5 Hz, a diurnal variation, with a maximum during noon and a minimum during night, has frequently been observed, Fig. 7.4 B. This cyclic variation, in which also a weekend effect can be seen, is supposed to be associated with cultural noise sources and just reflects the daily human activity (Bungum & Ringdahl, 1974). Only a few estimates of the statistical distribution of the noise level have been published. Mostly it is suggested that, provided corrections for the deterministic regular variations are made, a normal distribution is a valid approximation (Sax, 1968). Attempts have also been made to approximate the short-period noise by a nonstationary model (Tjöstheim, 1975a).

Sites close to the sea are generally more noisy than continental sites, and they also show a larger temporal variation in noise level. The temporal variation at the Eskdalemuir array in Scotland, for example, is about 20 dB, whereas it appears to be only about 6 dB at the Warra-

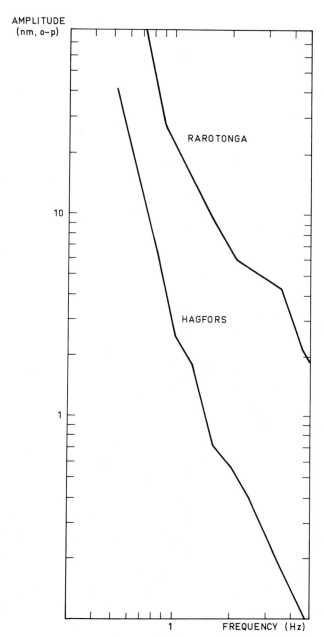

Fig. 7.5. Amplitude spectra of short-period vertical-component seismic noise recorded at Hagfors and on Rarotonga Island in the Pacific. The spectra are based on analog bandpass filtering. Both spectra were obtained in July of 1975.

munga array in Central Australia (Burch, 1968). The high noise level at stations close to the sea is clearly illustrated in Fig. 7.5, where noise spectra obtained at the Rarotonga Island in the Pacific Ocean and at the Hagfors Observatory are compared. Amplitude spectra obtained at different sites in North America show variations by almost 40 dB, the noise level for stations sited on bedrock being lower than that for stations sited on alluvium (Frantti, 1963). Extremely low and stable noise values have been reported for the Tamanarasset station in Algeria (Rocard, 1962). Comparative studies of noise on the ocean bottom and on land generally show significantly larger values at sea, although the spectra often are similar in shape (Bradner & Dodds, 1964; Bradner et al., 1965).

The reduction in noise with depth below the land surface has been studied in deep wells. It has been found that the depth dependence of noise is consistent with that of Rayleigh waves (Douze, 1967). The reduction, which can be about 10 dB for a depth of 1600 m, depends on several factors, such as noise frequency and geological structure, and is largest in unconsolidated rock. Due to interference, also the P-wave signals get smaller with increasing depth. Even in very deep wells no dramatic improvements in the signal-to-noise ratio have been observed (Douze, 1964).

7.2 DETECTION PROCESSES

In the foregoing section the main characteristics of seismic noise were outlined. The basic properties of P and Rayleigh waves were discussed in Chapter 4. The aim of the detection processes to be presented below is to explore differences between signals and noise in order to improve the detection capability. Slightly different techniques are applied depending on the type of seismic signal concerned. The first part of this section will deal with detection processes operating on a single recording channel and three-component recording at a single station. A description is then given of techniques for multichannel processing used at array stations. The treatment does not pretend to be complete, and only techniques of immediate relevance to the problems of detection seismology are discussed. For a comprehensive review of this field, the reader is referred to treatises on signal processing and statistical communication theory (Robinson, 1967; Van Trees, 1969; Beauchamp, 1973).

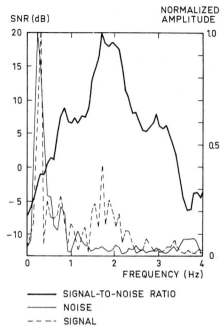

Fig. 7.6. Signal-to-noise ratio (SNR) as a function of the frequency of short-period signals at Hagfors from an earthquake in the Tibet–Indian Border region on 710130 at 201541 and with m_b(USGS) = 4.6. SNR is defined as the ratio of the amplitude in a time window covering the initial 10 s of the short-period signal to the amplitude in a 10-s noise sample preceding the signal onset.

7.2.1 Single stations

In single-channel recording signal detection is usually done visually. There is, so far, no method that computerizes "the procedure of the eye", which mainly attempts to detect changes in the amplitude level or in the frequency content of the records. Automatic detection has

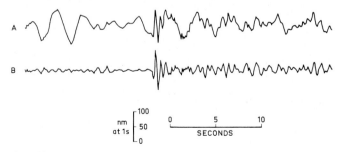

Fig. 7.7. The effect of bandpass filtering on the short-period signals at Hagfors from the earthquake in the Tibet–Indian Border region on 710130 (cf. Fig. 7.6). A–unfiltered. B–Filtered with a bandpass between 0.7 and 4 Hz.

been little used for single stations, although such a procedure is employed at the Seismic Research Observatories (SRO). The procedure is based on the ratio of the short-term average signal power to the long-term average signal power; if this ratio exceeds a preset threshold value, a detection is declared (Frasier, 1974).

Although signal detection is mostly based on visual inspection, there are automatic preprocessing methods to facilitate visual detection. These methods are based on different operational principles and are implemented by various kinds of filters. Some filters attempt to suppress the noise, while others enhance the signals. More optimal filters attempt to accomplish both effects. A few examples of such filters will be presented below.

The most commonly used filtering technique is based on information about both noise and signal. The spectral content of seismic signals usually differs from that of noise, and a comparison of the spectral components therefore suggests itself as a possible method to obtain optimum filtering. Figure 7.6 gives an example of the signal-to-noise ratio, SNR, as a function of the frequency of a short-period P-wave signal. The general shape of this curve, with a pronounced peak between 1.5 and 3.0 Hz, is quite typical of data obtained at Hagfors. A narrow bandpass filtering around the peak frequency provides an improved SNR, as is illustrated in Fig. 7.7. Bandpass filtering can also be used to separate interfering signals with different frequency content. This is illustrated by Fig. 7.8. Bandpass filtering has been found to operate very efficiently, and is also quite simple to implement for on-line processing. In digital recording systems the filter is usually designed as a recursive relation (Weichert, 1975 a); in analog recording systems ordinary electronic bandpass filters are used. Because of its simplicity and efficiency, bandpass filtering is the most widely used detection processing technique in seismology.

In cases where the waveform of a signal to be detected is known, so-called waveform filters can be utilized. This technique can be thought of as a cross-correlation between the signal to be filtered and a master waveform representing the known signal. The master signal can be either a synthetic signal or a previously recorded signal. Both kinds of master signal have been used for Rayleigh-wave detection (Capon & Greenfield, 1967; Toksöz, 1970). An example of a successful application of this method to Rayleigh-wave signals is shown in Fig. 7.9. Because of the large variability in the waveform of P signals, waveform

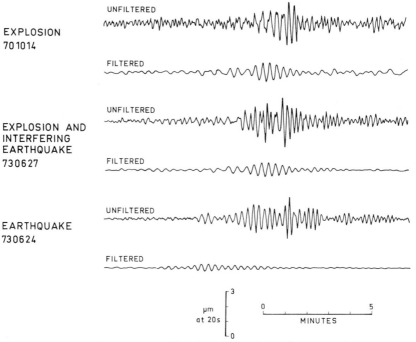

Fig. 7.8. Separation by bandpass filtering of interfering long-period signals from a Chinese atmospheric explosion and an earthquake recorded at the Hagfors Observatory. For comparison the records of undisturbed explosion and earthquake signals are also shown. The following source data were reported by the USGS:

Date	Origin time	Epicenter	Depth (km)	m_b
701014	072959	40.9N 89.4E	0	4.6
730627	035951	40.6N 89.5E	0	4.8
730627	034238	42.6N 145.8E	38	5.2
730624	200016	43.3N 146.8E	51	5.1

The bandpass filter for the band between 27 and 50 s effectively separates the explosion signal from the earthquake signal.

filtering is less suitable for these signals. It can, however, be helpful in a search for secondary *P*-wave phases, such as depth phases, where the primary *P* wave serves as master signal.

Noise-suppressing filters may be useful in cases where the character of the signals is only vaguely known. Such filters can also be valuable for estimating signal waveforms by minimizing the distorting effect of the noise. One type of noise-suppression filter predicts the noise from its past behavior and removes the predicted noise. These filters are often called prediction-error filters (Peacock & Treitel, 1969), and require to be successful a certain degree of stationarity of the noise. Short-period noise can usually be considered stationary over at

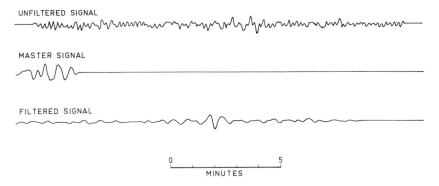

Fig. 7.9. Example of waveform filtering based on a long-period vertical-component record at NORSAR (Filson & Bungum, 1972). The master signal and the unfiltered signal are from presumed explosions in Eastern Kazakh, with the following source data according to the USGS:

	Date	Origin time	Epicenter	m_b
Master signal	710425	033258	49.8N 78.1E	5.9
Unfiltered signal	711009	060257	50.0N 77.7E	5.4

least a few hours. Even so, prediction-error filters are usually found to perform less efficiently than ordinary bandpass filters (Gjøystdahl & Husebye, 1972).

Another type of noise-suppression filter is the so-called notch filter, which essentially cuts off the frequency band of the noise peaks (Capon et al., 1968).

Differences in polarization between seismic waves and noise can also be utilized for SNR enhancement. For P waves, the particle motion coincides with the direction of propagation, as was mentioned in Chapter 4. P-wave polarization filters, which are nonlinear, and based on vertical as well as horizontal recording, have been designed by Shimshoni & Smith (1964), and by Montalbetti & Kanasewich (1970). Such filters might be useful for detection of secondary phases in short-period recording.

In conclusion, it is clear that among the filtering techniques presented above, bandpass filtering is the only one that has a widespread use for routine recording. The other types of filter have usually been applied only in special cases and on a nonroutine basis.

7.2.2 Array stations

In this section we discuss seismic arrays as a way to improve the signal-detection capability, i.e., a station consisting of several seismometers placed on the earth's surface in a predetermined pattern and with an

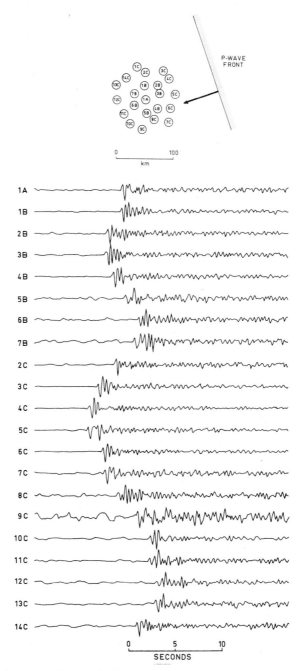

Fig. 7.10. Short-period signals from an underground nuclear explosion in the Eastern Kazakh test area on 750807 at 035700, propagating across NORSAR. The locations of the subarrays and the direction of the incoming waves are indicated.

extension usually not exceeding 200 km. The usefulness of an array is determined not only by the number of seismometers and their geometry, but also, to a great extent, by the way the data of the individual seismometers are combined. The array concept has been used for a long time in the field of geophysical exploration. It was not until the late 1950's that arrays began to be used also for detecting signals at teleseismic distances (Carpenter, 1965). The array concept is in fact quite general and is used in various applications of communication theory. A seismic array is analogous to, for example, a receiving antenna used in radar technology (cf. Chapter 6). It is therefore quite natural that some of the terminology of radar antennas has been adopted by array seismology (Robinson, 1967).

Here, we consider only horizontal arrays, i. e., arrays of seismometers placed on the surface of the earth. Three-dimensional arrays, with seismometers placed in deep holes, have been tested, but found to be of limited practical value (Broding et al., 1964).

Several techniques to combine the seismometer outputs of an array have been developed. We confine ourselves to two processing methods, called *beamforming* and *velocity filtering*, which have been of great practical value. Beamforming is so far the most widely used method and is standard technique in present medium-sized and large arrays, whereas velocity filtering has been applied mainly to smaller arrays.

Beamforming

Arrays based on beamforming can be used to carry out the following functions, which all are interrelated:

- to increase the signal-to-noise ratio for better signal detection,
- to separate interfering signals, and
- to estimate apparent velocities of seismic waves.

An example of a seismic wave front crossing a seismic array is shown in Fig. 7.10. The technique used to enhance the SNR is quite simple. After bringing the signals in phase by appropriate time delays, their straight sum is formed. If the noise is completely incoherent and the signals fully coherent across the array, delay and sum processing of N sensors in the array would theoretically provide an SNR gain of $\sqrt{N}$, as compared to a single seismometer. In practice, however, this theoretical gain is difficult to achieve, and the observed gain is usually several dB below the value of $\sqrt{N}$ (Green et al., 1965). The sum is

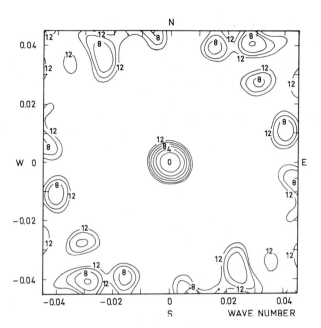

Fig. 7.11. Beam pattern for the long-period array at NORSAR (after Capon, 1971). The figures give the gain in dB below the main lobe at the origin. The relative locations of the instruments are shown in Fig. 7.10.

frequently called the *beam,* as the time delays are chosen so that the array is focused, "beamed", on some point on the earth's surface.

The capability of an array to separate two waves arriving simultaneously is determined by the number of seismometers, the array geometry, and the difference in directions of approach and wavelengths of the interfering waves (Lacoss, 1965). It is common practice in array seismology to describe a seismic wave by its horizontal *wave-number vector,* $\bar{k}$, which in turn characterizes the direction of approach and the wavelength (Kelly, 1964). For a given array configuration the separation capability can be mapped as a function of the relative wave-number vector. An example of such a mapping, which is usually called *beam pattern,* is shown in Fig. 7.11. The origin of the wave-number plane corresponds to the $\bar{k}$ value at which the array is phased. The array has its maximum gain at the origin of the beam pattern, which is called the *main lobe.* For effective detection and signal separation, the main lobe should be sharp and well separated from the secondary maxima, or *side lobes,* in the wave-number plane. Beamforming can be carried out also in the frequency-wave-number domain, $f-\bar{k}$, by estimating the spectral density of the wave field in the $f-\bar{k}$ domain from the outputs

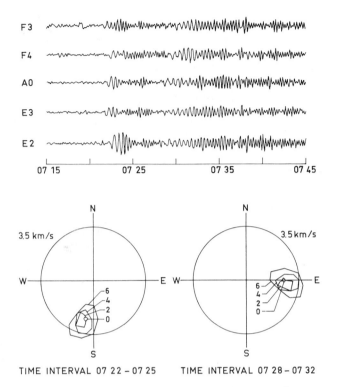

Fig. 7.12. The upper part of the figure shows interfering surface-wave signals at some subarrays of LASA (F3, F4, A0, E3, and E2) from earthquakes in the Easter Island and north of Ascension on 680625. The lower part shows wave-number spectra at 0.025 Hz. Contours of constant power are displayed in dB below the peak power at 0 dB. The circles correspond to phase velocities of 3.5 km/s. (After Capon & Evernden, 1971.)

of the individual seismometers (Capon, 1969 b; Capon & Evernden, 1971). Figure 7.12 gives an example of how interfering Rayleigh waves can be separated by a special high-resolution version of f–$\bar{k}$ processing.

For on-line detection with a seismic array, there is no prior knowledge of when and at what $\bar{k}$ values seismic waves will cross the array. In order to detect signals, one has therefore, by delaying the seismometer outputs correspondingly, to continuously tune the array towards $\bar{k}$ values of interest. Signal detection by beamforming automatically gives an estimate of the $\bar{k}$ value of the wave, which can be used to estimate the location of the epicenter of the associated seismic event. The location capability of array stations is further elaborated in Chapter 8.

The practical implementation of beamforming at the large-aperture arrays LASA and NORSAR is quite similar (Bungum et al., 1971 b;

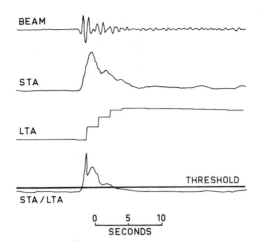

Fig. 7.13. Records of short-term average (STA), and long-term average (LTA), and the ratio STA/LTA for a beam at NORSAR of an earthquake in Tsinghai, China, on 700127 at 105900. (After Bungum et al., 1971 b.)

Dean, 1972). The analysis is carried out in digital form, and, after prefiltering, a narrow beam is steered at roughly 300 different velocities and azimuths corresponding to seismic regions and presumed explosion-test sites. A set of about 100 wider beams, covering almost the whole area of the globe from which the arrays can receive signals, is also included, to take care of unexpected seismic events in inactive regions. A detection is based on the ratio of two rectified average values of the beam, calculated on a sliding time window, as shown in Fig. 7.13. These averages are called long-term average, LTA, and short-term average, STA, corresponding to averaging times of 30 s and a few seconds, respectively. If the ratio STA/LTA in a certain time window exceeds a preset threshold, usually set to about 8 dB, the detection processor automatically flags a detection. Due to the statistical variations in the noise, a low value of the threshold will give a large number of false alarms. If, on the other hand, the threshold is set at a high level, there is a risk of weak signals passing undetected. This unavoidable trade-off between the number of false alarms and the detection capacity can be described in statistical terms by the so-called receiver-operating characteristics, a concept widely employed in statistical communication theory (Green & Swets, 1966).

The medium-aperture "UK type" array *Yellowknife* in Canada, which has recently been upgraded with a digital detection and recording system, uses a detection process slightly different from that used at large arrays (Weichert, 1975 a). The beam is formed from the logarithm of the individual sensor outputs (Weichert, 1975 b). There are a total of 121 beams at Yellowknife. The other "UK type" arrays are equipped with analog facilities, and the processing is based on a rather special technique (Whiteway, 1965). The signals from the seismometers of each leg are phased separately and the phased signals are then cross-correlated. The cross-correlation gives a sharper beam pattern and reduces the side-lobe effect, in particular for unwanted signals in phase along either of the legs.

If the signals are fully coherent and the noise is incoherent, beamforming should theoretically yield an SNR gain proportional to the square root of the number of sensors used. As was already mentioned in Chapter 4, the assumption about signal coherency does not seem to be valid at the large-aperture arrays, where large variations in both amplitude and wave form of signals across the array have been observed, cf. Fig. 7.10. This means that even if the signals are well recorded at the individual sensors, their phasing becomes difficult. The deteriorating effect of signal incoherency is illustrated by the fact that several of the outer subarrays at LASA could be removed without any significant change in the performance of the array. One way to cope with signal incoherency is to eliminate the Fourier phase components and sum just the spectral amplitude components of the signals (Lacoss & Kuster, 1970). In the time domain the phase elimination can be easily done by just squaring the signal amplitudes or taking their rectified values and then applying delay and summing on these modified traces (Ringdahl et al., 1971). Such modified beamforming procedures are usually called *incoherent beamforming*. At NORSAR the detection processor is supplemented with about 100 such "incoherent" beams.

Besides the techniques described above, which are mainly used for on-line purposes, other processing techniques based on more refined and detailed models of the noise and signal have been tested. In one model it is assumed that the signals are identical at the sensors, except for an amplitude factor, and the signals can then be estimated by least-squares techniques (Christofferson & Husebye, 1974). An even more detailed model assumes that the signal at each sensor can be expanded in an orthogonal set of functions the coefficients of which may differ from signal to signal (Christofferson & Jansson, 1969). So far, only subtle improvements by a few dB have been obtained by these more sophisticated techniques, as compared to standard beamforming and

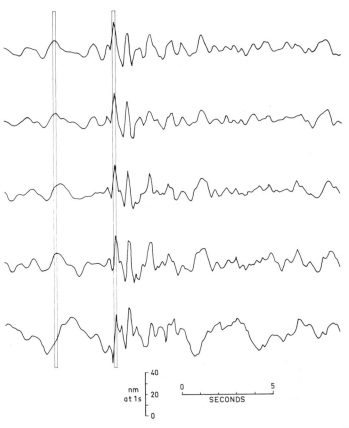

Fig. 7.14. Illustration of the difference in apparent velocity between teleseismic P signals and propagating noise. The two thin rectangles correspond to time windows of 125 ms. The P signals arrive within this window, whereas the coherent noise peaks span a larger time window. The records were taken by the velocity-filtering detector at Hagfors and come from an earthquake in the Kuril Islands on 701207 at 232754; magnitude m_b (USGS) = 4.8.

velocity filtering. In on-line processing, these detailed models have so far been of little practical use.

Velocity filtering

With close spacing of the seismometers, the P signals would be coherent also at higher frequencies. A close spacing of the seismometers would, however, mean that also the short-period noise would become coherent between the seismometers, and beamforming would no longer provide a gain proportional to the square root of the number of sensors. Teleseismic P waves and coherent noise do, however, have quite different

DETECTION

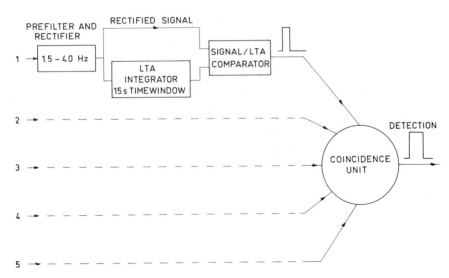

Fig. 7.15. Block diagram illustrating the detection logic of the velocity-filtering detector at Hagfors. At each of the five sensors of the array a rectified long-term average (LTA) is formed over a time interval of 15 s. The LTA is compared with the rectified signal, and when the signal-to-LTA ratio exceeds a preset threshold a pulse is emitted from the channel. The bandpass filters, the LTA integrator, and the signal/LTA comparator are identical in all five channels of the array. If all channels have emitted pulses, and if the time difference is less than 125 ms, corresponding to an apparent velocity of at least 8 km/s, a detection is defined by the velocity filter by the emission of a pulse from the coincidence unit.

propagating velocities. Teleseismic P waves travel across an array at speeds greater than 10 km/s, whereas coherent noise has a velocity of about 3 km/s. The utilization of this difference in array processing is called velocity filtering, the principle of which is illustrated in Fig. 7.14.

Velocity filtering is used at the Hagfors Observatory. The aperture of the velocity filtering array is only 1 km. The signal-detection process, which has been implemented by analog electronic equipment including bandpass prefilters, is summarized in the block diagram, Fig. 7.15 (Dahlman et al., 1971).

There is a difference in principle between signal detection by the velocity-filtering procedure used at Hagfors and that obtained by the beamforming procedure used elsewhere. A detection defined by the Hagfors array is based on a combination of detections at individual sensors, whereas a detection defined by a beamforming procedure is based on a single detection of a combination of the records from all the sensors in the array. In short, the Hagfors velocity filter combines decisions, and the beamforming arrays combine signals. Velocity filtering allows a very low threshold to be set for the signal/LTA ratio at each channel. The number of false alarms or noise triggers at the individual channels at Hagfors is roughly 10 000 per day. The number of noise triggers by the velocity filter as a whole is, however, only one or two per day. The capability of the automatic detector is comparable to that of an experienced seismologist basing his visual analysis on records which are bandpass-filtered through an optimal band. A velocity-filtering detector can sometimes detect signals which might be difficult to discern by visual inspection. For example, signals from both of the explosions referred to in Fig. 7.1 were automatically detected by the velocity-filtering detector at Hagfors.

In this chapter we have been discussing automatic techniques for signal processing and detection, techniques which in some cases are quite sophisticated. The automatic techniques have certainly been of great practical value by improving the seismological detection capability. Above all, they serve as valuable tools to be used by the seismologist in his analysis. It is, however, important to recognize that an experienced seismologist has a fundamental role to play when it comes to accepting or rejecting a signal detection. This is illustrated by the fact that analysts have the final say whether detections should be accepted or not also at the large and highly automated seismic arrays.

7.3 STATION DETECTION CAPABILITIES

In any detection process there is a trade-off between the number of detected signals and the number of false alarms. For conventional seismological stations the number of false alarms is generally considered to be so small that it can be ignored. However, due to the lack of an adequate population of reference events, it is very difficult to get adequate estimates of the detection and false-alarm probabilities of the most sensitive stations. This is because the reference data are dependent on observations of weak events by the stations whose detection capabilities are to be estimated. During the International Seismic Month, ISM, February 20 – March 19, 1972, there was compiled the most complete set of worldwide seismological data ever collected (Lacoss et al., 1974). This data set illustrates the number of associated and unassociated arrival times reported from some of the most sensitive stations. During the ISM, LASA reported 3881 and NORSAR 2297 arrival times, and of these were only 1358 (or 35 %) and 582 (or 25 %), respectively, associated with defined events. Corresponding figures for

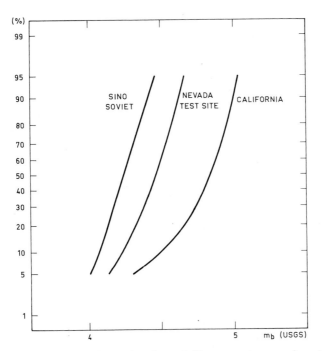

Fig. 7.16. Detection probabilities for three different regions as functions of m_b (USGS) for a single, short-period sensor at Hagfors. The curves were obtained by the following procedure. A bandpass providing maximum SNR was selected, and a linear relation between the logarithm of recorded amplitudes in the bandpass and m_b(USGS) was estimated. The statistical properties of the background noise of the bandpass and the linear relation were then used to calculate the detection probability.

the two medium-size arrays Yellowknife and Hagfors were 634 and 502 reported arrival times, respectively, of which 80 % and 71 %, respectively, could be associated with events. We do not claim that these figures reflect false-alarm rates, but rather that they show that a large number of reported arrival times cannot be associated with defined events, even when great efforts are made at defining events.

The problem of estimating the detection capability is less severe for those low-sensitivity stations which are located near a highly sensitive station or station network that can produce an adequate population of reference events. The problem is also less pronounced for long-period stations, as the detection capability of such stations generally is lower than that of short-period stations. The events detected and defined from short-period data can thus be used in estimating the detection thresholds of long-period stations.

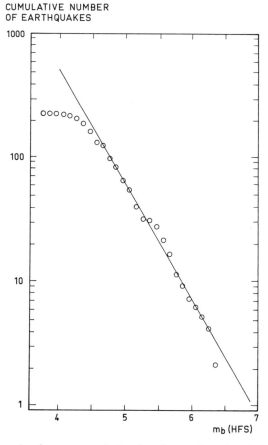

Fig. 7.17. Magnitude–frequency relationship for earthquakes defined by USGS in the USSR, detected at Hagfors. A linear relation is also indicated in the diagram. The deviation from linearity at lower magnitudes is an effect of reduced detection capability at Hagfors. The deviation could also partly be an effect of limited capability of USGS to define events in this area.

Two methods have been used more frequently than others to estimate the detection capability of seismic stations. The stringent way of defining the detection capability of a station is in terms of ground motion at the actual station. If we consider the noise amplitudes in different frequency bands and the SNR needed in the detection process, we can estimate the detection probabilities for different ground amplitudes at different frequencies (CCD/388, 1972). In converting these ground-motion values into threshold values for teleseismic signal detection, one also has to take into account the spectral content of the signals. Signals coming from events in different regions can have significantly different spectral contents at one and the same station. This means that it is a rather complicated process to convert ground-motion thresholds to, e. g., magnitude-detection thresholds for events in different parts of the world. Such station-specific magnitude thresholds might also be quite different for events in different regions, as is shown by Fig. 7.16.

The other method, which is used more extensively than the first-mentioned one, is based on the magnitude distribution of detected earthquakes. In this method a linear relationship between the logarithm of the cumulative number of events and the event magnitude is assumed, and the detection threshold is defined in relation to that line, as is shown in Fig. 7.17. Although in practice this method probably gives reasonable results, at least two objections to it can be raised. One is that there is no generally accepted physical explanation of such a linear relationship (Wyss, 1973). The other is that the estimated magnitude of an event is critically dependent on which stations are used in the magnitude computation. There might therefore be systematic differences in the magnitude scale between large events recorded by many stations and small events recorded by few stations only.

Thresholds for magnitude detection are specified at a certain probability, and are defined either as cumulative or incremental. For example, at the 90 % incremental threshold, signals from 90 % of the events with magnitudes equal to the threshold value are detected, whereas at the 90 % cumulative threshold signals are detected from 90 % of the events with magnitudes greater than the threshold value. A cumulative threshold depends on the magnitude distribution from which it is estimated and is therefore of less general interest than an incremental threshold. The cumulative threshold is always lower than the incremental; the difference depends on the actual slope of the recurrence line and is typically around 0.2–0.3 magnitude units.

The detection threshold depends on the background noise, which is discussed in Section 7.2. Due to the small time variation of the short-period noise, the detection capability of a short-period station is fairly constant throughout the year. The relatively larger variation of the long-period noise, on the other hand, causes a variation of about one magnitude in the detection capability of a long-period station. For long-period stations the problem of interfering signals is also important, as long-period signals may last for many hours, and in extreme cases for days. On the average, interfering signals have been observed

TABLE 7.2. *Detection threshold (90% incremental) for some short-period (SP) and long-period (LP) array stations.*

Station	Event region[a]	Thresholds m_b(USGS)	M_s	No. of events	Time period	Reference
ALPA	Kuril–Kamchatka		2.6			CCD/388 (1972)
Hagfors	US	4.9	4.0			Dahlman et al. (1971)
	USSR	4.2	3.5			
LASA	$\Delta=30-85°$	3.8		5600	1973	Dean (1975)
	$\Delta=30-85°$	4.0		4884	1971	CCD/388 (1972)
	Kuriles	4.2		42	1966	Sheppard (1968)
	Kuril–Kamchatka		3.1			CCD/388 (1972)
NORSAR	Iran–W. Russia	4.2		386	April '72–March '73	Bungum & Husebye (1974)
	Central Asia	4.0		524	April '72–March '73	Bungum & Husebye (1974)
	Japan–Kamchatka	4.1		841	April '72–March '73	Bungum & Husebye (1974)
	North America	4.7		114	Feb.–June 1972	Berteussen & Husebye (1972)
	Aleutian Islands	4.3		131	Feb.–June 1972	Berteussen & Husebye (1972)
	$\Delta=30-90°$	4.2		4335	April '72–March '73	Bungum & Husebye (1974)
	Kuril–Kamchatka		3.0			CCD/388 (1972)
Yellowknife	Asia	4.4		75	Jan.–April 1968	Anglin (1971)
	North America	4.2		155	Jan.–April 1968	Anglin (1971)
	Aleutian Islands	4.6		47	Jan.–April 1968	Anglin (1971)
	Kuril–Kamchatka	4.5		82	Jan.–April 1968	Anglin (1971)
	$\Delta=30-90°$	4.5		1126	Jan.–April 1968	Anglin (1971)

[a] Δ denotes epicentral distance.

for about 15 % of the events recorded by the VLPE stations (CCD/388, 1972).

Detection thresholds have been published for a limited number of stations only. Also for these stations the detection thresholds have been estimated in different ways, by using, e. g., different reference magnitudes, cumulative or incremental thresholds, etc. In Table 7.2 we have summarized available data on detection thresholds of various stations for some regions of interest in the CTB discussion. In an attempt to get a table that will allow some direct comparisons, data have been standardized. All short-period threshold values given in the table are expressed as 90 % incremental body-wave magnitudes, m_b (USGS). Threshold values published as cumulative have been converted to incremental by the addition of 0.2 magnitude units, and those given in the magnitude scale of some local station have been converted to m_b (USGS) by means of formulas usually given in the respective paper referred to. The long-period detection thresholds are defined for earthquakes only and are expressed as M_s, as defined in Chapter 4. The various calculations carried out to standardize the data might have introduced errors of one or two tenths of a magnitude, but this will not affect the overall picture.

NORSAR and Hagfors show strong regional variations in their detection capabilities. Both stations have a much higher detection capability for events in Asia and in the Japan–Kamchatka region than for events in North America. This depends on the significantly different characteristics of signals observed at Scandinavian stations from events in these two regions. Signals observed in Scandinavia from events in North America contain relatively less energy at high frequencies than do signals from events in the mainland of Asia and in the Japan–Kamchatka region; this is illustrated in Fig. 7.18.

LASA and NORSAR seem to have roughly the same sensivity of about m_b (USGS) = 4 for most areas, with the above mentioned exception for NORSAR. The long-period array stations ALPA, LASA, and NORSAR also seem to have roughly the same detection capability M_s of ca. 2.5 at 30 degrees distance (CCD/388, 1972).

The data presented in Table 7.2 have been used to estimate detection thresholds of the individual stations for explosions in the US and the USSR. These estimates are given in Table 7.3 as 90 % incremental thresholds, both in magnitudes and in hard-rock explosion yields. The yields, Y, were obtained from M_s values using the formula $M_s = {}^{10}\log Y$

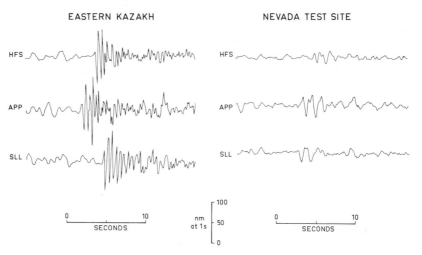

Fig. 7.18. Short-period vertical-component signals at Hagfors, demonstrating the regional variation in frequency content and amplitude. The signals originate from explosions at the Eastern Kazakh test area on 710525 at 040258 and at NTS on 710624 at 140000. In both cases the body-wave magnitude m_b (USGS) = 5.2, and the estimated yields are about 10 and 40 kt, respectively.

+2.0, and from m_b values using the relation m_b (USGS) = $0.9^{10}\log Y$ + 3.8 for US explosions, and m_b (USGS) = $0.9^{10}\log Y$ + 4.2 for USSR explosions. These yield relations are discussed further in Chapter 11. Table 7.3 shows that the estimated short-period detection capabilities are generally higher for explosions in the USSR than for explosions in the US. This difference is less pronounced for the long-period stations.

TABLE 7.3. *Detection thresholds (90% incremental) for hard-rock explosions in the US and the USSR for some short-period (SP) and long-period (LP) array stations.*

	US				USSR			
	SP threshold		LP threshold		SP threshold		LP threshold	
Station	m_b (USGS)	Yield (kt)	M_s	Yield (kt)	m_b (USGS)	Yield (kt)	M_s	Yield (kt)
ALPA			2.5	3			3.0	10
Hagfors	4.9	20	4.0	100	4.2	1	3.5	30
LASA					4.0	0.5	3.5	30
NOR-SAR	4.7	10	3.5	30	4.0	0.5	3.0	10
Yellow-knife	4.2	3			4.4	2		

TABLE 7.4. *Detection thresholds (90% incremental) for the 32-station ISM network.*

Station code	Station	m_b (ISM)	Station code	Station	m_b (ISM)
AFI	Afiamalu, W. Samoa	5.3	KBL	Kabul, Afghanistan	4.7
ASP	Alice Springs, Australia	4.8	KIC	Kosan Boka, Ivory Coast	4.8
BAG	Baguio City, Philippines	5.3	LAO	Lasa, Montana, USA	4.3
			MAT	Matsushiro, Japan	4.9
BDF	Brasilia, Brazil	4.8	MBC	Mould Bay, Canada	4.6
BLC	Baker Lake, Canada	4.8	NAO	Norsar, Norway	4.5
BNG	Bangui, C. Afr. Rep.	4.7	NUR	Nurmijärvi, Finland	4.8
BUL	Bulawayo, Rhodesia	4.9	PMG	Port Moresby, Papua	5.1
CHG	Chiengmai, Thailand	4.9	PNS	Penas, Bolivia	4.6
CLL	Collmberg, Germ. D.R.	5.0	QUE	Quetta, Pakistan	5.0
COL	College Outpost, Alaska	4.9	SHI	Shiraz, Iran	5.1
			SHL	Shillong, India	5.3
CTA	Charters Towers, Australia	5.0	SPA	South Pole, Antarctica	5.0
			SSF	Saint Saulge, France	4.9
EZN	Ezine, Turkey	5.2	TUC	Tucson, Arizona, USA	4.9
FBC	Frobisher Bay, Canada	4.9	UBO	Unita Basin, Utah, USA	4.4
GBA	Gauribidanur, India	5.0			
HFS	Hagfors, Sweden	4.6	YKA	Yellowknife, Canada	4.4

The 90% incremental short-period detection thresholds have been estimated for the 32 stations used in the ISM study and are listed in Table 7.4. These data are based on events located less than 90 degrees from the station concerned. It should be noted that the detection thresholds thus obtained are based on data for only one month, and are thus strongly dependent on the seismicity that occurred during that short time interval. The values for Hagfors Observatory, LASA, NORSAR, and Yellowknife for the ISM month do agree, however, quite well with those estimated from the larger data bases used in Table 7.2. About 80% of the stations in Table 7.4 have a threshold at or below $m_b = 5.0$, and the highest threshold value is $m_b = 5.3$. Among stations with high capabilities, we can mention, besides the four above arrays, the Unita Basin array and the single stations Mould Bay and Penas, with thresholds of $m_b = 4.4, 4.5,$ and 4.6, respectively.

7.4 NETWORK CAPABILITIES

A seismic station network may consist of both single stations and arrays. The detection capability of a network of short-period stations strictly means the capability to define seismic events from detected signals. A P-signal detection by an array the detection procedure of which is based on beamforming automatically defines an event, but usually the associated epicenter is not considered to be confident enough for weak events. Here it will be assumed that signal detection by at least four stations is needed to define an event. It can be noted that the degree of confidence of signal detection at the individual stations and the location of the stations relative to the event determine the confidence with which an event is defined. For the purpose of our simplified discussion, these effects are, however, neglected.

In detection seismology Rayleigh waves are used primarily to identify seismic events, as is discussed in Chapter 10. Theoretically, a detection by one station only would be sufficient. In practice, however, detection by several stations is desirable, mainly because a single Rayleigh-wave detection may be associated with more than one seismic event. The problems of associating a Rayleigh-wave detection with a located event have been pointed out by Filson (1974). The Rayleigh-wave detection capability of a network is here defined as the capability to detect at at least four stations Rayleigh-wave signals that can be adequately associated with a seismic event.

Below we distinguish between network detection capabilities that have been theoretically estimated and those that are observed. A theoretical estimate is an attempt to predict the capability of a network from estimates of the detection probabilities of the individual stations. An observed capability is obtained from a conducted, full-scale experiment. The performance of a network is determined not only by the capabilities of the individual stations but also by the efficiency of the analysis of the integrated data.

7.4.1 Short-period signals

Several studies of the theoretical detection capabilities of networks have been performed. Here we discuss the results of Basham & Whitham (1970), which are quite representative. Basham & Whitham analyzed data obtained in response to a resolution by the UN General Assembly (UN, 1969) requesting that the member nations state to what extent they would make available for international exchange seismic data obtained within their territories. There was a return from 75 member nations, 45 of which supplied information on the characteristics of about 300 stations and arrays from which they would be prepared

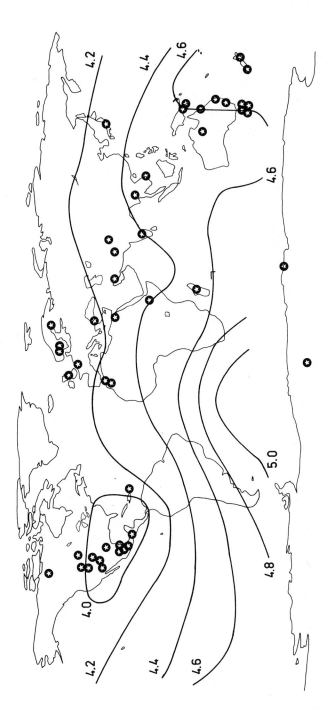

Fig. 7.19. Short-period detection thresholds for a worldwide network consisting of 46 stations, as defined by Basham & Whitham (1970). The thresholds are given as the body-wave magnitude, m_b, for which short-period signals of earthquakes and explosions would be detected with 90% probability by at least four stations in the network.

to supply data on an international exchange basis. Basham & Whitham found that the number of stations could be reduced to a network of 46 well distributed stations without significantly reducing the detection capability. From estimates of the incremental detection capabilities of the individual stations, they calculated for each point of the earth's surface the magnitude threshold for which there was a certain probability of P-wave detection by at least a certain number of stations. By calculating these magnitudes for an appropriately selected set of points on the earth's surface, the global contours of the magnitude threshold could be outlined for the entire earth.

Figure 7.19 delineates the magnitude threshold at which P waves from a shallow earthquake would be detected with 90% probability by at least four stations in the 46-station network. Some interesting features can be noted. There seems to be a systematic difference between the northern and southern hemispheres by about half a unit of magnitude, threshold values being lower in the northern hemisphere. The magnitude threshold seems to be at its minimum of 4.0 for North America and is slightly larger, about 4.2, for Eurasia. The highest threshold value, about 5.0, is obtained in the southernmost part of the Atlantic Ocean and Antarctica. This variation in magnitude threshold reflects the geographical distribution and the sensitivity of the seismic stations of the network. If the number of stations required to define an event is doubled, which leads to more confident event definition, the detection threshold becomes roughly half a magnitude larger. Finally it can be mentioned that other theoretical studies on network detection capabilities have yielded results comparable to those quoted above, also for networks consisting of somewhat different stations. In a study by Romney (1971) of a network of 9 arrays and 15 single stations it was found that this network would provide detection capabilities similar to those of the above mentioned 46-station network. There are some simplifications applied in this type of theoretical procedure for assessing network capabilities. First, the regional variability in the detection capability of the individual stations is not taken into account. Secondly, the criterion of detections by at least four or eight stations to define a seismic event is rather arbitrary.

The other way of estimating network capabilities is by conducting full-scale experiments. The ISM experiment (Lacoss et al., 1974) mentioned in Section 7.3 is an example of this. It is interesting to note that a reduction of the number of stations from almost 200 to 32 in

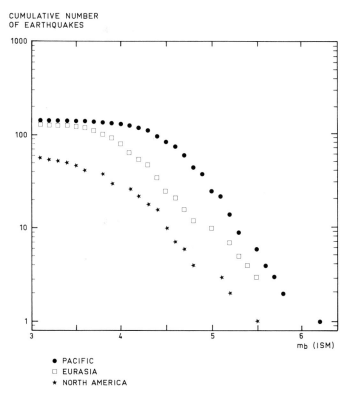

Fig. 7.20. Cumulative distributions of the body-wave magnitude, m_b(ISM), defined during the International Seismic Month for earthquakes in Eurasia, North America, and the Pacific.

the ISM experiment did not significantly change the results. There is a significant difference in the way of defining events between the ISM experiment and the earlier mentioned theoretical evaluation by Basham & Whitham (1970). In the ISM experiment advantage was taken of the capabilities of the seismic arrays in defining seismic events. Figure 7.20 illustrates the results obtained for the Eurasian continent. The magnitude—frequency curve defines a threshold between m_b = 4.0 and 4.3, in agreement with the detection thresholds estimated by Basham & Whitham (1970) and Romney (1971). Similar values have also been obtained from independent data sets of events in the USSR (Dahlman et al., 1974; Austegard, 1974). Two other regions, the North American continent and the Pacific, have been searched for events in the ISM list. The cumulative magnitude curves for these regions are compared in Fig. 7.20 with the Eurasian data. These curves give some indication of what can be expected in terms of detection thresholds. The linear

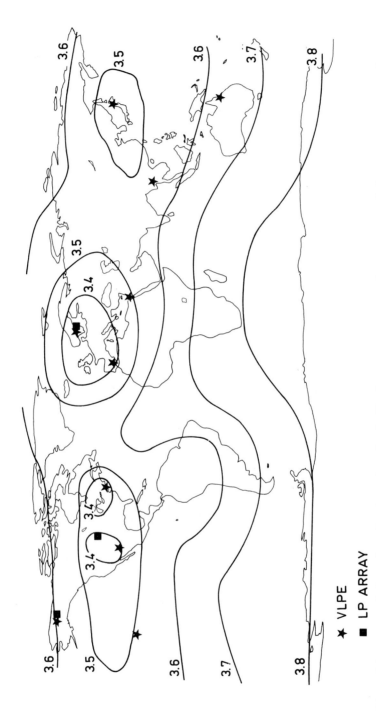

Fig. 7.21. Long-period earthquake detection thresholds for a seismic network consisting of the ALPA, LASA, and NORSAR arrays and ten VLPE stations, as defined by ARPA (1973). The thresholds are given as the surface-wave magnitude, M_s, for which long-period signals of shallow earthquakes would be detected with 90% probability by at least four stations in the network.

trends taper off in the magnitude intervals 4.0–4.5 and 4.5–5.0 for the North American and Pacific data, respectively. These intervals again agree quite well with the theoretical values shown in the contour plot in Fig. 7.19.

7.4.2 Long-period signals

In the discussion of networks for detection of Rayleigh waves, only stations and arrays equipped with high-gain instrumentation are considered. The theoretical capability of a network consisting of the large-aperture arrays ALPA, LASA, and NORSAR together with a 10-station VLPE network is shown in Fig. 7.21 (ARPA, 1973). The calculations behind this presentation are based on average noise values typical of the winter season in the 17–23 s band and on the assumption of an SNR for detection of at least 4 dB at each station. The contours in Fig. 7.21 correspond to at least four stations detecting signals with 90 % probability. As for the P waves, there is a systematic, hemispherical difference, with a threshold as low as $M_s = 3.4$, in some parts of the northern hemisphere. The threshold is going up to about 3.8 in the southernmost part of the southern hemisphere. This difference between the northern and southern hemispheres is similar to the above mentioned difference in the capability for short-period signals, for which the m_b threshold ranges from 4.0 to 5.0. This long-period network consists of only 13 stations, as compared to 46 and 32 stations, respectively, in the two short-period networks discussed above. An increased number of long-period stations would probably make the Rayleigh-wave detection capability less sensitive also to signal interference and large noise-level variations, effects not considered in the calculation of the contours in Fig. 7.21.

Data from the ISM experiment have also been used to estimate the long-period detection capability (Filson, 1974). The network studied by Filson consisted of large and medium-size arrays, the VLPE stations, the Canadian Seismograph Network, and a number of selected standard stations of the WWSSN. Data were used from a total of 46 stations, which were rather uniformly distributed in North America, Europe, and Southern Asia, but with only a few stations in the southern hemisphere. To regard surface waves as detected, it was required that at least two long-period detections be associated with an event. The 90 % incremental thresholds were estimated to be about $m_b = 4.8$ for shallow events in Asia, and about $m_b = 5.5$ worldwide. For earthquakes these body-wave magnitudes correspond roughly to $M_s = 3.7$ and $M_s = 4.3$, respectively. The observed threshold values are somewhat higher than those based on the theoretical calculations mentioned above. It was pointed out earlier that the Rayleigh-wave detection capability is sensitive to signal interference and rapidly changing noise levels. Estimates based on data of only one month may be biased by these factors.

8. EVENT DEFINITION AND LOCATION

Event definition is the process by which a seismic event is born from seismic records. The definition of strong events recorded at a large number of stations is fairly trivial, whereas the definition of events from weak signals recorded at a few stations might be a delicate task. Event definition is usually done implicitly in the event location, but is a fundamental process of its own specifying the events for further considerations.

By the *location* of a seismic event we mean the estimation of its latitude, longitude, depth, and origin time. Throughout the years the location of seismic events has been an important task for seismologists, both to map seismic areas and to estimate with high accuracy the locations of individual earthquakes. From a geophysical point of view, accurate event location is important for the detailed study of the seismicity within special regions, e. g. for geophysical interpretation and earthquake prediction. Nuclear explosions with known positions have given seismologists an opportunity to test the accuracy of epicenter location and to revise travel-time curves. The travel-time curves used for event location (Jeffreys & Bullen, 1967; Herrin et al., 1968) and their accuracy are discussed in Chapter 4.

In the context of seismic monitoring of a CTB an interesting question might be to see whether an event is located within the borders of a certain country. More accurate event location might be needed to tell whether an event is within a certain test area, as is discussed in the Threshold Test Ban Treaty, or to pinpoint a certain area for further studies, e. g. by satellite data.

In this chapter we will briefly present ways to estimate event locations and discuss the achievable accuracy and the applicability of the different methods. We have found the problem of depth estimation to be so important both for test-ban monitoring and for general geophysical considerations that a special chapter (Chapter 9) is devoted to this subject. We have not found that the accurate estimation of the origin time of an event is of any particular interest for the applications considered in this book, but rather that the time accuracy of a few seconds routinely achieved in epicenter location is satisfactory. The estimation of origin time is therefore not discussed.

8.1 LOCATION BY AN ARRAY STATION

If from one point on the earth we want to locate a seismic event taking place at another point, we have to know the distance and the azimuth from that point to the place of the event. The distance can be estimated either from the time differences between the arrival of different waves, e.g. P, S, and Rayleigh waves at our observation point, or from the apparent velocity of the P waves. The apparent velocity, which is the inverse of the derivative of the travel-time curve, $dT/d\Delta$, can be converted into epicentral distance, as is discussed in Chapter 4. In theory the accuracy of this epicentral-distance estimate depends on the slope of the $dT/d\Delta$–distance curve, which together with its derivative is shown in Fig. 4.11, but in practice local travel-time variations play a more significant role. The azimuth to an event can be estimated either from the direction of arrival of the incoming waves or from the amplitude ratio of the horizontal components. The accuracy of this last method is low, and the method has not gained much practical use.

To estimate the apparent velocity and the arrival direction of an incoming wave, at least three seismometers are needed arranged in a two-dimensional and not too narrow pattern. Seismic-array stations, where several seismometers are distributed over a limited area, provide the necessary data for estimating both the apparent velocity and the azimuth of the incoming waves, primarily the P waves. The location accuracy depends on the distances between the sensors and on the accuracy in estimating the arrival-time differences between the seismometers. If the latter accuracy is constant, the location accuracy will increase with the distance between the seismometers. The signal coherency does, however, decrease with increasing seismometer spacing, which in turn decreases the accuracy of the time estimates. The rate of decrease in coherency depends on local geophysical conditions and on the signal frequency, as is further discussed in Chapter 4. In

EVENT DEFINITION AND LOCATION 173

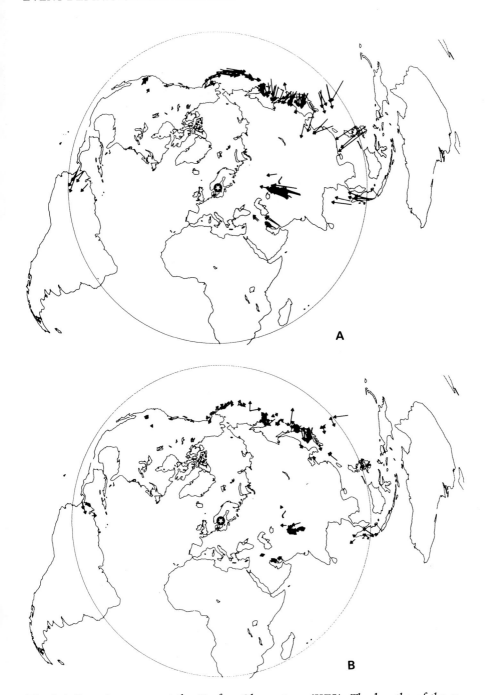

Fig. 8.1. Location errors at the Hagfors Observatory (HFS). The lengths of the arrows show the difference between USGS (arrow butt) and HFS (arrowhead) epicenter estimates. A—Uncorrected. B—Corrected for systematic bias. Radius of circle 10 000 km.

practice, one has to compromise between large seismometer spacing and high signal coherency, where also the method of analysis employed must be considered. Beamforming seems to require higher signal coherency and thus smaller seismometer spacing than does visual estimation of time differences. Most of the array stations have apertures between 20 and 100 km, see Chapter 6.

Several studies of the location capability of seismic-array stations have been carried out which show that there is always systematic bias in array-station estimates of azimuth and distance, giving systematic mislocation of events. This is illustrated in Fig. 8.1A, where the differences between US Geological Survey (USGS) epicenter estimates and uncorrected locations made at Hagfors Observatory are shown. The average difference between the USGS and the array locations is about 500 km, and differences up to 1000 km can be observed. These figures are also typical of uncorrected location differences observed at other array stations (Noponen, 1974; Weichert, 1975 a). The differences between observed and theoretical $dT/d\Delta$ values at the three Scandinavian array stations, Hagfors, Helsinki, and NORSAR, show rather similar variations with event location (Noponen, 1974). This illustrates that location bias is caused not only by local geological heterogeneities below the array station but also by more large-scale effects in the upper mantle.

The systematic part of the array-mislocation vectors can be corrected by time corrections estimated from calibration events well located by data from other seismological stations. The HFS locations corrected for the systematic bias are shown in Fig. 8.1B. The remaining errors are random and cannot be eliminated by calibration events.

In Table 8.1 we have summarized reported estimates of the location accuracy for four array stations, Hagfors (HFS), LASA, NORSAR, and Yellowknife (YKA) for seismic events in some regions of interest in the CTB discussion. All differences are calculated relative to the USGS locations, and corrections for systematic bias are applied. There is a substantial variation both between the array stations and between the different source regions. The values given in Table 8.1 do not, however, represent extreme values; NORSAR, for example, has a 90 % location error of more than 800 km for events in Central America and on the Mid-Atlantic ridge (Bungum & Husebye, 1974). The error in azimuth is usually somewhat less than the error in distance. Theoretical considerations by Shlien & Toksöz (1973) suggest that the errors

TABLE 8.1. *Location errors of array stations.*

Station	Source region[a]	Location difference (90% of the events) (km)	Distance difference S.D. (km)	Azimuth difference S.D. (degrees)	Reference
HFS	Δ=30–90°	620	290	1.9	
	Central Asia	430	220	1.2	
LASA	Eastern Kazakh	350	250	1.2	Estimated from LASA Bulletin
NORSAR	Central Asia	270	160	0.7	Bungum & Huseby (1974)
	W. North America	310	100	1.2	
	Δ=30–90°	490	290	1.5	
YKA	Δ≈65°		200	0.7	Weichert (1975a)

[a] Δ denotes epicentral distance.

in estimating distances should indeed be larger than those in azimuths. Attempts have been made at the Hagfors Observatory to reduce the distance errors in event locations by using Rayleigh-wave data. Using regionalized, well-calibrated Rayleigh-wave dispersion curves, the distances can be estimated with an accuracy of the order of 100 km. Von Seggern (1972) showed that, if adequate calibration data are available, NTS explosions can be located with an accuracy of 10–30 km using surface-wave data from 5 to 10 stations.

We can conclude that an array station can locate a seismic event with an expected accuracy of a few hundred kilometers. The exact figure may differ somewhat between arrays and between source regions. This accuracy is low compared to what may be achieved using a network of globally distributed stations. A single array, however, represents an independent national capacity to detect, define and locate seismic events rapidly and with an accuracy that in most cases is adequate to decide whether the event deserves further attention or not.

8.2 LOCATION BY A NETWORK OF STATIONS

The most common way to locate seismic events is to use P-wave arrival-time data from many stations distributed around the epicenter of the event. The epicenter is then estimated as that location which gives the least travel-time residuals. The computations are usually iterative, consisting of solving a system of linearized equations by a least-squares procedure, with one equation for each station (Bolt, 1960, 1970). The computations, which are rather extensive when data from many stations are used, are routinely carried out on computers. The location errors obtained in this way are considerably smaller than those obtained by array stations. The errors contain one systematic and one random part. The random part depends on the number of recording stations involved, on their distribution, and on the accuracy of the arrival-time estimates. The errors given in seismological bulletins are only estimates of this statistical part. Such statistical error estimates are sometimes given as confidence regions around the estimated epicenter (Flinn, 1965). For large events recorded at many stations these random errors can be rather small. The real location errors might, however, be considerably larger than the estimated random errors, as will be discussed below.

The systematic location errors depend on the difference between the used travel-time model and the real earth, and can be disclosed only by calibration events with known epicenters. To increase the location accuracy and reduce the systematic location errors, computational methods have been developed whereby locations are estimated simultaneously for a number of seismic events in a region (Douglas, 1967; Douglas & Lilwall, 1972). In the computation, travel-time corrections are estimated for each of the recording stations used, and a substantial part of the systematic errors is compensated by these station corrections, especially in those cases where the stations are well distributed in azimuth around the epicenter. The location accuracy increases further if the position of at least one of the events within the area is known. The positions of the other events in the area are in such cases estimated relative to this reference event.

8.2.1 Location of explosions in the US

The large number of nuclear explosions with announced positions at NTS provide an opportunity to study the accuracy of seismological location methods. To illustrate the achievable location accuracy under

EVENT DEFINITION AND LOCATION 177

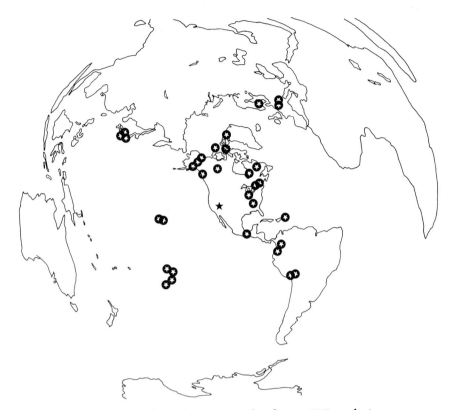

Fig. 8.2. Seismological stations used to locate NTS explosions.

different conditions, we conducted a numerical experiment using arrival times from 19 explosions with announced positions, reported from all or from a large number of the 32 seismological stations shown in Fig. 8.2. These stations were selected so as to provide as large a coverage as possible in azimuth and distance in the distance interval of from 2000 to 10 000 km. The locations, which were estimated one by one using arrival times from 13 to 32 stations and a standard computer program for event location, are shown together with the announced positions in Fig. 8.3A. This figure clearly shows a systematic location bias of 15 km.

The 19 explosions were also located using various subsets of the 32 stations. The estimated locations relative to the announced positions are in Fig. 8.4 shown for the estimates based on all available stations, i.e. the locations shown in Fig. 8.3, and also for estimates based on 8–10 stations, either well distributed in azimuth, or located within a 90° azimuth sector from NTS. The figure clearly illustrates the well-

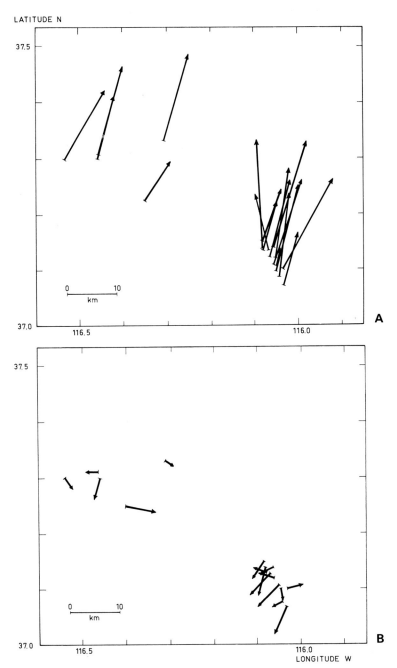

Fig. 8.3. Locations of NTS explosions relative to their announced positions. The arrowheads show the estimated locations. A—Estimation one by one. B—Joint epicenter estimation, two explosions fixed at their announced positions. One of the explosions used in the computation was located outside the area shown in the figure.

EVENT DEFINITION AND LOCATION 179

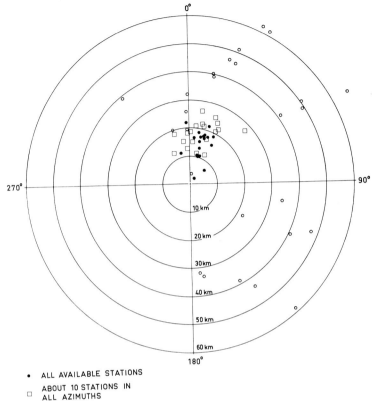

Fig. 8.4. Locations of NTS explosions estimated with different station configurations. Origin: announced locations.

known fact that a good azimuthal distribution of stations around the epicenter is essential for obtaining accurate location estimates. We can also see that there is no pronounced difference between the estimates carried out using 8—10 azimuthally well distributed stations and those based on 13—32 stations.

In Fig. 8.5 we have also illustrated the random location errors, corrected for the systematic bias, obtained from all available stations and from 8—10 stations, either well distributed in azimuth or within an azimuth sector of 90°, respectively. For estimates based on all available stations, 90 % of the remaining errors are within 10 km. Ninety percent of the random location errors for 8—10 well distributed stations are less than 12 km, compared to 57 km for 8—10 stations within an azimuth sector of 90°. These location errors are only about one

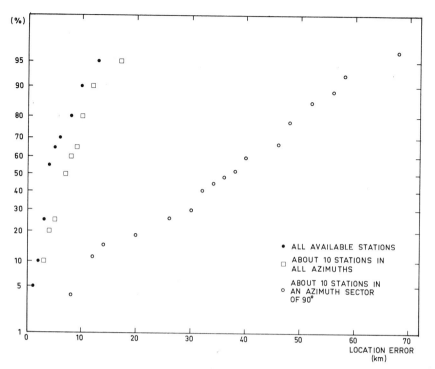

Fig. 8.5. Distribution of location errors for different station configurations after correcting for systematic bias.

tenth of the errors earlier discussed for location by an array station.

In the Threshold Test Ban Treaty it is agreed that epicenter data for two nuclear explosions at each test site would be exchanged between the US and the USSR. To see what location accuracies can be achieved when the positions of two events at a test site are known, we have used a joint epicenter-estimation program, designed by Slunga (1975), to estimate the locations of 17 of the 19 explosions, utilizing the known position of the other two explosions and arrival-time data from the stations referred to earlier. The estimated locations are shown together with the announced positions in Fig. 8.3B. No systematic bias can be seen, and 90 % of the errors are less than 6 km, which is of the same order of magnitude as the random errors observed when the events are located one by one.

The location accuracies obtained here are considerably better than those reported for NTS explosions by Blamey & Gibbs (1968). Ninety percent of the location errors they reported were less than 30 km when the explosions were located one by one and when the systematic

EVENT DEFINITION AND LOCATION

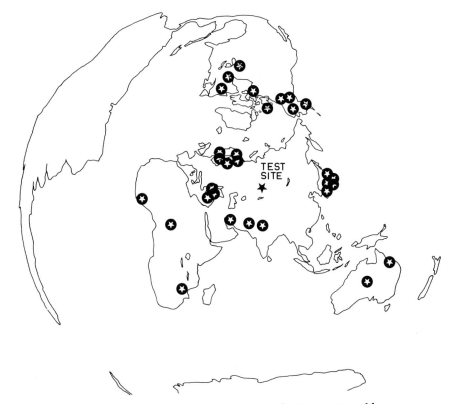

Fig. 8.6. Stations used to locate explosions in the Eastern Kazakh test area.

errors were removed. The corresponding figure when the explosions were located by a joint epicenter-determination method, fixing one explosion to its announced position, was 20 km. Blamey & Gibbs do not specify the seismological station network they used.

Another exampel of a most carefully studied calibration event is the "Long Shot" underground nuclear explosion in the Aleutian Islands. The difference between the true location and that estimated by seismological data from more than 300 globally distributed stations was 22 kilometers. The 90 % confidence region for that estimate had a major axis of only 8 kilometers and did not cover the explosion point. The estimate also gave an explosion depth of 58 km, whereas the explosion was fired at a depth of only 0.7 km (Lambert et al., 1969). This strong divergence between real and computed epicenters might be exceptional, as the Aleutian Islands are part of a complicated geological structure with large heterogeneities, but it does illustrate that the real location error can be considerably larger than the statistical estimate.

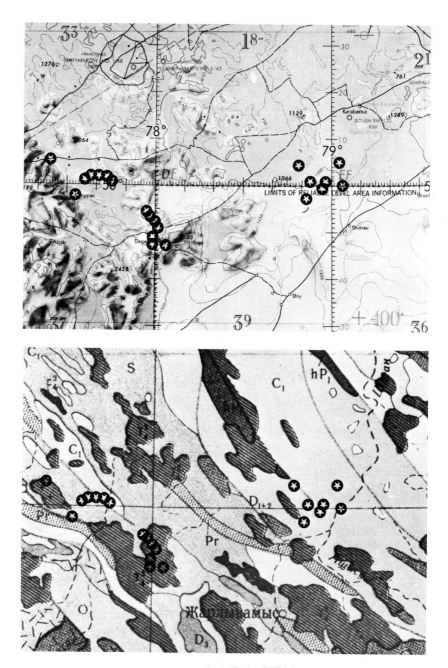

Fig. 8.7. Estimated locations of explosions in the Eastern Kazakh test area. A—Topographic map (Operational Navigation Chart ONCE-6, Aeronautical Chart and Information, US Air Force). The elevation is given in feet. B—Geological map (Map No. 15, Ministry of Geology and Mineral Resources, USSR). The following notations are used: C_1 = Carboniferous. D_{1+2}, D_3 = Devonian. O = Ordovician. Pr = Proterozoic. γ_4^3 = Paleozoic. ϵ_4^3 = Feldspathic.

8.2.2 Location of explosions in the USSR

The locations of about 20 nuclear explosions conducted in the Eastern Kazakh test area in 1971–1975 have been estimated using data from 30 seismological stations the distribution of which is shown in Fig. 8.6. The explosions were selected to cover all three test sites, and the stations to give as good a coverage as possible both in azimuth and in distance, for distances between 2000 and 10 000 km. The estimated locations are shown in Fig. 8.7. The true locations of these explosions are unknown to us, however, since the station coverage of the Eastern Kazakh test area is better than that of NTS, it seems reasonable to assume that the relative location errors for Eastern Kazakh are of the same order of magnitude as, or possibly somewhat smaller than, those obtained for the NTS explosions (Fig. 8.4). All the locations may, however, have been offset by an unknown quantity (note that for the NTS explosions this systematic error was 15 km).

Figure 8.7 clearly shows that the Eastern Kazakh test area consists of three separate test sites, one eastern, one western, and one central. The central test site is the one most frequently used. The estimated locations are in Fig. 8.7A shown on a topographic map. The central test site is in an area with a pronounced hill reaching an altitude of about 1000 m. The western test site is on a plain, and the eastern one on slightly lower land characterized by salt lakes and marshes. A geological map, Fig. 8.7B, suggests that the central and western sites contain granite and silurian strata. The eastern test site seems to be sited on carboniferous strata.

Marshall (1972) has demonstrated that the locations of several explosions in the USSR, estimated from arrival-time data obtained from carefully analyzed records from the four array stations in Australia, Canada, India, and Scotland, are in substantial agreement with those reported by USGS. The average difference between the USGS locations and those made from the four-array-station data is only 25 km for eleven explosions in the USSR.

8.2.3 Combined use of array stations

Above we have discussed location estimates based only on arrival times. Array stations can, however, report not only arrival times but also estimates of apparent velocities and azimuths. Although the uncertainty in the location estimate from apparent velocity and azimuth

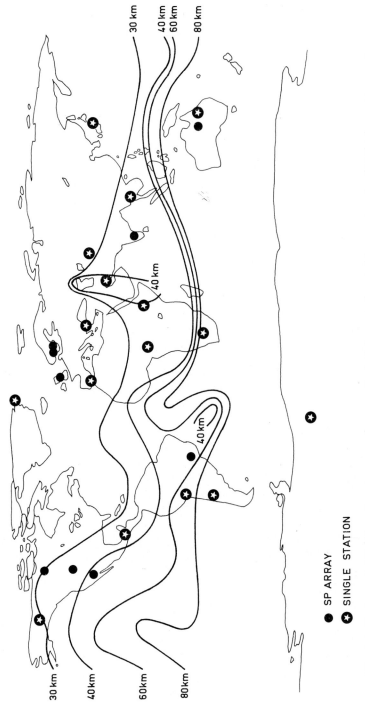

Fig. 8.8. Estimated global location accuracy for magnitude-4.5 events located by a network of 25 high-sensitive stations.

obtained at one array is large compared to that obtained by a network of globally distributed stations, locations based on data from a limited number of arrays are of interest for those weak events which are recorded only at the most sensitive stations. Computer programs for estimating event locations based on arrival times, apparent velocities, and azimuths from several stations have been developed (Julian, 1973b). The only study so far published on multiarray location is a preliminary evaluation of the joint LASA and NORSAR location capability (Gjøystdal et al., 1973). This study showed a minor decrease in location error when data from the two arrays were combined. The average difference between the joint locations and those of USGS for 14 events in the West Indies was 130 km, compared to 160 km for LASA alone and 310 km for NORSAR alone.

More work has to be done to find efficient ways to combine array estimates of azimuth and velocity with arrival-time data from arrays and other stations. We are convinced, however, that the combined use of such data will be of great importance in the future to define and locate weak events which would otherwise remain unreported.

8.2.4 Estimates of network capabilities

Several attempts have been made to estimate the location capability of global networks of seismic stations (Evernden, 1969a, 1971a, Romney, 1971). The capability is usually defined in terms of confidence regions which take into account stochastic but not systematic location errors.

In evaluating the location capability of a station network, one has to consider the detection capability of the network, which determines its applicability, i.e., the lowest explosion yield the network can locate. The location capability for any network will also decrease with decreasing event strength. Signals from weak events may pass undetected at some stations. A decreased signal-to-noise ratio will also give less accurate arrival-time estimates. Romney (1971) presented an estimate of the location capability of a network of 25 globally distributed, highly sensitive stations, most of which are operating today. The positions of the stations are shown in Fig. 8.8, together with the average radius of the estimated 95 % confidence location ellipse for events with magnitude m_b =4.5. This figure shows that the location accuracy in the USSR and the main part of Eurasia is almost the same as in eastern US, and that the accuracy is considerably lower

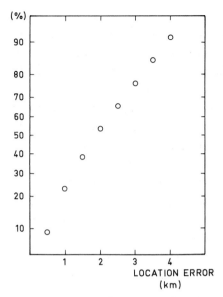

Fig. 8.9. Distribution of location error for NTS explosions estimated by a network of ca. 15 local stations. Other NTS explosions with announced positions were used as calibration events.

in the Pacific. Such rather general estimates of location accuracy must be considered as order-of-magnitude estimates only. The experimentally estimated stochastic location errors discussed above are, however, significantly smaller than those presented by Romney.

8.3 EVENT LOCATION AT SHORT DISTANCES

During the discussion of a CTB it has been suggested to utilize seismological stations located in seismic areas to detect and locate seismic events at short distances. Today, national local networks of seismological stations monitor the local seismicity in certain areas, such as Japan, South-western USSR, and Western US. Seismic waves recorded at distances of a few hundred kilometers have propagated through the earth's crust or the uppermost part of the mantle, where the structure varies considerably from one region to another, as is discussed in Chapter 4. Local travel-time models must therefore be used for event location at these distances, and the location accuracy depends to a larger extent at these short distances than at teleseismic distances on calibration explosions.

High location accuracy can be achieved when data on many explosions from stations at short distances are available to construct a

EVENT DEFINITION AND LOCATION   187

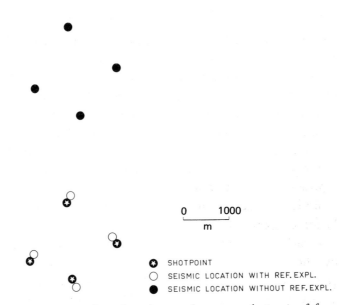

Fig. 8.10. Estimated locations (with and without reference explosions) of four closely spaced chemical explosions from data obtained at three stations within an azimuth of 90° and at distances of from 200 to 400 km.

fairly detailed travel-time model of the area. Slunga (1973) has used reported P and S arrival times for 13 NTS explosions obtained at about 15 stations at distances of between 200 and 4000 km from NTS to compute station corrections and a special travel-time model. Location errors observed when this model and the estimated station corrections were used to locate NTS explosions are shown in Fig. 8.9. Ninety percent of the location errors are less than 4 km. These very small errors must be regarded as the best that can be achieved from reported arrival times when good calibration data and observations from close-in stations are available.

Another experiment with closely spaced chemical explosions, recorded at three temporary seismological stations at distances of between 200 and 400 km, was conducted in Sweden to estimate the location accuracy achievable at such distances. The locations were estimated using a joint epicenter-determination method modified for local events (Slunga, 1975). As for the US explosions described earlier, two computations, with and without master events, respectively, were made. The locations using master events gave an error of 100 to 200 meters, Fig. 8.10. When master events were not used, a systematic error of about 5 km was observed.

8.4 INTERNATIONAL DATA EXCHANGE

An extensive exchange of seismological bulletins is traditionally going on between seismological observatories and institutes around the world. These bulletins usually contain observed arrival times, primarily for P waves, and to some extent also amplitude and period data as well as information on later phases. Seismological bulletins are issued with a delay ranging from one day to several months or years. Although the analysis of seismic records is carried out rather routinely at the participating stations, the production of these bulletins taken together represents a considerable effort.

Bulletin data are utilized by some organizations and institutes for locating seismic events routinely.

A rather fast and comprehensive event-location summary is published by the US Geological Survey, USGS. The USGS preliminary *Earthquake Data Report* is normally distributed within 4—5 months after an event and has since the early 1960's been extensively used throughout the world as a main reference for event definition and location. USGS also issues a monthly listing of *Preliminary Determination of Epicenters*, with somewhat longer delay. Event locations are also made by the International Seismological Centre, ISC, which was reorganized in 1970, and which is now supported by institutes in about 25 countries. The event-location summaries from ISC are published with a delay of two years, the delay being due partly to the ambition to include all possible information. Despite the considerable number of seismic events reported by USGS and ISC, many events detected at teleseismic stations are left unreported. During the earlier discussed International Seismic Month, 996 events were defined and located. USGS reported only 353, or 35 % of these events; the corresponding figure for ISC was 515 (52 %). ISC reported 299 additional events not included in the ISM list, most of which were based on close-in stations. ISC utilized all the arrival-time data supplied, which means that hundreds of stations might be included in the location of a strong event, which fact does not increase the accuracy of the location estimate, but only the volume of the bulletin. Although ISC is a good manifestation of international cooperation, its present operative philosophy, resulting in long delays and little new information, makes the ISC bulletin inadequate for operative purposes. The present ISC output therefore seems to be of no interest in a CTB monitoring.

Event locations are also routinely carried out by the Institute of

Physics of the Earth, Moscow, using data mainly from stations in the USSR, by the recently instituted Centre Séismologique Européo-Mediterranéen, Strasbourg, focusing on European earthquakes, and by the Hagfors Observatory using data from about 50 globally distributed stations. Array locations are also promptly reported by the array stations Hagfors, NORSAR, and Yellowknife. The Blacknest Data Analysis Center is reporting locations of selected events. Locations of local earthquakes estimated from rather dense local networks are reported by a number of institutes, e.g. the Japan Meterological Agency for Japanese earthquakes and the Seismological Observatory in Wellington for New Zealand earthquakes.

Event-location estimates today are usually based on data from a large number of stations having quite different sensitivities. The stations also report with a varying degree of accuracy and reliability, and with rather different time delays. From the preceding discussion it may be concluded that it is not possible to increase the teleseismic location accuracy substantially beyond what can be achieved by 10 or 20 well distributed stations. As far as event location is concerned it is thus of little advantage to use more than about 20 stations. For a routine monitoring operation it is important to have stations with high operative reliability, where data are carefully analyzed and promptly reported. Such a global network might well be combined with local station networks monitoring the seismicities in areas of special interest. We think that a special network of selected stations, fulfilling the above discussed criteria, should be used not only to locate events for monitoring a CTB but also for establishing a more adequate reference base of earthquakes for seismological purposes in general.

In summary we regard the achievable location accuracy of about 10 km for seismic events, corresponding to fully contained explosions in hardrock of about 1 kt, as adequate for monitoring a CTB without on-site inspection. The stations needed to achieve this are, with few exceptions, operating today. The routine analyses and the regularity in data submission have to be substantially improved at some stations. The main undertaking to create a system capable of producing adequate locations for monitoring a CTB is thus not the establishment of a large number of new stations, although some new stations might be needed, but rather to analyze and report carefully the data recorded at the highly sensitive stations operating today. Other operative monitoring requirements are discussed in Chapter 15.

9. DEPTH ESTIMATION

In Chapter 4 it is concluded that only a small minority of earthquakes are shallower than 10 km. The depth distribution does, however, vary regionally. Along transform faults the focal depths seldom go below 20 km, whereas in convergence zones earthquakes can occur at depths greater than 100 km. The significance of this regional variation is illustrated by Fig. 9.1, in which the depth distributions of earthquakes in the US and the USSR are compared.

Present drilling techniques do not allow a penetration deeper than about 10 km for the emplacement of nuclear explosions. Although there seems to be sufficient knowledge about how to go a few kilometers deeper, one cannot foresee that any really significant improvement in present drilling capabilities will be made in the near future (Press & Siever, 1974).

The practical limit that is thus set to the depth of nuclear explosions and the factual depth distribution of earthquakes present a possibility to identify by depth estimates a large portion of the earthquakes. It is also clear from Fig. 9.1 that the prospects of identifying seismic events by depth estimates are significantly different for different regions. The data in the figure show that 40 % of the earthquakes in the USSR have an estimated depth of more than 100 km, whereas virtually no earthquake in the US is deeper than 30 km.

Even if the difference in depth between explosions and earthquakes appears to be large, it must be noted that the accuracy of the present depth-estimation methods is far from sufficient for unambiguous discrimination between earthquakes and explosions based on focal depth. Another problem with focal depth also stresses the significance of effective depth estimation for seismic discrimination purposes. As discussed in Chapters 4 and 10, the signatures of seismic signals from a deep earth-

DEPTH ESTIMATION

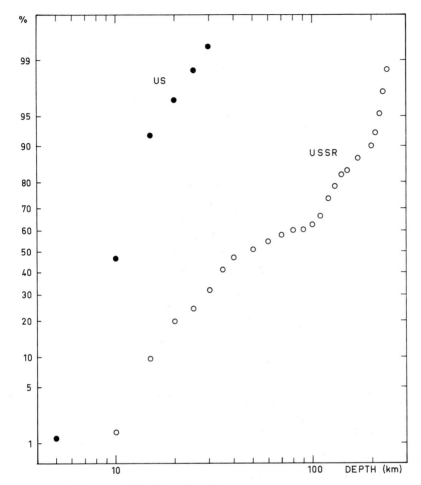

Fig. 9.1. Distribution of focal depths of earthquakes in the US in 1972 and in the USSR in 1971–1972, as compiled by Israelson et al. (1976), Dahlman et al. (1974), and Austegard (1974). The data in the figure are based on 204 US and 174 USSR events. Events with depths restricted to the normal value, 34 in the US and 51 in the USSR, are not included.

quake (at depths below about 70 km) frequently appear quite explosion-like. Without knowledge of the focal depth, and using the signal signatures, one may take such an earthquake for an explosion.

Precise depth estimates also play an important role in the interpretation of tectonic processes discussed in Chapter 4. Billington & Isacks (1975) used precise depth estimates to outline deep fault planes in subducting lithospheric zones. Accurate hypocenters are important also in the field of earthquake prediction.

In the presentation of depth-estimation methods below, we also dis-

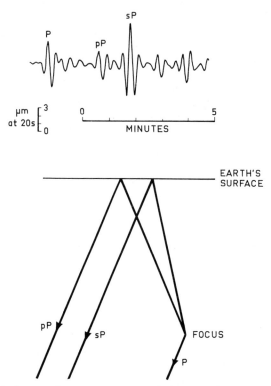

Fig. 9.2. Long-period vertical-component record showing the P, pP, and sP phases and their travel paths in the source region. The record was obtained at Hagfors from an earthquake in the Sea of Okhotsk on 730728 at 200636.0 occurring at latitude 50.5°N and longitude 148.8°E; depth 592 km, m_b (USGS)=5.5.

cuss their applicability, accuracy, and utility for seismic monitoring. Several methods have been deployed over the years, and some have been used on a routine basis, but little has been published in quantitative terms about the operational performance of these methods. We will focus on techniques used in the routine work by national and international seismic services, and we will distinguish between methods based on travel times and methods based on signal wave forms. In the presentation we also distinguish between local methods and teleseismic methods. Finally we summarize and discuss potential improvements obtainable by the combination of different methods.

9.1 SURFACE-REFLECTED PHASES

Seismic body waves travelling directly upward from the focus are at the earth's surface reflected back into the earth again and then follow a path similar to that of the direct body wave, as illustrated by Fig. 9.2.

DEPTH ESTIMATION

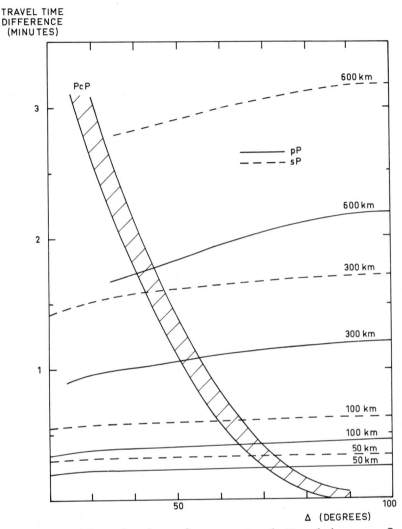

Fig. 9.3. Time delays of surface reflections (pP and sP) and the core reflection, PcP, relative to the direct P wave as functions of epicentral distance, Δ, and focal depth, according to the travel-time table of Jeffreys & Bullen (1967). The band for PcP covers the depth interval of 0–100 km. (After Yamamoto, 1974.)

The surface-reflected waves can be compared with echos of sound waves. A surface-reflected wave is denoted *pP* or *sP*, depending on the character of the upgoing waves. For example, *sP* says that an upgoing *S* wave is converted into a *P* wave at the reflection. It should be mentioned that also waves of types *sS* and *pS* are generated, but in detection seismology only reflected waves which arrive at the receiver as *P* waves are of practical interest. The time delays of *pP* and *sP* are

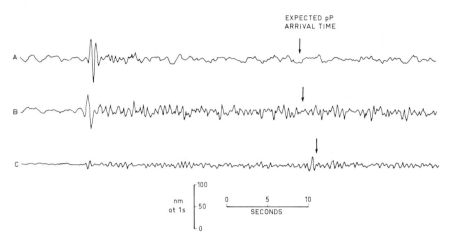

Fig. 9.4. Short-period records obtained at Hagfors of three closely spaced, deep earthquakes in the Tadzhik–Sinkiang border region with the following source data according to USGS:

	Date	Origin time	Epicenter	Depth (km)	m_b
A	720206	185124	38.3N 73.7E	113	4.7
B	720406	223837	38.4N 73.5E	118	5.0
C	710408	183338	38.4N 73.3E	129	5.0

The expected arrival times for pP are indicated in the figure. The pP phase can be clearly seen only in trace C. Note also the large difference between P codas in the three signals.

chiefly functions of the focal depth, and are virtually independent of the epicentral distance for events shallower than 100 km, as shown in Fig. 9.3.

When the *pP* or the *sP* phase, or both, can be found in the seismic records, it is easy to estimate the focal depth from the curves in Fig. 9.3. The accuracy of such an estimate is mainly determined by the reading error of the time difference. With a reasonable signal-to-noise ratio of the reflected signal, as for example in Fig. 9.2, the total depth error is within ± 5 km and thus quite small.

It can be quite difficult, however, to detect any secondary signals at all in some events, and one must also verify that detected secondary signals really are surface reflections. Figure 9.4 illustrates this problem of detecting secondary signals. It shows in the lower record the ideal case where the reflected *pP* signals stand out clearly. In the upper two records, however, secondary signals can hardly be seen at all. The theoretically predicted arrival times of *pP* are indicated in the figure. The three signals in Fig. 9.4 are associated with three deep and rather closely spaced earthquakes in Central Asia. The difficulty of detecting secon-

DEPTH ESTIMATION

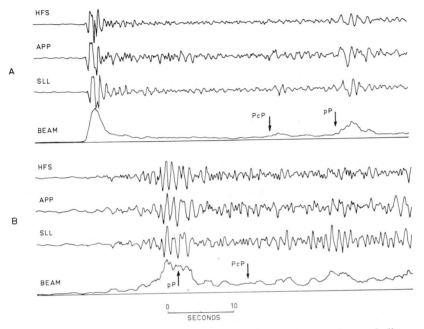

Fig. 9.5. Short-period records at Hagfors from one deep and one shallow earthquake in the Aleutian Islands, with the following source data according to USGS:

	Date	Origin time	Epicenter	Depth (km)	m_b
A	710405	090448	53.4N 170.6W	153	5.8
B	710207	022928	51.4N 176.7W	50	6.0

The signals at the three Hagfors substations, HFS, APP, and SLL, were aligned, and a beam based on rectified signals was formed. The expected arrival times of pP and PcP are indicated in the figure. Both these phases are clearly seen in the top records.

dary phases can be due to several factors. The reflected signals may be hidden in the coda generated by scattering or multipathing. The reflected phases may also be severely attenuated at the source, if there is a strongly absorbing geologic structure between the focus of the earthquake and the surface. In some earthquakes the radiation pattern can be such that only little energy is radiated directly upwards (Douglas, 1975). As shown by Fig. 9.5, detection of pP and sP is usually more difficult from a very shallow earthquake, above 20 km, than from a deep one, below 100 km; this is partly due to interference with reverberations in the source crust.

Some of the difficulties of detecting secondary phases by single-channel recording may be overcome by signal processing, as is discussed by Landers (1974). For three-component recording, experiments have also been made with the polarization filters mentioned in Chapter 7.

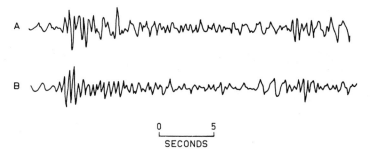

Fig. 9.6. Short-period records at Hagfors of two closely spaced underground nuclear explosions in China, with the following source data:

	Date	Origin time	Epicenter	$m_b(HFS)$
A	690922	161459	41.4N 88.3E	5.2
B	751027	010004	41.4N 88.3E	5.0

On both records the initial P signal is followed 21 sec later by a clear secondary phase.

Encouraging results from this application have been reported by Basham & Ellis (1969). Seismic arrays facilitate detection of secondary phases. Promising results have been reported from Yellowknife, where surface reflections of events close to the detection threshold were detected during the International Seismic Month (Weichert et al., 1975).

Secondary phases which fit the time difference of PcP cannot be used for depth estimation, and have to be discarded. The PcP traveltime difference is principally a function of the epicental distance and nearly independent of the focal depth, as can be seen in Fig. 9.3. At arrays one can, however, theoretically discriminate between PcP and pP or sP according to differences in slowness. There are also secondary phases the origin of which is difficult to explain. A case in point is shown in Fig. 9.6 displaying short-period records at Hagfors from two underground nuclear explosions in China. Twenty-one seconds after the initial P phase, there is a clear secondary phase. A similar phase has been reported at many other stations and arrays. It has been suggested that part of the direct downgoing P wave close to the focus is reflected backwards to the surface, and after hitting the surface is reflected back into the earth again (Davies & Frasier, 1970). Strong phases have also been reported to arrive between P and pP from some deep earthquakes (Lin & Filson, 1974). These phases are denoted pdP and are interpreted as reflections at boundaries above the source, at depth d.

9.1.1 Case studies

It is clear that the possibility to make depth estimations based on surface reflections depends on the magnitude and focal depth of the event, on regional effects, and on the method of signal processing and analysis of the records. As shown by Fig. 9.5, it is usually more difficult to detect pP and sP from a very shallow earthquake, above 20 km, than from a deep one, below 100 km. Only few systematic studies have been carried out considering all these aspects. Here we discuss mainly case studies made by Lacoss (1969) and Yamamoto (1974).

The study by Lacoss (1969) is based on records at LASA from about 150 events in Eurasia. It was found that a secondary phase could be detected in about 60 % of the events. Detection was made by an analyst picking the phases visually from the beam at LASA. About 80 % of the picks could be associated with pP or sP. This means that it was possible to identify pP or sP in about half the number of events studied. Three presumed explosions were among the events where unassociated secondary phases appeared. To make readings taken at a single station less subjective, one can combine picks made by several analysts in independent readings of the same records. It has been found that the combined results by three or four analysts may eliminate several unassociated phases, without significantly changing the number of associated phases (Yamamoto, 1974). More confident readings can also be achieved by using data from more than one station. Observed readings of secondary phases at the two arrays LASA and NORSAR for the events of a Eurasian event sample agreed in about 50 % of the cases (Lacoss & Lande, 1969). Several methods have been proposed to combine readings of secondary phases at several stations. A method proposed by Keilis-Borok et al. (1972) is essentially a search for the depth that maximizes the number of agreeing phases. The method is graphically illustrated in Fig. 9.7. Starting at zero depth, one proceeds in reasonable steps down to 700 km, and for each step one counts the number of fitting phases. The depth range showing the maximum number of agreeing phases defines the depth. Using WWSSN data, Yamamoto (1974) applied this method to about 300 Japanese earthquakes. To accept a depth solution the maximum number of agreeing phases must not be too small. Yamamoto required at least three agreeing phases. For comparison it can be noted that the depth-estimation procedure employed by USGS requires two agreeing phases. The confidence of a depth solution is, however,

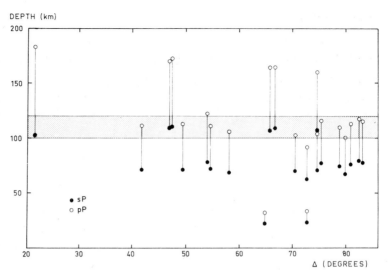

Fig. 9.7. Procedure for depth estimation from secondary arrivals using multistation data. For each arrival the depths corresponding to pP and sP are calculated. The depth interval maximizing the number of fitting arrivals is indicated in the diagram. The basic data have been taken from Yamamoto (1974) and represent readings from an earthquake in the Japanese region on 730214.

determined not only by the number of agreeing phases, but also by the number of unassociated phases. The ratio of the number of agreeing to the number of non-agreeing phases can therefore be a more appropriate measure of the confidence of the depth estimate. For the Japanese events it was found that the method can be applied to more than 70 % of the events with body-wave magnitude m_b(USGS) larger than 5.0. An even higher applicability was found for deep events, below 100 km.

Regional variations in the number of reported depth phases were observed by Yamamoto, who found that USGS determines depth from depth phases for about 40 % of the events in Japan with m_b of at least 5, whereas the corresponding figure for an unregionalized, worldwide sample is only about 15 %.

Another interesting result obtained by Yamamoto is the finding that data from only about 15 stations gave results similar to those obtained by using data from more than 100 stations. This probably reflects both the variation in the station policies in regard to the reporting of secondary phases and the fact that a rather limited number of well distributed stations conducting a close analysis of their data are almost as efficient as the present worldwide network.

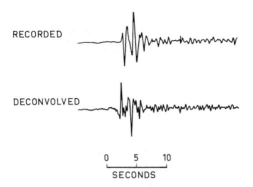

Fig. 9.8 Illustrating an attempt to isolate the surface reflection of an explosion P signal by spiked filtering. The top record is the unprocessed signal, and the bottom record is the deconvolved signal. The data were obtained at Yellowknife from an explosion near Bukhara in the USSR on 680521. (After Douglas et al., 1972.)

9.1.2 Explosions

As a rule of thumb, the firing depth (in meters) of a fully contained nuclear explosion varies with the yield, Y (in kilotons) as $160 \, Y^{1/3}$. The P-wave velocity above the shot point can usually be assumed to be at least 3 km/s. This means that for explosions with yields less than 1 Mt, the reflection pP is delayed by only about 1 s or less and becomes mixed up with the direct P itself. Several techniques to resolve the surface reflection have been devised and tested. Some of these techniques operate in the time domain (Frasier, 1972; Douglas et al., 1972a) and attempt to remove the effects of the attenuation in the mantle and of the instrument response on the recorded signals. Fig. 9.8 gives an example of an explosion signal together with such a *deconvolved* signal, on which a secondary phase can be clearly seen. Other methods operate in the frequency domain of the signals. It is assumed that the pP is similar to P except for an amplitude factor. Then there should theoretically be a scalloping of the amplitude spectrum of the combined P signals. This scalloping can be detected by forming the spectrum of the spectrum, a so-called cepstrum (Cohen, 1970). Estimates of the delay time of pP made by different methods seem to yield concordant results (Frasier, 1972). However, because of the shallowness of explosion depths compared to the wavelength of the waves involved, one may question the validity of the simple, elastic surface-reflection model. The overburden at the shot point is also subject to a significant upheaval and an associated slapdown. A separate phase, which may be associated with this slapdown, has also been reported in some cases

(Frasier, 1972; Bakun & Johnson, 1973; King et al., 1974). So far, little systematic and routine analysis has been carried out to estimate the depths of explosions by surface reflections.

9.2 P-ARRIVAL TIMES

The surface-reflection method discussed above is considered to be the most effective method for depth estimation. Quite often, however, it is impossible to apply this method, in which case one may attempt using teleseismic P-arrival times of a seismic network to estimate the depth along with the epicenter and the origin time. In brief, one fits the observed arrival times to a travel-time model of the earth, using a least-squares procedure. The observed arrival times are functions of the epicenter, the depth, and the origin time. Since teleseismic P waves are radiated only in a small solid angle, it becomes difficult to separate the influences of origin time and depth from each other. This means that in cases where only a small number of arrival times are available, one can run into problems with the convergence of the least-squares procedure. In such cases, the depth is usually by convention restricted to 33 km, which means that no actual estimate has been obtained. For a worldwide sample of events defined by the USGS, and with $m_b \geqslant 5$, about 50 % of the depth estimates were based on arrival times and about 35 % were restricted (Yamamoto, 1974). The remaining 15 % were based on surface reflections. Depths obtained by standard travel-time models are considered to be less accurate than those based on surface reflections. For example, the nuclear explosions "Longshot" on the Aleutian Islands, discussed in Chapter 8, not only had its epicenter misplaced by 22 km, but was also estimated to have taken place at a depth of 58 km (Lambert et al., 1969).

There are other examples of explosions in the USSR as well as in the US which by standard travel-time models have been placed at depths around 20 km (Austegard, 1974). Depths estimated from arrival times are considered to have an uncertainty of ±50 km (Basham & Ellis, 1969). The large uncertainties are due to regional variation of the seismic velocities and to lateral inhomogeneities. The error is also a function of the number of reporting stations and their distribution relative to the epicenter and of the actual depth itself.

By refining the standard travel-time model, the uncertainties can be reduced. The so-called joint hypocenter-determination method, where the travel-time model is estimated jointly with a number of hypocenters,

DEPTH ESTIMATION

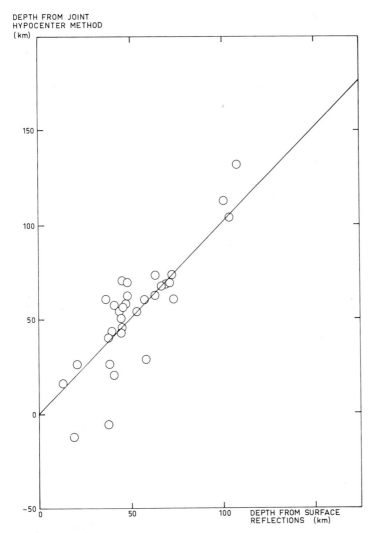

Fig. 9.9. Comparison of depths estimated by the joint hypocenter method and by surface reflections. The data represent 32 earthquakes occurring in and near Japan during 1973. (After Slunga, 1975.)

is discussed in Chapter 8. Here we only briefly describe one such model, the one constructed by Slunga (1975). In this the travel time is modelled by a second-order polynomial of the epicentral distance, provided the event occurs at a certain reference depth. The coefficients of the polynomial may differ from region to region. Special corrections are included in the travel-time formula to provide the travel time from an event at a depth different from the reference depth. These corrections are functions of the slowness, the depth itself, and the velocity structure at the source, which latter is assumed to be known. The accuracy of the depth estimates by this model has turned out to be improved by a factor of almost two as compared to the accuracy of estimates by a standard earth model. Fig. 9.9 compares depth estimates based on the joint hypocenter method with estimates based on surface reflections. The estimates agree within 15 km for most of the events, with somewhat higher differences for the most shallow earthquakes of the data set. The joint hypocenter method utilizes master events, and events with depths estimated from depth phases may well be used as such master events.

Even more refined models utilizing ray tracing have been proposed (Jacobs, 1970; Julian, 1970a). By such models details of the velocity structure, including lateral variations, can be taken into account. The basic principle of ray propagation is that seismic rays travel along that path which minimizes travel time. Starting from a hypocenter estimate obtained by standard travel times, one can trace the rays from there and compare the observed and the computed arrival times. The differences can be used to correct the observed arrival times for lateral effects, and a relocation can be made. This procedure can be repeated until the differences are negligible. Ray tracing is computationally complicated, requiring adequate velocity models, and has so far not been much used in routine work.

9.3 NEAR DISTANCES

In this section we will discuss methods based on seismic recording at close distances, from a few kilometers out to several hundred km. As is mentioned in Chapter 6, the possibility of operating seismic stations in this distance range has been considered for the monitoring of a CTB. Depth estimates based on close-in seismic data are currently carried out on a routine basis by local national networks to monitor local seismicity, for example, in Western US and in Japan (Lee et al., 1972). These estimates use the arrival times of local P and S phases, the travel paths of which are schematically indicated in Fig. 9.10 for a simple, layered crustal structure. Some of the existing standard depth-estimation methods using local data have been summarized by Greensfelder (1965). The most widely used method estimates the hypocenter and the origin time by minimizing the residuals between observed and theoretically calculated arrivals, as described in Chapter 8. The calculations are usually based on some horizontal layer model, and the parameters of such a model are sometimes obtained from controlled experiments with chem-

DEPTH ESTIMATION

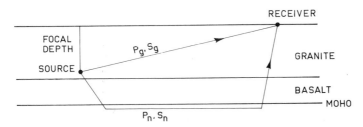

Fig. 9.10. Schematic travel paths of local phases.

ical explosions. To be able to separate the focal depth from the origin time, it is important to have data on waves radiated both upwards and downwards from the source. For local networks this means that the epicenter of the event must be located within the network. The epicentral distance of the nearest station should be less than the focal depth, and the distance of the furthest station large enough to record waves refracted downwards. With such a network it is sufficient to use the first arrivals of the direct wave through the granitic layer, P_g, at the stations to get accurate depth estimates. The accuracy of such depth estimates for earthquakes in California with magnitudes $M_L \leqslant 1.5$ is between 2 and 5 km (Lee et al., 1972). For events occurring outside the network, the S phases can be used to control the origin time. If the S-to-P-velocity ratio is constant throughout the traversed medium, S and P waves will propagate along the same path. A plot of S-arrival times against P-arrival times at a number of stations will then yield the origin time (Evernden, 1969b, 1975). In practice it can, however, be difficult to accurately read the S waves from the seismograms.

It becomes more difficult to make a hypocenter and origin-time estimation with a sparse local network, since a small number or a total lack of direct arrivals removes much of the depth control (Crampin, 1970). Slightly modified procedures to minimize arrival-time residuals have been designed for such cases. Epicenters are calculated for focal depths restrained at fixed intervals. The depth and the corresponding epicenter having the least root-mean-square residuals are selected as the solution. This type of procedure, with a depth increment of 10 km, is used by the Japan Meteorological Agency (JMA) for routine evaluation of local seismicity (Ichikawa & Mochizuki, 1971). JMA depth estimates usually agree within 10 km with estimates obtained from surface reflections (Yamamoto, 1974).

Ray tracing, which was discussed for teleseismic waves above, has

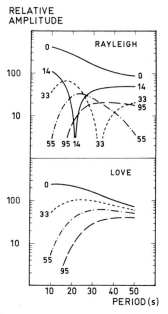

Fig. 9.11. Theoretical amplitude spectra of Rayleigh and Love waves for different depths (given in km in the figure). The earthquake source is modelled as a vertical strike slip fault with a step-time function, and the Gutenberg model is used for the crust and the upper mantle. (After Tsai & Aki, 1970a.)

also been applied to data obtained at local distances to account for lateral heterogeneous structures (Engdahl, 1973).

Finally it should be pointed out that local arrival-time data combined with teleseismic P-wave arrival times can provide quite accurate and well controlled depth estimates. Local S-P-arrival-time data can be used to control the estimate of the origin time; as indicated above, the depth is then estimated from the teleseismic P waves (Evernden, 1975).

9.4 LONG-PERIOD SIGNALS

The depth-estimation methods discussed so far are based only on arrival times of seismic signals. In this section and the next one methods utilizing the signal shape are discussed. Much attention has recently been paid to surface waves and the shape of their amplitude spectra (Massé et al., 1973). Theory shows that for periods between 10 and 50 s the shape of the Rayleigh-wave spectrum is rather sensitive to the depth (Tsai & Aki, 1970a). Fig. 9.11 gives examples of theoretical Rayleigh-wave and Love-wave spectra. The most striking feature of the Rayleigh-wave spectrum is a pronounced hole, the period of which changes with

DEPTH ESTIMATION

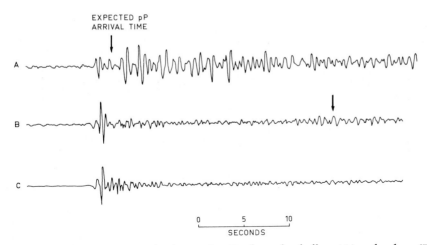

Fig. 9.12. Short-period records obtained at Hagfors of a shallow (A) and a deep (B) earthquake in the Afghanistan–USSR border region, and of a presumed explosion (C) in Eastern Kazakh, with the following source data according to the USGS:

	Date	Origin time	Epicenter	Depth (km)	m_b
A	710210	233143	38.7N 70.6E	30	4.8
B	700510	123233	36.1N 71.1E	121	4.7
C	700906	040257	49.8N 78.1E	0	5.6

depth. The shape of the surface-wave spectrum depends, however, not only on focal depth but also on the focal mechanism and the properties of the crust and the mantle. This means that to be able to estimate the depth from Rayleigh-wave spectra one needs accurate knowledge of the structure of the medium and of the parameters of the earthquake source. The Rayleigh-wave spectral method has been applied to very shallow Mid Ocean Ridge earthquakes (Weidner & Aki, 1973). By utilizing both amplitude and initial-phase spectra, they calculated the errors of the depth estimates to be within a few km. Without precise knowledge of structure and source, such a high accuracy is not likely to be obtained. It has also been suggested that uncertainties about the shear velocities in the depth range of 50–500 km will reduce the usefulness of the surface-wave spectral method (North, 1974). The requirement of knowledge about the source mechanism makes surface-wave spectral methods of small practical interest for seismic discrimination. So far, these methods have been applied only to a few earthquakes, all with body-wave magnitudes larger than 5.5–6.

Not only the shape of the Rayleigh-wave spectrum but also the overall level are significantly changed with depth, as shown in Fig. 9.11, and for a given body-wave magnitude m_b the corresponding surface-

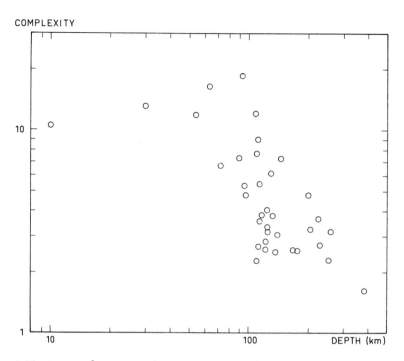

Fig. 9.13. A wave-form–complexity parameter of short-period P signals obtained at Hagfors, plotted against focal depth for 36 earthquakes in the Afghanistan–USSR border region in 1970–1973; magnitude range $4.0 \leqslant m_b(USGS) \leqslant 5.2$. (For a definition of the parameter, see Chapter 10.)

wave magnitude M_s decreases with depth (Austegard, 1974). The implications of this effect are further elaborated in Chapter 10. A depth effect has been observed for the ratio of the amplitudes of the shear phase SS and the Rayleigh wave from deep earthquakes in Central Asia recorded at Hagfors (Austegard, 1974).

9.5 SHORT-PERIOD SIGNALS

Among seismologists it has long been recognized that short-period P signals from deep earthquakes frequently differ significantly from short-period P signals from shallow earthquakes. The quite simple signal of the deep earthquake reproduced in Fig. 9.12 (record B) differs considerably from the complex signal of the shallow earthquake (record A). In fact, the deep-earthquake signal looks very much like the explosion signal (record C) shown for comparison in Fig. 9.12. The depth effect may well be partly due to the fact that the signals of a deep earthquake traverse the earth's crust only at the receiver end. However, the crust

DEPTH ESTIMATION

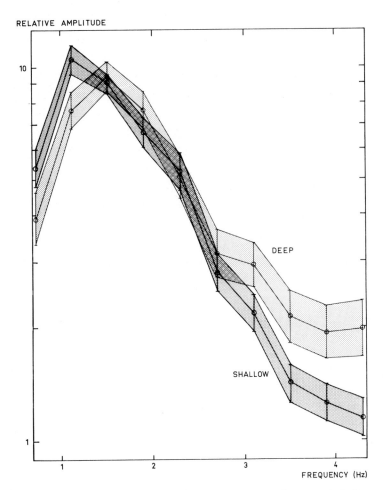

Fig. 9.14. 99.9 % confidence intervals for average spectra of deep and shallow earthquakes in Eurasia. The data are based on records at Hagfors of 260 and 115 earthquakes with depths below and above 100 km, respectively. Events with restricted focal depth have been omitted. Before averaging, the spectra have been corrected for the strength of the events. The spectra have not been corrected for instrumental response.

alone does not shape the P signal, as is seen from the explosion signal, which is quite simple although it has a travel path similar to that of a shallow-earthquake signal. The change in wave form with depth can be utilized for depth estimation. This is demonstrated by Fig. 9.13, which shows a parameter expressing the degree of wave-form complexity as a function of focal depth for earthquakes in the Afghanistan–USSR border region. It can be noted that little has, so far, been done to use this type of parameter for depth estimation in a routine manner.

The spectral content of short-period signals depends also on focal depth, as shown by the average spectra for deep and shallow earthquakes in Fig. 9.14. Preliminary attempts have been made to separate heuristically between deep and shallow earthquakes and nuclear underground explosions by using a rather detailed description of the P signals (Israelson, 1972; Pisarenko & Poplavskii, 1972). In the study by Israelson (1972) attempts were made to distinguish between earthquakes below and above 70 km, without estimating the actual depths of the earthquakes. In a test of this procedure on a Eurasian data set, about 60 % of the deep earthquakes were correctly classified.

Over the years there have been considerable discussions on the nature of deep earthquakes and on the question of whether they differ fundamentally from the shallow ones (Knopoff & Randall, 1973; Randall & Knopoff, 1973). One has mainly been concerned with the character of the focal mechanism, which for shallow earthquakes usually is described by the double-couple mechanism presented in Chapter 5. For at least some deep earthquakes there are indications that, besides the double-couple component, also a significant constituent of pure compression occurs, which can be associated with phase changes that may occur at considerable depths (Dziewonski & Gilbert, 1974).

9.6 COMPARISON AND COMBINATIONS

According to our presentation above, there are two principal methods for depth estimation which are used routinely by national and international seismic services. These methods are based respectively on the surface reflections and on the arrival times of short-period P waves. In cases where no estimate can be provided by these methods, the depth is usually restricted to a normal depth of 33 km. It should be stressed that a restricted depth is artificial, and the depth might not even be shallow. Unrestricted estimates reported by different seismic services are mostly in close agreement, as shown in Fig. 9.15, in which depths published by the USGS and by the Institute of Physics of the Earth in Moscow are compared.

Although the two principal depth-estimation methods have been in use for a long time, one has only a vague, qualitative idea of their performance. Surprisingly few data have been published which describe the effectiveness and applicability of these methods. It is therefore not very clear how the depth-estimation methods operate on explosion data. At present, explosions are mostly restricted to zero depth, with

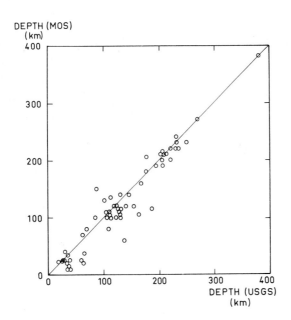

Fig. 9.15. Comparison of depths reported by the USGS and by the Institute of Physics of the Earth (MOS) in Moscow for earthquakes in the USSR in 1971 and 1972. (After Austegard, 1974.)

no actual estimation being done, because they are either announced as such or, for other reasons, presumed to be explosions.

Some of the methods presented above have so far been applied only to a small number of events. With more extensive testing and further refinements, some of them may become of great practical value. This could, for example, be the case with the methods based on the wave form of the short-period P-wave signals.

To ensure a wide applicability of a depth-estimation method it is important that it uses short-period P signals. Methods based on long-period signals will only be of limited value for routine analysis. Occasionally they may be helpful for precise estimates of depths of very shallow reference earthquakes.

Even if new, improved methods for depth estimation are developed in the future, it is unlikely that one single method will make other methods obsolete. The characteristics of seismic signals are so variable that probably only a combination of several methods, based on multistation data, will lead to improved depth estimation. Very few studies have been conducted on these problems, and we confine ourselves here to stress the importance of multistation data for depth estimation. Data can be combined at different levels: raw seismic records, measurements obtained from the records, or different depth estimates from different methods. Rather than analyzing the records at each station independently, it may be more effective, when picking the surface reflections, to make a joint analysis of a number of records obtained at different stations. Combining actual records from several stations, however, requires an effective coordination of data, and can from a practical point of view only be carried out for selected events. In our opinion it would be desirable, therefore, that various depth parameters be extracted from the records of several, highly sensitive stations around the globe. Such data should then be included in the routine data exchange. Appropriate methods to utilize such multistation depth data could also be developed at some existing data center or at a center-to-be established for CTB monitoring.

10. IDENTIFICATION

The central question in the discussion of an adequate monitoring of a CTB is that of confidently identifying seismic events, i. e., to judge whether observed seismic signals are generated by an underground nuclear explosion or by an earthquake. The estimation of event locations and depths, parameters of main importance for the identification of seismic events, is discussed in Chapters 8 and 9. Events occurring in areas where nuclear explosions can be excluded a priori or occurring at depths below those which can be reached by drilling can safely be classified as earthquakes. In this chapter we therefore present identification methods to be used in those cases where an event cannot be confidently identified on the basis of its location and depth.

In this context, two aspects of the identification methods have to be considered: their *efficiency*, i. e., their ability to separate explosions and earthquakes, and their *applicability*, i. e., the minimum event strength to which they can be applied. Generally, there is a trade-off between efficiency and applicability, such that the most efficient identification methods are applicable only to larger events, and that the efficiency of a given method decreases with decreasing event strength. In Chapter 9 we discuss the difficulties encountered when estimating the depth of a weak event. In this chapter we encounter more examples of how the difficulties of event identification depend on event strength.

In this chapter, we deal with event identification by means of dynamic parameters obtained from short-period and long-period seismic signals, and begin with a brief discussion of statistical aspects of event identification. The concept of identification curves is introduced, where the probability of correctly identifying an explosion is described as a function of the probability of mistaking an earthquake for an explosion. This has turned out to be a convenient way of describing observed identification data, although one must be careful when extrapolating such functions outside the event population on which they are based. The discrimination capability estimated in a case study can be critically dependent on the sample of events used, and this applies in particular to explosion populations.

A comparison of the amplitudes of short-period and long-period signals is the basis of the so-called $m_b(M_s)$ discriminant, which is the identification method that has gained the most widespread acceptance. Parameters based solely on short-period data have also been suggested as valuable tools for event identification. Such parameters generally provide less separation between explosion and earthquake samples than does the $m_b(M_s)$ discriminant, but they can, on the other hand, be applied to weaker events. A few studies of identification by other seismic signals will also be reviewed. Combination of either several identification parameters obtained at one station, or values of a single identification parameter obtained at several stations, significantly increases the identification capability. Few studies on the combination of identification data from several stations have been published so far, the main reason being that present seismological stations do not routinely evaluate and report identification data.

Most studies presented here are case studies of data selected for the purpose of the study. The results of such studies are therefore only examples of what can be achieved, and are not necessarily representative of present monitoring capabilities. Most of the studies include earthquakes from regions outside the territories of present or potential nuclear countries. In order to illustrate a monitoring situation, we therefore discuss also a few operative studies, where all seismic events detected within the US and the USSR during a certain time period were investigated. However, these studies do not utilize all available seismic data, and are not representative of the globally achievable capability. They rather illustrate an operative assessment of seismic data for test-ban control, a point that to a large extent has been overlooked in earlier technical discussions.

10.1 STATISTICAL ASPECTS

In this section we deal with some statistical aspects of seismic identification. The presentation is rather formal and general, but to understand the subsequent presentation in a general way, it is not necessary

IDENTIFICATION

to have a full understanding of all the details presented here. The less mathematically oriented reader may disregard this section and proceed to the next one, which deals with applications.

First of all it should be made clear, that it is not always possible even for an experienced analyst to discriminate between explosions and earthquakes by visual inspection of recorded signals only. Therefore, one has to extract from the seismic signals characteristics that can be made the subject of statistical analysis. An example of such a characteristic is wave-form complexity, which expresses the degree of complexity of a seismic signal. The precise definition of complexity is given in Section 10.3. Signal characteristics will in the following be called *discriminants*. They can usually be described by a numerical value, which makes it possible to set down in an objective and quantified way differences and similarities between explosion signals and earthquake signals. Discriminants can be one-dimensional or multidimensional. Usually, multidimensional discriminants are reduced to one-dimensional in a manner which is described later on in this section. To begin with, it is therefore sufficient to consider only one-dimensional discriminants.

It is quite helpful to study the efficiency of a discriminant in terms of decision theory (Ericsson, 1970). Seismic events can be misclassified in two ways. First, an earthquake can be mistaken for an explosion. This type of error is called a *false alarm*. Secondly, an explosion can be mistaken for an earthquake. Rather than talk about missed explosions we will here describe the effectiveness of a discriminant by giving the probability that an explosion really is classified as such. This probability is here called *deterrence*. The concepts of false alarm and deterrence are analogous to the concepts of false alarm and detection probability introduced in Chapter 7. One can assume, also, that the explosion values for a given discriminant, D, are on the average smaller than the earthquake values. If they are not, one can readily transform the data to conform with this requirement, for example by changing their sign. It then seems reasonable to apply the following classification rule to an observed value of D and a chosen threshold, T:

if $D \leqslant T$, the event is classified as an explosion,
if $D > T$, the event is classified as an earthquake.

The false-alarm and deterrence probabilities can be used to describe in a convenient way the effectiveness of a discriminant (Ericsson, 1970). Let us assume that the probability distribution functions F_Q and F_X

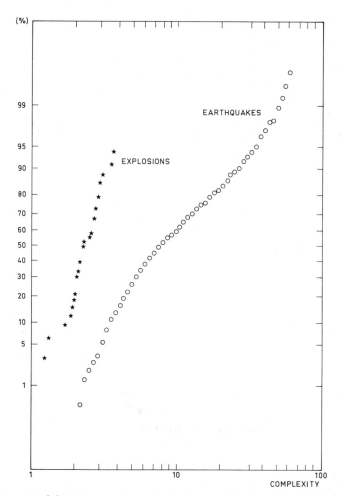

Fig. 10.1. Empirical distribution functions for a waveform-complexity discriminant. The explosion sample contained 31 events in the Eastern Kazakh test area, and the earthquake sample 535 shallow events in Eurasia.

of D for earthquakes and explosions, respectively, are known. Figure 10.1 shows the empirical distribution functions F_Q and F_X of a waveform-complexity discriminant. The data on which Fig. 10.1 is based were obtained at the Hagfors Observatory from Eurasian events; they are further discussed in Section 10.3. The false-alarm and deterrence probabilities, FA and DE, respectively, can be written:

$$FA = P(D \leq T \text{ if earthquake}) = F_Q(T),$$
$$DE = P(D \leq T \text{ if explosion}) = F_X(T).$$

For a given value of FA, say, 1 %, one can in principle obtain the value of the threshold, T, as:

IDENTIFICATION

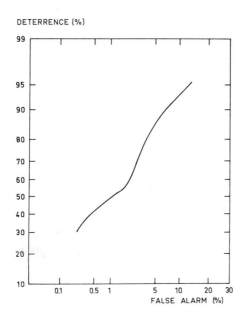

Fig. 10.2. Identification curve estimated from the empirical distribution functions in Fig. 10.1.

$$T = F_Q^{-1}(FA).$$

This value of T can then be used to calculate the corresponding deterrence probability, and we can formally write:

$$DE = F_X(F_Q^{-1}(FA)).$$

This relation between the false-alarm and deterrence probabilities has been called the *identification curve*, and is here denoted IC (Ericsson, 1970). It is analogous to the receiver-operating characteristics used in statistical communication theory (Robinson, 1967). Figure 10.2 gives the IC for the empirical distribution functions in Fig. 10.1. At a false-alarm probability of 1 %, the waveform-complexity discriminant in Fig. 10.1 provides a deterrence probability of about 50 %. The closer to the upper left-hand corner of the diagram an IC is located, the greater is the effectiveness of the discriminant.

In practice the true distribution functions F_X and F_Q are not known, and have therefore to be estimated or replaced by corresponding empirical distributions, as in Fig. 10.2. Quite frequently, normal distributions have been used as approximations to F_X and F_Q (Ericsson, 1971a; Weichert & Basham, 1973). The occurrence of earthquakes can be considered as a stochastic process the parameters of which can be estimated. The empirical distribution corresponding to F_Q can therefore be con-

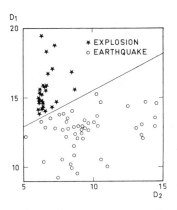

Fig. 10.3. Scatter diagram of two discriminants, D_1 (third moment of the frequency) and D_2 (waveform complexity), of short-period signals from a sample of Eurasian events obtained by Anglin (1972).

sidered as a well-defined approximation to F_Q itself. A similar interpretation in the case of explosions cannot easily be made. Because of this, an *IC* describes the separation between explosions and earthquakes for the actual data sample only. The use of *IC* should therefore be limited to qualitative comparisons of the effectiveness of different discriminants. For such applications, however, *IC* is a convenient tool, which in a standardized way shows the relative separation between particular explosion and earthquake samples.

From a statistical point of view there is a fundamental difference between F_X and F_Q; an alternative way of classifying seismic events could therefore be based on the hypothesis that earthquakes can be modelled by some kind of statistical process which cannot easily be changed by human activity. Explosions, on the other hand, are entirely controlled by human decision. The problem of seismic event identification is then formulated in terms of identifying earthquakes as such, considering non-earthquake-like events from case to case. Few such studies have, however, been conducted so far.

There is so far no single discriminant which in all situations is superior to other discriminants. This means that combinations of several discriminants generally will improve the discrimination. Figure 10.3 illustrates what can be achieved by combining two discriminants, D_1 and D_2, which describe different characteristics of the short-period *P*-wave signal. When applied individually, either discriminant fails to separate explosions and earthquakes. When considered jointly, however, they provide full separation, as shown by the straight line drawn in the figure.

This means that a linear combination, L, of discriminants, of the form

$$L(D_1, D_2) = W_1 D_1 + W_2 D_2,$$

where W_1 and W_2 are appropriately chosen weights, is sufficient to provide complete separation of the samples.

If the number of discriminants to be combined into a linear combination is large, the weights, W, cannot be obtained directly by fitting a straight line, as in the two-dimensional case. The most widely used method to calculate the weights is based on the assumption that the multidimensional discriminants are normally distributed (Booker & Mitronovas, 1964). It is usually also assumed that explosion data and earthquake data are scattered in a similar manner around their means in the multidimensional discriminant space. In statistical terms this means that the covariance matrices for explosions and earthquakes are equal. This assumption usually overestimates the scatter of the explosion data, since the covariance matrix is based mainly on earthquake data, and since one should expect less scatter of explosion data than of earthquake data. The difference in scatter for the one-dimensional case is illustrated by Fig. 10.1. Moreover, in practice comparatively few explosion data are available to provide reasonable estimates of the covariance matrices for explosions.

It can be shown that the discriminant which maximizes the deterrence probability for a given false-alarm probability is a quadratic form of the individual components, D_i, of the discriminant $\overline{D}$ (Ferguson, 1967). If the covariance matrices are equal, this quadratic form reduces to a linear combination:

$$L(\overline{D}) = \sum_i W_i L_i .$$

Linear weights have also been derived for the case where no restriction on the covariances with respect to equality is made (Anderson & Bahadur, 1962). These weights have so far been tested only for a few cases, but have not resulted in any significant improvement (Ericsson, 1973). The weights of multidimensional discriminants are usually estimated from data samples. The number of observations in the samples must, however, be appreciably larger than the number of dimensions of the discriminant; otherwise one might easily overestimate the capability of the discriminant (Sammon et al., 1970; Shumway & Blandford, 1970).

The applicability of a multidimensional discriminant is limited by

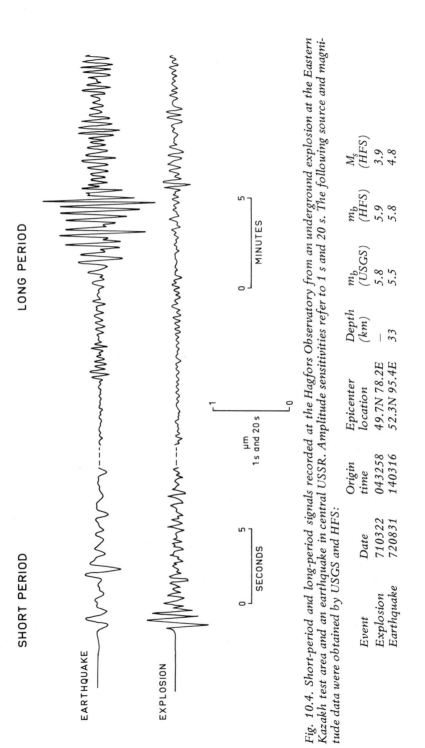

Fig. 10.4. Short-period and long-period signals recorded at the Hagfors Observatory from an underground explosion at the Eastern Kazakh test area and an earthquake in central USSR. Amplitude sensitivities refer to 1 s and 20 s. The following source and magnitude data were obtained by USGS and HFS:

Event	Date	Origin time	Epicenter location	Depth (km)	m_b (USGS)	m_b (HFS)	M_s (HFS)
Explosion	710322	043258	49.7N 78.2E	—	5.8	5.9	3.9
Earthquake	720831	140316	52.3N 95.4E	33	5.5	5.8	4.8

the component which is most difficult to detect. The number of dimensions can be reduced by just deleting a component for which data are missing and forming a new linear combination from the remaining components. Such a reduction cannot be carried too far, however, as is demonstrated by Fig. 10.3. If one of the two discriminants D_1 and D_2 is missing, one has to accept a significant reduction in the efficiency. For some discriminants it may be possible to say whether their values are below a certain threshold, such as one imposed by noise, even if the discriminant values themselves cannot be estimated. When forming the linear combination in such a case, one can use the upper limit instead of the actual value. The technique of utilizing upper-limit estimates is called *verification by negative evidence,* and has been successfully applied to the so-called $m_b(M_s)$ discriminant (Elvers, 1974). The application of this technique is discussed in Section 10.2.

10.2 THE $m_b(M_s)$ DISCRIMINANT

The $m_b(M_s)$ discriminant is the discriminant which has been most extensively discussed for use in the identification of seismic signals from earthquakes and explosions. The discriminant compares the short-period P-wave amplitude at a period of about 1 s with the long-period surface-wave amplitude at a period of about 20 s. The pronounced difference between explosions and earthquakes in respect to the ratio of P-wave to Rayleigh-wave strengths is illustrated in Fig. 10.4, which shows P and Rayleigh waves from an underground explosion and an earthquake in the USSR recorded at the Hagfors Observatory. The P waves from these two events have roughly the same amplitudes, but the Rayleigh-wave amplitudes from the earthquake are about five times as large as those from the explosion. Relative amplitude spectra of the short-period and long-period signals for these two events are shown in Fig. 10.5.

Formally, the $m_b(M_s)$ discriminant compares the short-period P-wave magnitude, m_b, with the long-period surface-wave magnitude, M_s. Usually, the M_s values are estimated from Rayleigh waves, but Love waves have also been used. In Chapter 4 we describe formulas for the calculation of body-wave and surface-wave magnitudes, and also discuss the ambiguity inherent in such calculations. In most of the studies of the $m_b(M_s)$ discriminant discussed in this chapter, the m_b values are either single-station measurements or values reported by the USGS. Especially for low-magnitude events, the USGS magni-

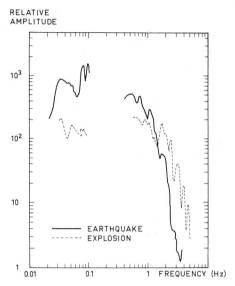

Fig. 10.5. Relative amplitude spectra of the short-period and long-period signals from the same events as in Fig. 10.4.

tude might be inconsistent, as data from different stations may have been used for different events.

To illustrate the variation in $m_b(M_s)$ data for the various types of event, we show in Fig. 10.6 data obtained at the Hagfors Observatory from underground and atmospheric nuclear explosions and from shallow and deep earthquakes. The underground explosions are all from the USSR, and the atmospheric explosions are from the Chinese test site at Lop Nor. The earthquakes are all located in the USSR, most of them in the southern part of the country. The focal depths of the shallow earthquakes are estimated to have been less than 30 km, and those of the deep earthquakes to have exceeded 100 km. The figure shows that the $m_b(M_s)$ values for the underground explosions differ from those for shallow earthquakes. The degree of separation that can be achieved between these two kinds of event is discussed later in this chapter. Figure 10.6 also shows that the $m_b(M_s)$ values observed for some of the deep earthquakes are close to those observed for underground explosions. This illustrates the importance of having depth-estimation methods that allow us to estimate the depths of such deep events with a high degree of confidence. Otherwise, they might appear explosion-like on the $m_b(M_s)$ diagram. The figure further demonstrates the large difference between $m_b(M_s)$ values for atmospheric and those

IDENTIFICATION

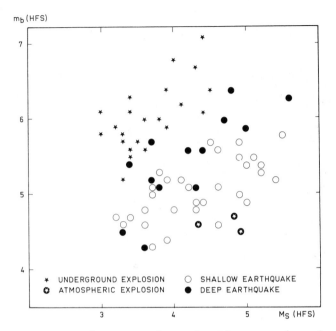

Fig. 10.6. Observations of $m_b(M_s)$ at the Hagfors Observatory for underground and atmospheric nuclear explosions and shallow and deep earthquakes. All events were located in the USSR, with the exception of the atmospheric explosions, which occurred in China.

for underground explosions. The seismic source functions for those two kinds of explosion are also quite different. Seismic waves from an atmospheric explosion are generated by the atmospheric shockwave hitting the ground over a considerable area near the explosion point, and also by resonance coupling to the earth of the atmospheric wave propagating outward from the explosion point. Such sources are more extended both in time and in space than are those of underground explosions (see Chapters 3 and 5). Since atmospheric explosions can be adequately monitored by sampling the radioactivity of the atmosphere and by satellite-based detection equipment (Dickinson & Tamackin, 1965; Singer, 1965; SIPRI, 1972), we do not pursue the matter of identifying such explosions any further.

Several attempts have been made to explain the differences in the $m_b(M_s)$ discriminant observed between underground explosions and earthquakes. It has been suggested that these differences are due to differences in source dimension, source time function, source mechanism, and focal depth, and to a combination of these (cf. Chapter 5). The source dimension of an earthquake is supposed to be larger than

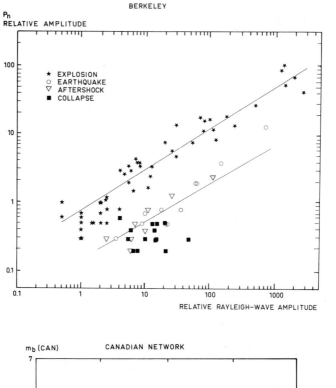

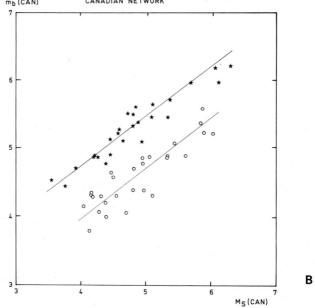

Fig. 10.7. The $m_b(M_s)$ diagram for North American explosions and earthquakes obtained in case studies using data from (A)–Berkeley, (B)–Canadian Network, (C)–Ogdensburg, and (D)–Hagfors (cf. Table 10.1). The linear trend lines for the

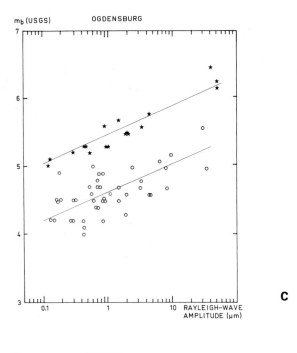

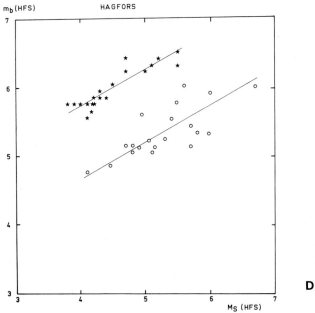

earthquakes are least-square fits, whereas those for the explosions are fitted to be parallel with the earthquake lines. Symbols shown in Fig. 10.7 A apply also to B–D.

that of an underground explosion generating the same amount of P-wave energy. Let us compare the source dimension of an explosion with that of an earthquake, both having an m_b value of 6. For an explosion of this strength (which corresponds to a yield of the order of 100 kt), the diameter of the elastic sphere which can be regarded as its seismic source is about 1 km. For an earthquake of this strength, on the other hand, the fault has a length of the order of 10 km (see Chapter 5).

Various source-time functions have also been suggested as an explanation of the observed $m_b(M_s)$ differences. The rise time of an $m_b = 6$ explosion is of the order of 0.5 s, whereas the rupture time of the fault of an earthquake with the same m_b value is about ten times as long. These large differences both in time and in space might well explain the large M_s values generally observed for the earthquakes.

Theoretical calculations have shown that the $m_b(M_s)$ values for a point source may vary considerably depending on the source mechanism (Douglas et al., 1972b). To illustrate such influences, $m_b(M_s)$ values measured at short distances from 14 collapse events at the Nevada Test Site are shown in Fig. 10.7A together with values from nearby underground explosions and earthquakes. The $m_b(M_s)$ values for collapse events are in general agreement with those observed for nearby earthquakes, but quite different from those of explosions. Since the source dimension of a collapse is of the same order as that of an explosion, the observed difference in $m_b(M_s)$ could be due to different source mechanisms and source-time functions.

The relative $m_b(M_s)$ values for deep earthquakes differ, as can be seen in Fig. 10.6, from those for events at shallow depths. Deep earthquakes usually have $m_b(M_s)$ values close to those observed for underground explosions. The influence of depth on the excitation of body and surface waves has been subject to theoretical considerations. A large difference between $m_b(M_s)$ values for explosions and those for earthquakes occurring very close to the earth's surface (depth 1—2 km) has been anticipated, but for earthquakes deeper than 10 km it should vanish (Douglas et al., 1971). This very strong depth dependence is, however, not supported by experimental data. Observations so far show that there is appreciable separation between explosions and earthquake down to depths of about 50 km.

The transmission properties of the earth between source and receiving station significantly influence the $m_b(M_s)$ values for both explosions

TABLE 10.1. Case studies of $m_b(M_s)$ discriminants.

	Explosions			Earthquakes					Deterrence probability for 1% false alarm (%)	Reference
Region	No.	Magnitude range		Region	No.	Magnitude range	Time period	Station		
NTS	51	M_L 4.0–6.3		Nevada region	34 incl. 22 collapses and aftershocks	M_L 3.8–6.5	1966–1971	Berkeley	87	McEvilly & Peppin (1972)
NTS, New Mexico	26	m_b(CAN) 4.5–6.3 M_s(CAN) 3.5–6.3		SW North America	28	m_b(CAN) 3.8–5.7 M_s(CAN) 4.0–6.0	1965–1968	Canadian network	94	Basham (1969)
NTS, Colorado	18	m_b(USGS) 5.0–6.5		SW North America	44	m_b(USGS) 3.9–5.6	1968–1970	Ogdensburg	98	Savino et al. (1971)
NTS	20	m_b(HFS) 5.5–6.5		SW North America	20	m_b(HFS) 4.8–6.0	1970–1974	Hagfors	>99	
Kazakh, W. USSR, Nov. Zemlya	29	m_b(USGS) 4.8–6.2 M_s(WWSSN) 3.0–4.4		Eurasia	83	m_b(USGS) 4.1–5.6 M_s(WWSSN) 3.1–6.7	1968–1970	42 Eurasian WWSSN stations	90	Marshall & Basham (1972)
Kazakh, W. USSR, Nov. Zemlya	46	m_b(USGS) 4.4–6.9 M_s(NAO) 2.4–5.5		Central Asia	45	m_b(USGS) 4.1–6.0 M_s(NAO) 3.1–6.4	1971–1974	NORSAR	96	Bungum & Tjöstheim (1976)
Kazakh, W. USSR, Nov. Zemlya	25	m_b(USGS) 5.1–6.7 M_s(ALPA) 2.7–4.6		Central Asia	66	m_b(USGS) 3.5–6.5 M_s(ALPA) 2.6–6.5	1971–1974	ALPA	98	Strauss (1973)
Kazakh, W. USSR	27	m_b(HFS) 5.2–7.1		Mainland USSR	127	m_b(HFS) 3.9–6.4	1969–1975	Hagfors	97	

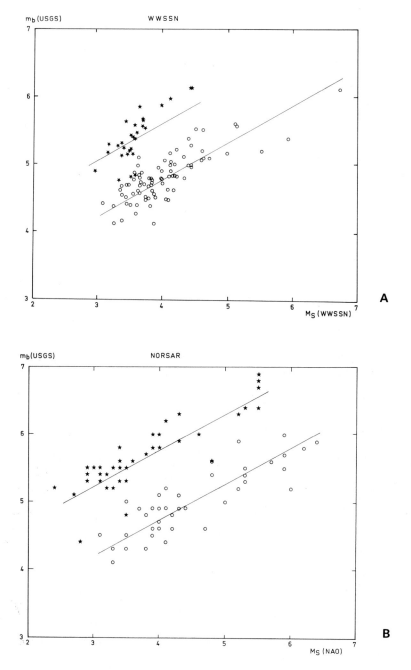

Fig. 10.8. $m_b(M_s)$ data for Eurasian explosions and earthquakes in four case studies using data from (A)–WWSSN, (B)–NORSAR, (C)–ALPA, and (D)–Hagfors. The trend lines were obtained as described in the legend to Fig. 10.7.

IDENTIFICATION

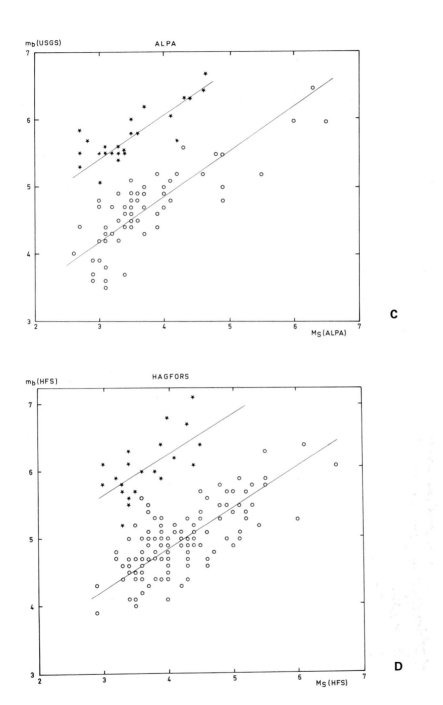

Fig. 10.8, C & D. (See legend on p. 226.)

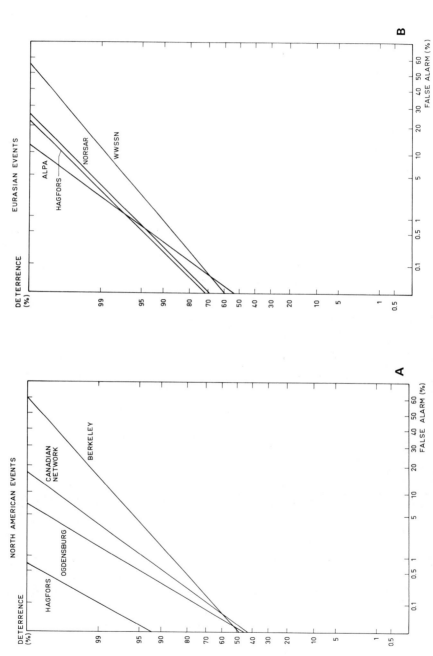

Fig. 10.9. Identification curves describing the relative efficiency of the case studies in (A)—Fig. 10.7, and (B)—Fig. 10.8 in terms of false alarm and deterrence. The empirical distribution functions for the discriminants were approximated by normal distributions.

IDENTIFICATION

and earthquakes. Strong lateral variations have been observed, and the $m_b(M_s)$ discriminant should therefore only be applied to earthquakes and explosions in one and the same region.

The degree of separation between earthquakes and explosions can change from region to region, but a generally valid observation seems to be that shallow earthquakes and underground explosions, with only few exceptions, exhibit different $m_b(M_s)$ ratios. The parameters that influence the $m_b(M_s)$ relation also seem to be understood, but we are still far from being able to predict in detail the $m_b(M_s)$ relation valid for explosions in new, untested areas.

10.2.1 Observations of $m_b(M_s)$

Throughout the years a large number of studies of event identification by means of the $m_b(M_s)$ discriminant have been published, for example by Basham (1969), Savino et al. (1971), Filson & Bungum (1972), Landers (1972), Marshall & Basham (1972), McEvilly & Peppin (1972), Lambert & Becker (1973), Strauss (1973), Evernden (1975), and Nuttli & Kim (1975).

Our intention here is to briefly summarize and evaluate some of these studies, without trying to cover everything written. Studies based on few data, and studies where data from significantly different regions are merged, i. e., where data from explosions in one part of the globe are compared with data from earthquakes in another part, are not considered. We will discuss eight studies in all, four of which were based on explosions and earthquakes in Western North America, and four on explosions in the USSR and on Eurasian earthquakes.

These studies are summarized in Table 10.1, and the $m_b(M_s)$ data are shown in Figs. 10.7 and 10.8. For each data sample a straight line has been fitted through the $m_b(M_s)$ data for the earthquakes, and a line having the same slope has then been fitted through the explosion data. For each data sample, identification curves have been calculated in the way discussed in the previous section of this chapter. These identification curves, which are shown in Fig. 10.9, facilitate a comparison of the degrees of separation between $m_b(M_s)$ data from explosions and earthquakes obtained in the different studies. Deterrence probabilities obtained at a false-alarm probability of 1 % are also given in Table 10.1.

North American events

The four studies of North American events are almost entirely based on explosions at Nevada Test Site and earthquakes in the southwestern part of North America. Most of the epicenters are within an area with a diameter of a few hundred kilometers; some of the events are common to two or more studies.

From the diagrams in Fig. 10.7 it can be seen that there is no overlap at all between the explosion and earthquake samples studied. The identification curves in Fig. 10.9 also indicate for all four studies a considerable separation of the two kinds of event. Some doubts have earlier been expressed as to whether the $m_b(M_s)$ discriminant would maintain its discrimination capability also for weak events (SIPRI, 1968). It is therefore of special interest to notice in Fig. 10.7A the complete separation for weak events down to a magnitude M_L of ca. 4.0, which magnitude corresponds to an explosion in hard rock with a yield of the order of 1 kt (cf. Table 10.1). As these four studies cover a considerable number of events and involve stations at different distances it is reasonable to conclude that the $m_b(M_s)$ discriminant applied to events in the southwestern parts of North America provides a high degree of separation between explosions and earthquakes.

The four studies referred to are case studies, and do not necessarily reflect achievable operative capability. Nevertheless, they are of interest, since they indicate how far down in event strength $m_b(M_s)$ data can be applied. In the study using local data from stations at distances of about 500 km from the epicenters, Rayleigh waves with ground amplitudes of the order of 50 nm, or local magnitudes, M_L, of about 4, were included (McEvilly & Peppin, 1972). This corresponds to explosion yields of the order of 1 kt in hard rock. Canadian M_s data have been reported from explosions with magnitudes down to m_b(CAN) = 4.5, which limit corresponds to a hard-rock yield of about 10 kt. M_s data obtained at the Ogdensburg station, at roughly the same distance from the region of events in question as some Canadian stations, have been reported from explosions with m_b(USGS) down to about 5.0, which corresponds to a yield of about 30 kt in hard rock (Ericsson, 1971b). This illustrates that, although VLPE stations are well able to record long-period surface waves from earthquakes, they have a rather limited capability to record explosion signals. M_s data obtained at the Hagfors Observatory, at a distance of about 75 degrees, could be applied down to m_b(HFS) = 5.5, corresponding to a hard-rock yield of about 75 kt.

These figures illustrate that, although the ability to separate explosions and earthquakes by the $m_b(M_s)$ discriminant is similar in all studies presented here, the applicability of $m_b(M_s)$ is quite variable, depending on the ability to obtain the necessary data. In these studies the lower limit to the applicability ranged from hard-rock yields of the order of 1 kt for data recorded by local stations, up to yields of the order of 100 kt for data from a station at large distance and on another continent.

Eurasian events

The seismicity of the USSR has been discussed in Chapter 4. In one of the four $m_b(M_s)$ studies of Eurasian events discussed here, only earthquakes occurring within the USSR borders were utilized, most of them having been located in the southern border region of the Union. In three other studies, also events from adjacent seismic areas were used. Nuclear explosions have been carried out in the USSR not only at the two main test sites in Eastern Kazakh and at Novaya Zemlya, but also at a number of sites in western USSR. The individual studies have used somewhat different explosion populations. The regions where the explosions were conducted are given in Table 10.1.

All $m_b(M_s)$ data were obtained by stations located outside the USSR, which means that close-in data corresponding to those used by McEvilly & Peppin (1972) for the North American events are not available for the USSR explosions. The $m_b(M_s)$ data for the studies presented in Fig. 10.8 show a good separation between the explosion and earthquake samples. The estimated identification curves, Fig. 10.9, are quite similar to those for the North American events.

In the CCD there have been technical discussions about a sample of earthquakes in the Tibet–Assam region $m_b(M_s)$ data for which are close to those observed for explosions in the USSR (CCD/388, 1972; CCD/399, 1973; CCD/407, 1973). The earthquakes referred to in the US working paper CCD/388 (1972) were discussed in relation to $m_b(M_s)$ data obtained by Marshall & Basham (1972) from Eurasian explosions. The $m_b(M_s)$ data for these two event populations are shown in Fig. 10.10. For reasons of consistency, we have from the Marshall–Basham study selected the Eastern Kazakh explosions and the earthquakes occurring in central Asia. The locations of these events and those in the US working paper are shown in Fig. 10.11. The separation between these populations of explosions and earthquakes (Fig. 10.10) is smaller

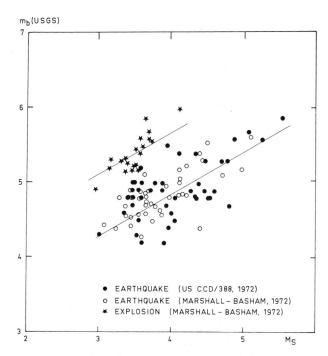

Fig. 10.10. The $m_b(M_s)$ values for Eurasian events studied in the US working paper CCD/388 (1972) and by Marshall & Basham (1972). The events were located as shown in Fig. 10.11.

than in the earlier presented studies. In a Japanese working paper (CCD/399, 1973) the consistency of the short-period magnitude estimates was discussed, and by using other m_b values, a somewhat better separation was achieved. M_s values from three of the earthquakes discussed in the US working paper were also given in a paper by Landers (1972); the two magnitudes differ for all three events by up to 0.5 magnitude units. This discussion does show not only the importance of using consistently calculated values of m_b and M_s for event identification, but also that the $m_b(M_s)$ discriminant has to be applied on regionalized data, as the $m_b(M_s)$ ratio may vary regionally, and particularly in geophysically extreme places. We can, however, conclude that the $m_b(M_s)$ method, when applied in a consistent way, will give also in Eurasia a very high degree of separation between explosions and earthquakes, despite the fact that here the events are spread over a large continent. In the study by Marshall & Basham (1972), explosions with m_b (USGS) down to 4.8, corresponding to a hard-rock yield of about 5 kt, were included. The $m_b(M_s)$ data on these

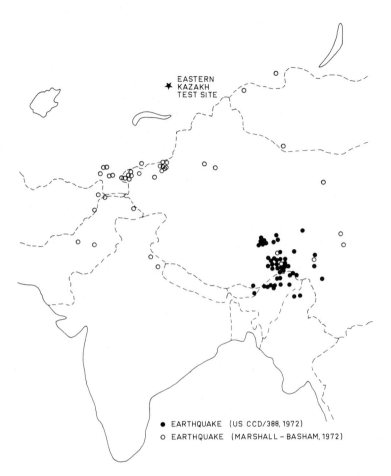

Fig. 10.11. Locations of the Eurasian events studied in the US working paper CCD/ 388 (1972) and by Marshall & Basham (1972).

weak explosions were, however, mixed with those on earthquakes. It also seems difficult to confidently detect surface waves from such weak explosions by conventional long-period equipment. The NORSAR $m_b(M_s)$ data referred to were applied to explosions in western USSR down to m_b(USGS) of 4.4, corresponding to a hard-rock yield of about 3 kt. The Hagfors $m_b(M_s)$ data were obtained from explosions with m_b(HFS) down to 5.2 in central Kazakh, which corresponds to a yield in hard rock of 10 kt. Finally, the ALPA $m_b(M_s)$ data were reported down to m_b(USGS) = 5.1, which also corresponds to a hard-rock yield of about 10 kt.

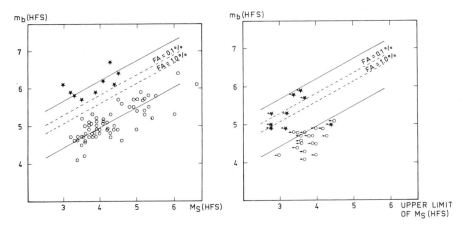

Fig. 10.12. Verification by negative evidence of USSR events in 1971 (Dahlman et al., 1974). The diagram at left shows $m_b(M_s)$ values obtained at Hagfors, where both m_b and M_s could be calculated. The diagram at right shows data for m_b versus estimates of upper limits to M_s indicated by arrows. The dashed lines are estimates of false-alarm probabilities, FA, of 0.1 and 1 %.

Applicability

The applicability of the $m_b(M_s)$ method is limited by the surface-wave detection capability. The applicability figures achieved from the case studies presented here are in general agreement with the long-period detection thresholds mentioned in Chapter 7. The detection capability for long-period signals can be significantly reduced by interference, particularly as long-period signals have long duration and thus easily overlap and mix. A study of the number of mixed signals at the VLPE and long-period array stations indicates that about 15 % of the events have mixed long-period signals (CCD/388, 1972). Array processing can separate some of these signals, and some signals can be separated using a network of globally distributed stations. Surface waves from very large earthquakes do, however, severely disturb all long-period stations on the globe for a considerable time period after the origin time.

In the method called verification by negative evidence the $m_b(M_s)$ discriminant is used to identify explosions in cases where no surface waves have been detected (Elvers, 1974). The noise amplitude at the expected arrival time of the surface waves is used as an estimate of the upper limit to M_s. In Fig. 10.12 we illustrate this method by means of $m_b(M_s)$ data obtained at the Hagfors Observatory from explosions and earthquakes in the USSR in 1971. The diagram at left contains the

IDENTIFICATION

events from which both m_b and M_s data were obtained at Hagfors. In the diagram are drawn the 0.1 % and 1 % false-alarm lines, these should be interpreted to mean that there is a probability of 0.1 % and 1 %, respectively, that $m_b(M_s)$ data from the actual earthquake population will fall above these lines. The diagram at right contains those events for which an m_b value is available together with an upper-limit estimate of M_s. We can see that in this sample six additional explosions can be identified with a probability of <1 % of making a false alarm about an earthquake. In this example, the applicability of the $m_b(M_s)$ method is thus increased by about half an m_b unit, or down to m_b (HFS)=5.0, corresponding to a hard-rock yield of about 5 kt in Eastern Kazakh.

We conclude that in a large number of studies, based on data obtained at many different stations from explosions and earthquakes in different parts of the world, the $m_b(M_s)$ discriminant has proved to provide very good separation between explosions and earthquakes. It has also been shown that this separation is attainable also for weak events, that is, events equivalent to explosion yields in hard rock down to about 1 kt. The applicability of the $m_b(M_s)$ method is determined by the ability to detect long-period surface waves. The present detection capability of the most sensitive of the existing long-period stations, located at teleseismic distances from the explosions but on the same continent, is equivalent to hard-rock explosions with a yield of about 10 kt. In cases where data from close-in stations are available or where the method of identification by negative evidence is applied, the $m_b(M_s)$ method can be applied to still weaker explosions.

10.3 SHORT-PERIOD DISCRIMINANTS

In Chapter 7 it is shown that the most effective way to detect seismic sources at teleseismic distances is to use the frequency band 0.5–5 Hz to detect the short-period part of vertical-component P waves. It is quite natural that efforts already in the early days of detection seismology were devoted to developing discriminants based entirely on the short-period P waves (Thirlaway, 1965). Figure 10.13 shows P-wave signals from an earthquake and an explosion, where the explosion signal looks pulse-like and quite simple, whereas the earthquake signal appears to be of much longer duration and more complex. Besides the differences in wave-form complexity, the two signals in Fig. 10.13 also exhibit differences in spectral content. It can be seen that the characteristic signal period is shorter for the explosion than for the earthquake.

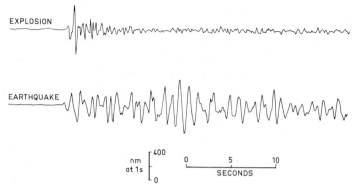

Fig. 10.13. Short-period vertical-component signals obtained at the Hagfors Observatory from a shallow earthquake and an underground explosion, both occurring in the USSR. According to USGS, the events had the following source data:

Event	Date	Origin time	Epicenter location	Depth (km)	m_b
Explosion	700906	040257	49.8N 78.1E	0	5.6
Earthquake	700605	045306	42.5N 78.8E	20	6.0

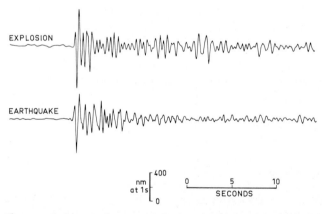

Fig. 10.14. Short-period vertical-component signals obtained at the Hagfors Observatory from the San Fernando earthquake in California and from the explosion "Handley" at the Nevada Test Site, according to the USGS having the following source data:

Event	Date	Origin time	Epicenter location	Depth (km)	m_b
Explosion	700326	190000	37.3N 116.5W	0	6.5
Earthquake	710209	140042	34.4N 118.4W	13	6.2

The wave-form complexity and the spectrum of the short-period signals are the characteristics most widely used for discrimination. Figure 10.14 shows explosion and earthquake signals from another region. In

IDENTIFICATION

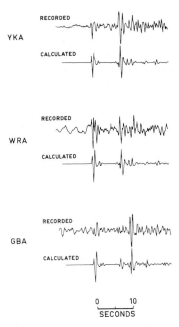

Fig. 10.15. Modelling of short-period signals recorded at the three UK-type arrays YKA, WRA, and GBA from an event in Eastern Kazakh on 690501 at 040008, site 44.0N 77.9E, m_b(USGS) = 4.9. In the UK working paper CCD/440 (1974) the event was considered to have been an earthquake.

this case there is no clearly seen difference in wave-form complexity nor in characteristic period between explosion and earthquake. In a simple way, the four signals shown in Figs. 10.13 and 10.14 set the whole scene of discrimination by short-period signals. In many cases there seem to be large differences between explosions and earthquakes, but there are also several cases where it is difficult to see any differences at all. Attempts have been made to explain by various physical models the large variability that short-period signals exhibit. Such models, some of which have already been outlined in Chapter 5, describe the source, the crust, and the mantle by various parameters, and have been used to match computed and observed signals. Figure 10.15 shows a case where a fairly successful match was obtained. Signal matching has so far been conducted only in a few cases, and few data are available about its capabilities. This, so to say, physical approach to the discrimination problem does however look promising. In the meantime, however, one has to base discrimination by short-period signals on heuristically derived discriminants, and describe their performance in statistical terms. The remainder of this chapter is dealing with this approach.

Discriminants can be one-dimensional or multidimensional, i. e. characterize the short-period signals by one or several numbers. As is discussed in Chapter 6, several array stations provide signals in digital form. The most general discriminant for such digital recording would be the whole set of values sampled from short-period signals. Sampling rates of 10 or 20 Hz would, however, lead to far too many dimensions, since there is usually only a limited number of explosions available for analysis. One must therefore reduce the number of dimensions somehow, without losing the inherent differences between explosions and earthquakes.

Most of the short-period discriminants that have been used are related to the wave-form complexity and the frequency content. In the following we therefore consider discriminants in the time domain and in the frequency domain separately and start the presentation of time-domain discriminants.

10.3.1. The time domain

The majority of discriminants utilizing the signatures of the short-period signals themselves can be related to the concept of complexity, which essentially involves comparison of the amplitudes of the initial part of the short-period signal with those of the succeeding coda. One definition of the complexity, C, can be written as

$$C = \frac{\int_{t_1}^{t_2} s^2(t)\,dt}{\int_{t_0}^{t_1} s^2(t)\,dt},$$

where $s(t)$ denotes the signal amplitude as a function of time, t. The signal $s(t)$ in the formula varies with the type of recording station employed. In the case studies based on records at HFS, $s(t)$ corresponds to an appropriately bandpass-filtered single-sensor signal (Israelson, 1972). At LASA, $s(t)$ represents the beam (Kelly, 1968); the original definition of C at the UK-type array stations was based on a smoothed version of the cross-correlation between the two beams of the orthogonal legs of the arrays (Marshall & Key, 1973). The complexity of a signal can also be calculated manually from analog records, using the definitions by Bormann (1972).

A detailed study of different definitions of C by Kelly (1968) suggests that the effectiveness of C does not depend critically on the exact values of t_1 and t_2. Usually, t_0 is taken to be the onset of the short-

IDENTIFICATION

period signal, and typical values of the interval length vary between 2 and 5 s for $t_1 - t_0$, and between 25 and 35 s for $t_2 - t_1$.

The multidimensional time-domain discriminants represent a more detailed description of the wave forms of the P signals than the complexity, C. A ten-dimensional vector, $\bar{S}$, describing the relative amplitudes during the initial ten seconds after the onset has been used by Israelson (1972). The components S_i of this vector are defined by

$$S_i = \frac{\int_{i-1}^{i} |s(t)| dt}{\int_{i=1}^{10} |s(t)| dt}, \quad i = 1, 2, ..., 10.$$

The dimension of $\bar{S}$ can be reduced to unity by forming a linear combination, as discussed in Section 10.1:

$$L(\bar{S}) = \sum_{i=1}^{10} W_i \log S_i.$$

The coefficient W_i can be calculated to provide maximum separation between explosions and earthquakes for each individual data sample. The reason for using a logarithmic transformation of S_i is that it improves the separation and that the logarithms of the observed values of S_i fit a normal distribution better than do the S_i values themselves. Since the coefficients W_i have to be calculated from a given data sample, this type of discriminant can be applied only to fairly large and representative data bases.

Other definitions of the signal vector than $\bar{S}$ can be applied to analog recording as well (Kedrov, 1971).

Autoregressive models have also been used to represent P signals. Tjöstheim (1975b) has estimated the coefficients a_i of the relation:

$$s(t) + \sum_{i=1}^{h} a_i s(t-i) = n(t),$$

where $s(t-i)$ is the signal $s(t)$ shifted in time, and $n(t)$ denotes an artificial process of white noise, i. e. a process with constant amplitude spectrum. Tjöstheim found that $h = 3$ gives a reasonable fit to the relation, and used the coefficients a_i as discriminants.

Noponen (1975) has introduced a time-domain discriminant calculated from the deconvolved signal, i. e. the signal corrected for the response of the seismometer. This discriminant is called *pulse width*, and is defined as the ratio of the area to the maximum amplitude of the deconvolved signal.

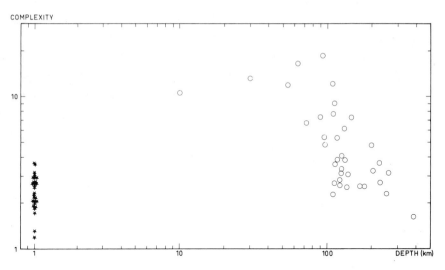

Fig. 10.16. Wave-form complexity obtained at the Hagfors Observatory versus focal depth for earthquakes in the USSR–Afghanistan border region. Data from explosions in the Eastern Kazakh test area are shown for comparison. The explosions have been placed at a depth of 1 km in the figure.

Time-domain discriminants are on the average different for explosions and earthquakes, as illustrated by the data for wave form complexity in Fig. 10.1. The effectiveness and applicability will be discussed in relation to the case studies in Section 10.3.3. Here we only outline the variation in time-domain discriminants caused by factors other than the source type. Since the effects of these factors are similar regardless of the exact definition of the time-domain discriminant, they are discussed in terms of complexity. Complexities for shallow earthquakes often increase with increasing magnitude. This can be an effect of the increased dimension of the earthquake source region (see Chapter 5). Explosion complexities, however, do not change significantly with magnitude. The complexity is usually a function of the focal depth, as can be seen from Fig. 10.16. For earthquakes at depths larger than about 70 km the complexity is frequently quite explosion-like. The simplicity of the short-period signals from deep earthquakes may be due partly to the source mechanism and partly to the fact that there is little interaction with the crust or with the surface at the epicenter. Signals from explosions and earthquakes at distances less than about 30 degrees are usually quite complex, probably due to multipathing (see Chapter 4). At distances beyond about 80 degrees, the *PcP* phase may interfere quite strongly with the *P* coda. This limits the applicability of time-domain discriminants to a distance range of from 30 to 80 degrees. The complexity discriminant is also subject to regional variation. The complexity values observed for explosions at most of the test sites in Eurasia, North America, North Africa, and South Pacific are all quite small, as one would expect, but there are two notable exceptions, i.e., the values for Novaya Zemlya and the Aleutian Islands. Explosions at the main test site on Novaya Zemlya consistently give complex and earthquake-like short-period signals at many seismic stations in North America; the signals at most stations in Europe are, however, simple and quite explosion-like (Douglas et al., 1973). A similar effect was observed for the "Longshot" explosion in the Aleutian Islands. Attempts have been made to relate the observed irregularities of "Longshot" to lateral velocity contrasts in the descending lithospheric plate beneath the Aleutians (Davies & Julian, 1972; Jacobs, 1972). More recently it has been proposed by Douglas et al. (1973) that the large complexity of some explosions is due to multipathing with differential absorption. They suggest that the initial P pulse travels along a path through the mantle with high nonelastic absorption, whereas the coda is due to waves travelling along a path with low absorption.

10.3.2 *The frequency domain*

It is well known that a signal in the time domain can be represented by its Fourier transform in the frequency domain. The amplitude component, $A(f)$, of the Fourier transform, as a function of frequency, f, has been widely used in searching for effective discriminants. The phase component shows a strongly irregular behavior, and is difficult to interpret, and has therefore been less extensively studied. From digital records a discrete version of the amplitude spectrum can readily be obtained by established, discrete Fourier-transform algorithms (Beauchamp, 1973). Such spectral calculations are usually carried out over a record length of from 5 to 20 s. $A(f)$ is contaminated with seismic noise, which has to be corrected for. As the short-period noise usually is stationary during at least several minutes, one can use noise samples from the record just preceding the short-period signal for this correction. At an array it is usually not practical to estimate $A(f)$ from the beam signal, as the high-frequency part of the short-period signals is cancelled out in the beamforming, as pointed out in Chapters 4 och 7. An average from the signal spectra of the individual sensors usually provides a more effective estimate of the high-frequency part of the short-period frequency band. In analog recording, $A(f)$ can be approximated by the amplitude of narrow passbands (Miyamura & Hori, 1972). Here we emphasize discriminants based on digital recording, but the conclusions are to a large extent applicable also to discriminants obtained from analog recording. The discriminants derived from $A(f)$ mostly attempt to describe the high-frequency content of the short-period band.

The dominant period of the short-period signal presents itself immediately as a discriminant that can be obtained easily also from analog records. This period corresponds roughly to the frequency for which $A(f)$ has its largest value. The dominant period can be measured also on weak signals, but gives a rather incomplete characterization of the spectral content. One early discriminant based on $A(f)$ compared the amplitudes in one low-frequency and one high-frequency band, and has therefore been called *spectral ratio, SR* (Kelly, 1968). It can be written as:

$$SR = \frac{\int_{h_1}^{h_2} A(f) df}{\int_{l_1}^{l_2} A(f) df}.$$

The integration limits h_1 and h_2 of the high-frequency and l_1 and l_2 of the low-frequency bands are different for different stations, and are usually determined by comparing average spectra of explosions and earthquakes. Typical bands that have been used for LASA data are 1.45–1.95 and 0.35–0.85 Hz (Kelly, 1968). By comparing average spectra of explosions and earthquakes it has been found that the separation was improved by using several bands to form a combined spectral ratio (Filson, 1969).

There are also spectral parameters expressing the relative high-frequency content without considering the detailed amplitude spectra. One such parameter, which has been widely used, is the *third moment of frequency, TMF,* defined as (Weichert, 1971):

$$TMF = \left\{ \frac{\int_0^5 f^3 A(f) df}{\int_0^5 A(f) df} \right\}^{\frac{1}{3}},$$

where f is expressed in Hz. This formula puts high weights on the high-frequency components. Therefore, explosions usually yield large *TMF* values. *TMF* has no firm theoretical basis and is a somewhat arbitrary

IDENTIFICATION

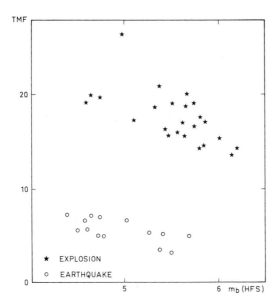

Fig. 10.17. Plot of the third moment of frequency, TMF, versus m_b for explosions in Eastern Kazakh and earthquakes in the Kirgiz–Sinkiang region. Data are based on records obtained at the Hagfors Observatory.

or heuristic parameter; nevertheless, it has turned out to be a quite useful discriminant.

Discriminants based on data obtained from the whole spectrum, and therefore more rich in information, have also been tested. Dividing the whole spectral range of interest into n narrow bands of width $b_k - b_{k+1}$, one can characterize the spectrum by a vector $\bar{A}$ with components A_i defined by (Israelson, 1972):

$$A_i = \frac{\int_{b_{i-1}}^{b_i} A(f) df}{\int_{b_0}^{b_n} A(f) df}, \qquad i = 1, 2, ..., n.$$

This spectral amplitude vector is analogous to the signal vector $\bar{S}$ in the time domain, and is for discrimination purposes usually treated in a similar way as the latter. The spectral discriminants generally seem to provide a larger separation between explosions and earthquakes than do the time-domain discriminants. The effectiveness and applicability of these two types of discriminant will be discussed in the next section.

The spectral discriminants do not depend only on the nature of the seismic event (explosion or earthquake), but also on several other factors. We therefore conclude this section with a brief discussion of the

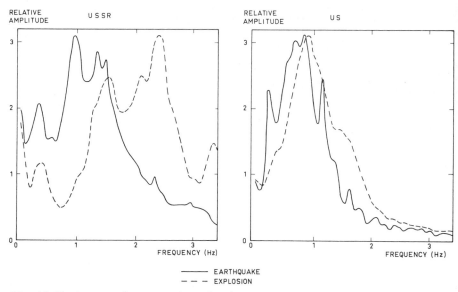

Fig. 10.18. Average short-period amplitude spectra for explosions and earthquakes in the US and the USSR, based on data from the Hagfors Observatory. Each spectrum in the figure is the average of about 20 events.

effects of some of these factors. Both for explosions and earthquakes, the frequency-domain discriminants usually are functions of the source strength. This is to be expected from seismic source models (Aki, 1972; Filson & Frasier, 1972). The larger the source the smaller the proportion of energy radiated at high frequencies. This effect, which is about the same for explosions and earthquakes, can be corrected for in many cases. Figure 10.17 gives an example of the effect of event magnitude on the *TMF* discriminant. Deep earthquakes often have explosion-like values of the frequency-domain discriminant. This is probably partly due to the fact that P waves from deep earthquakes traverse the crust only at the receiver end of their path. Systematic large-scale regional variations in the spectral content of P waves have also been observed. Signals recorded at the Hagfors Observatory from North American seismic events, whether explosions or earthquakes, contain, as illustrated in Fig. 10.18, generally much less high-frequency components than do signals from events in Central Asia (Israelson, 1971). This means that, to provide effective discrimination, spectral discriminants have to be tailored to each specific region and each seismic station. The regional dependence of spectral discriminants can be quite strong also on a fine scale. Rapid changes over just 100–200 km have been observed for explosions and earthquakes across the Eurasian continent (Israelson,

IDENTIFICATION

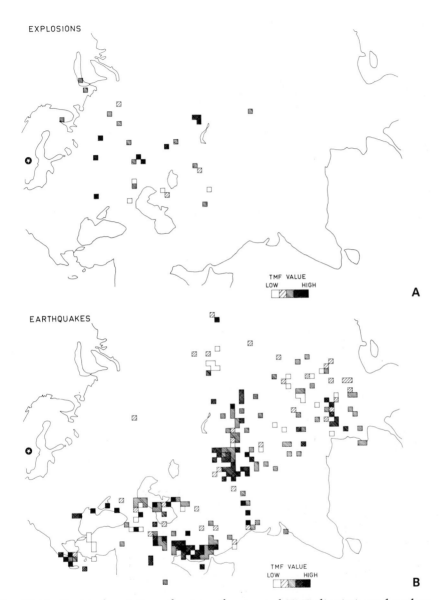

Fig. 10.19. Regional variation of a strength-corrected TMF discriminant based on data obtained at the Hagfors Observatory from (A)—explosions, and (B)—earthquakes in Eurasia. The figure shows the variation relative to the average value for explosions and earthquakes, respectively; however, explosions and earthquakes are not compared, as their average values are different. The data represent regional averages taken over areas of 1 deg. by 1 deg., and are classified on a five-grade scale from "low" to "high".

1975). An example of this fine-scale variation is given in Fig. 10.19 showing *TMF* values obtained at the Hagfors Observatory for about 700

TABLE 10.2. Case studies of short-period discriminants.

	Explosions			Earthquakes						Discriminants	Deterrence probability for 1% false alarm (%)	Reference
Region	No.	Magnitude range		Region	No.	Magnitude range	Time period	Station				
NTS	49	m_b(YKA) 4.4–5.8		Gulf of California	59	m_b(YKA) 4.2–5.8	1966–1970	YKA		Modified moment of frequency	69	Manchee (1972)
NTS	29	m_b(USGS) 4.6–6.5		North America	37	m_b(USGS) 4.7–6.5	1969–1972	HFS		Complexity and TMF	25	Israelson & Wägner (1973)
Sino-Soviet	67	m_b(USGS) 4.5–6.0		Eurasia	242	m_b(USGS) 4.1–6.3	1965–1972	YKA		Complexity and TMF	34	Anglin (1972)
Sino-Soviet	42	m_b(USGS) 4.4–6.3		Eurasia	508	m_b(USGS) 3.3–6.1	1969–1974	HFS		Complexity and TMF	61	Israelson (1975)
Sino-Soviet	48	m_b(USGS) 4.8–6.9		Eurasia	63	m_b(USGS) 4.8–6.0	1966–1972	TSK		Spectral magnitudes at 1 and 2 Hz	58	Miyamura & Hori (1972)
Soviet	40	m_b(USGS) 4.4–6.8		Eurasia	45	m_b(USGS) 4.1–6.0	1971–1974	NORSAR		Autoregressive coefficients	59	Tjöstheim (1975b)

earthquakes and 50 explosions in Eurasia. Large spectral differences between explosions in Western and Eastern Kazakh have also been observed at other stations, e. g. Tsukuba in Japan (Miyamura & Hori, 1972) and Yellowknife in Canada (Anglin, 1972). This indicates that effects near the source can have a significant influence on the frequency-domain discriminants. Clear spectral differences between explosions and earthquakes at virtually one and the same epicenter have been demonstrated for several cases (Lambert et al., 1969; Bakun & Johnson, 1970; Murphy & Hewlett, 1975).

10.3.3 Case studies

In order to illustrate the performance of short-period discriminants we present here data of a few case studies. It should be noted that, because of the strong regional dependence of short-period discriminants, the results pertain strictly to these particular case studies only. Projections to the performance of short-period discriminants in previously untested areas should be made with great care.

The amount of actual data on short-period discriminants published so far is much smaller than the amount published on the $m_b(M_s)$ discriminant. In addition, most case studies on short-period discriminants are based on only a rather small number of events. Moreover, the data seldom represent a systematic search for events in a given region and over a given period of time. Here we will briefly discuss the results of six studies, two of which deal with North American and four with Eurasian events. The pertinent data of these studies are summarized in Table 10.2. The discriminants were obtained from digital data recorded at three arrays, Hagfors Observatory, NORSAR, and Yellowknife, and from analog data recorded at the station of Tsukuba, Japan. The basic discriminant data are shown graphically in Fig. 10.20. Somewhat different discriminants have been calculated at the different stations. Complexity and *TMF* were calculated at Hagfors Observatory and at Yellowknife, the *TMF* parameters of Hagfors Observatory being corrected also for the magnitude of the event. For the North American events, however, there are no complexity values for Yellowknife, because of the small epicentral distances. For the NORSAR data the two highest autoregressive coefficients were defined and plotted against each other. The discriminant at Tsukuba consists of the estimated magnitudes from signals at 1 Hz and 2 Hz. Linear discriminants of the type described in Section 10.1 have been formed to calculate identification curves—these

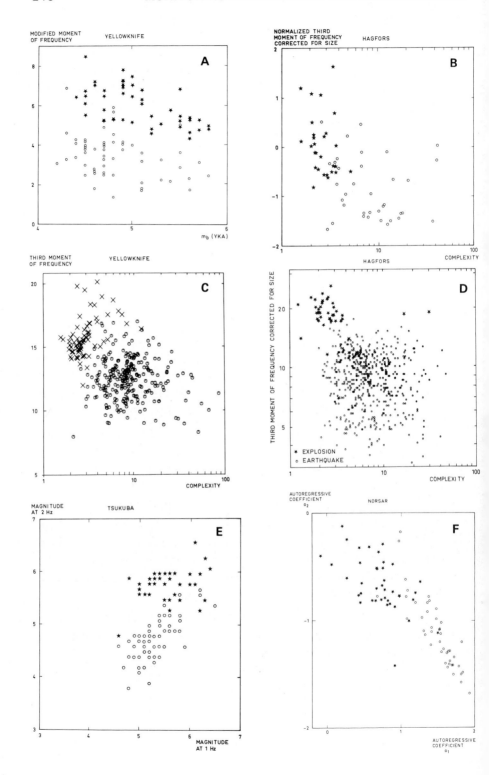

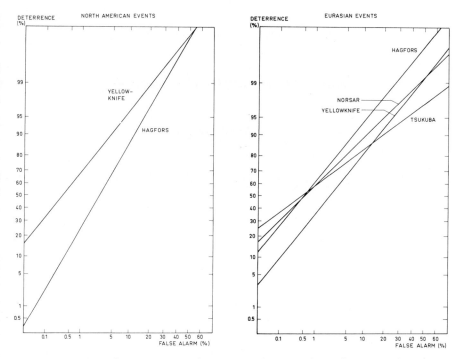

Fig. 10.21. Identification curves for case studies based on short-period discriminants and the data shown in Fig. 10.20.

are shown in Fig. 10.21. It is important to note that all studies are based on different events. Only the samples of Eurasian events for Hagfors Observatory and Yellowknife can be regarded as large.

The events used in the studies of North America and Eurasia differ in various ways. The explosions in North America are essentially concentrated to the Nevada Test Site, whereas explosions on the Eurasian continent are distributed between about twenty widely separated places. Also the Eurasian earthquakes are more scattered geographically than the North American ones, as was discussed in Section 10.2. The higher seismic activity in Eurasia also gives a larger number of earthquakes in that region.

The identification curves shown in Fig. 10.21 suggest that the short-period discriminants at Hagfors are more effective in separating explosions and earthquakes in Eurasia than in separating events in North

Fig. 10.20. Scatter diagrams of short-period discriminants for two case studies of North American events, based on data at (A)—Yellowknife, and (B)—Hagfors, and for four case studies of Eurasian events, based on data at (C)—Yellowknife, (D)—Hagfors, (E)—Tsukuba, and (F)—NORSAR.

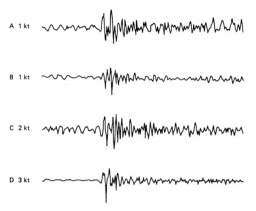

Fig. 10.22. Short-period seismic signals obtained at the Hagfors Observatory from explosions at the Eastern Kazakh test site, having the following source data according to the USGS:

Signal	Date	Origin time	Epicenter location	m_b
A	720706	010258	49.7N 78.0E	4.4
B	691229	040158	49.7N 78.2E	4.6
C	740625	035658	49.9N 78.1E	4.7
D	740416	055302	50.0N 78.8E	4.8

The yield values in kilotons refer to estimated hard-rock yields (cf. Chapter 11).

America. This may be due to the severe attenuation of high-frequency short-period signals recorded at Hagfors from events on the North American continent. A comparison between the IC in Figs. 10.9 and 10.21 shows that the $m_b(M_s)$ discriminant would provide a larger separation between explosions and earthquakes than would short-period discriminants. This was confirmed by comparisons of the methods applied to identical event samples (Dahlman et al., 1974).

Although short-period discriminants do not separate explosions and earthquakes as well as does the $m_b(M_s)$ discriminant, they are still quite useful for seismic discrimination. The short-period discriminants have wider applicability than the $m_b(M_s)$ discriminant. Table 10.2 shows that short-period identification parameters could be obtained at Yellowknife and Hagfors for NTS explosions down to body-wave magnitudes, m_b(YKA)=4.4 and m_b(USGS)=4.6, respectively. This corresponds to yields in tuff of the order of 2–3 kt and 5 kt, respectively. Short-period identification parameters for explosions in the USSR were obtained at Hagfors down to m_b(HFS)=4.6, and at NORSAR down to about m_b(USGS)=4.4. These magnitudes correspond to a yield in hard rock of about 1 kt. Figure 10.22 shows the short-period records obtained at the

IDENTIFICATION

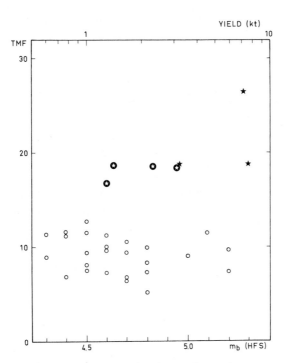

Fig. 10.23. Plot of TMF versus m_b for low-yield explosions in Eastern Kazakh (stars) and low-magnitude earthquakes in Central Asia (open circles). The four explosions of Fig. 10.22 are represented by open stars in filled circles. The yield scale refers to estimated hard-rock yields (cf. Chapter 11).

Hagfors Observatory from four USSR explosions with yields estimated to be between 1 and 3 kt in hard rock. The TMF identification parameters for these explosions and some other low-yield explosions are compared in Fig. 10.23 with data from low-magnitude earthquakes. This figure demonstrates that good separation between explosions and earthquakes is maintained also for weak events. A signal-to-noise ratio of about 6–8 dB is usually needed for calculating a short-period discriminant. This means that the threshold of applicability is approximately 0.3–0.4 magnitudes above the detection threshold for P waves. Moreover, short-period discriminants are not as sensitive as the $m_b(M_s)$ discriminant to noise variations and signal interference.

10.4 OTHER DISCRIMINANTS

Besides the $m_b(M_s)$ and short-period discriminants discussed above, a variety of other discriminants have been proposed and tested over the years. Most of these discriminants are generally applicable to large events only. We therefore confine ourselves to a brief summary of these discriminants.

Polarity of initial motion

In the technical discussions of the experts in Geneva in 1958 the *polarity discriminant* was believed to provide the solution to the problem of seismic discrimination. This discriminant is based on differences in polarity of the initial motion of P waves. As mentioned in the discussion of simplified seismic models, Chapter 5, an explosion source is assumed to be completely symmetric and to radiate compressive initial motion in all directions. Theoretically the radiation from certain source models of earthquakes exhibits, on the other hand, two lobes of compressive motion and two lobes of rarefactional motion. It was soon realized that these radiation patterns put severe limitations to the usefulness of the polarity discriminant (De Noyer, 1963). First of all, some observed explosion polarities proved to be rarefactions (Enescu et al., 1973). Moreover, it is often difficult or even impossible to discern the initial motion from weak events. The method also requires an adequate azimuthal coverage of seismic stations, which for many events is difficult to achieve.

Corner frequency

In Chapter 5 it was mentioned that the far-field amplitude spectra of most seismic source models are quite simple. The spectra are constant from zero frequency up to the so-called corner frequency, above which the amplitude drops with frequency f as f^{-2} or f^{-3}. Theoretically, the corner frequency is higher for an explosion than for an earthquake having the same amplitude at zero frequency. At teleseismic distances the corner frequency can be estimated from the amplitude spectra of long-period and short-period P-wave signals. Observed corner frequencies differ by a factor of 3—10 between explosions and earthquakes. So far, only small data samples of large events have been analyzed in this respect (Wyss et al., 1971; Noponen, 1973). The usefulness of corner frequency as a discriminant is probably limited to large seismic events.

Broad-band P waves

Spectral differences between explosions and earthquakes can be clearly seen on broad-band P-wave records, as is illustrated by Fig. 10.24.

IDENTIFICATION

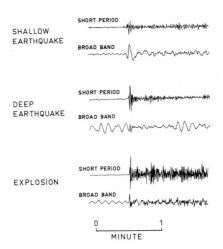

Fig. 10.24. A comparison of short-period and broadband signals from a shallow earthquake, a deep earthquake, and an explosion, having the following source data according to the USGS:

Event	Date	Origin time	Epicenter location	Depth (km)	m_b
Shallow e.q.	710518	224444	64.0N 146.1E	33	5.8
Deep e.q.	720120	113629	36.4N 70.7E	213	6.0
Explosion	701014	055957	73.3N 55.1E	0	6.7

The records were obtained with the broadband array near Blacknest, UK (Marshall et al., 1972).

Whereas there is no apparent difference between the short-period records from the deep earthquake and those from the explosion, the frequency contents of the corresponding broad-band records from the two events differ clearly. In many cases, as in the example shown in Fig. 10.24, also the differences between broad-band records of explosions and those of shallow earthquakes are so clear, that discrimination is obtained simply by measuring the period of the P wave (Burton, 1975b). Although discrimination based on broad-band recording is so far limited to large events, this discriminant can serve as a valuable tool for counteracting some evasion methods, as will be further discussed in Chapter 13 (Marshall & Hurley, 1975).

Love waves

Theoretically, an explosion in a perfectly elastic medium should not radiate any Love waves, whereas an earthquake should generate such waves in certain directions. This means that Love waves, from a theoretical point of view, should be well suited for identification of explosions by the technique of negative evidence presented in Section 10.2. However, Love waves have in fact been observed from some explosions. It is usually assumed that such Love waves are associated with strain release, and that the amount of Love-wave radiation emitted depends on the strength or depth of the explosion (Aki & Tsai, 1972). In certain earthquake focal models the radiation patterns of Love and Rayleigh waves are complementary to one another so that at azimuths where Rayleigh-wave radiation is at a minimum, Love-wave radiation has a maximum, and vice versa. This could be utilized to improve the $m_b(M_s)$ discriminant by basing M_s on the maximum of the Love- and Rayleigh-wave amplitudes (Lacoss, 1971). So far little has been published on discrimination based on Love waves (CCD/484, 1976), which fact may partly be due to the difficulty to record weak explosion signals on horizontal-component, long-period seismometers. The VLPE instruments (Savino et al., 1971) and the planned SRO network have, however, rearoused interest in Love-wave discriminants.

S waves

Generally, explosions generate less shear-wave energy than do earthquakes (Willis et al., 1963; Ryall, 1970). Only a few attempts have been made to utilize S waves recorded at teleseismic distances for the purpose of discrimination (Briscoe & Sheppard, 1966). This is due to the difficulty of detecting S waves from weak seismic events. For ALPA, however, a fairly high detecting capability is reported to have been obtained, the threshold for 90 % probability of S-wave detection of earthquakes in the Kuril–Kamchatka region was about m_b=4.5 (Strauss, 1973). In cases where discriminants based on S-wave amplitudes or amplitude-period ratios can be applied, the observed values differ by a factor of about 15 (Evernden, 1969b). Theoretically, an explosion source in a perfectly elastic medium should not radiate any shear waves, whereas earthquakes should do so. This means that classification based on S-wave discriminants could be carried out by negative evidence. Highly sensitive horizontal-component measurements combined with the negative-evidence technique could thus improve the applicability of discriminants based on S waves.

Discriminants at short distances

This section is concluded with a brief presentation of discriminants based on recording at distances of from a few kilometers to 2000 km.

IDENTIFICATION 255

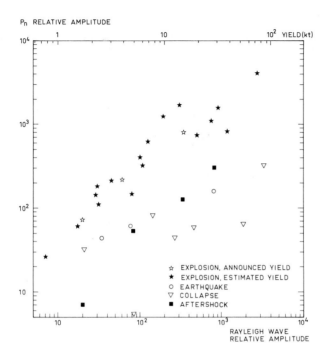

Fig. 10.25 P_n and Rayleigh-wave amplitude data from events at and close to NTS, obtained at the KANAB station in Utah, USA, at a distance of about 400 km from NTS (after Peppin & McEvilly, 1974). The yield scale is based on officially announced and estimated yields (cf. Chapter 11).

Such discriminants can be applied to events in the low-magnitude range and are of particular interest in the context of close-in stations mentioned in Chapter 6. Data on local discriminants have so far been published only for explosions at the Nevada Test Site. As seismic waves are sensitive to local geophysical conditions, one must be careful to extrapolate the results from the Nevada Test Site to other regions.

We have already mentioned that close-in studies of the $m_b(M_s)$ discriminant at distances less than 500 km separated explosions and earthquakes down to $M_L=3.5$ (Peppin & McEvilly, 1974). Figure 10.25 shows the result for two of the discriminants studied by Peppin & McEvilly. Spectral ratios for local P phases similar to the short-period discriminants for teleseismic distances have also been tested. They gave complete separation for events at the Nevada Test Site down to about $M_L=3.5$ corresponding to explosions in tuff with a yield of about 1 kt (Bakun & Johnson, 1970). At local distances S waves are more easy to detect, and they can be used for discrimination also of weak explosions. One S-wave discriminant is defined as the ratio of P-wave to S-wave amplitudes. This method has been applied to events with magnitudes below 3 at distances of about 50 km. More detailed descriptions of the local signals, similar to the multidimensional signal vectors for short-period signals, have also been used. The components represent the amplitudes at different time intervals, corresponding to different phases of P, S, and short-period surface waves (Booker & Mitronovas, 1964).

Among discriminants so far discussed in this section, those based on local data seem to have the greatest practical utility in the context of monitoring a CTB. Local discriminants are of particular interest for monitoring systems with local stations, which could provide discrimination of weak events in areas of special interest, e. g. where the geological conditions are suitable for the use of evasion methods.

10.5 MULTISTATION DISCRIMINANTS

The different seismic discriminants presented in the preceding sections usually express different properties of the seismic signals. A combination of several discriminants may therefore improve the discrimination (Bungum & Tjöstheim, 1976). The Tibetan earthquakes mentioned in Section 10.2.1, which appeared explosion-like in the $m_b(M_s)$ diagram, had short-period discriminant values at Hagfors Observatory which were quite typical of earthquakes.

In the same way, a combination of discriminants obtained at seismic stations widely separated in direction and distance from the source can improve discrimination by utilizing the difference in symmetry between explosion sources and earthquake sources. Multistation discriminants could also reduce some of the variation in the discriminants caused by propagation effects.

The $m_b(M_s)$ discriminant is often based on multistation data, as can be seen from Table 10.1. Multistation m_b and M_s values are usually calculated as direct mean values of observations at several stations. Station corrections accounting for peculiarities associated with the receiving conditions at the different stations are sometimes applied. The stations from which the m_b value is estimated are usually not the same as those on which the M_s value is based. Individual m_b or M_s values deviating with more than about half a magnitude are usually excluded from the mean values. New mean values are calculated, until all abnormal observations have been rejected. This way of eliminating extreme m_b and M_s values may well degrade the discrimination capability.

IDENTIFICATION

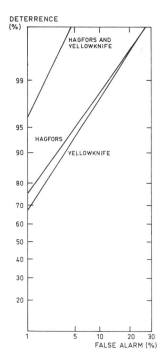

Fig. 10.26. Identification curves for short-period discriminants obtained at Hagfors and Yellowknife separately and jointly. (After Anglin & Israelson, 1973).

A case study of $m_b(M_s)$ observations at two different and widely separated stations was conducted by Israelson & Yamamoto (1974). The results indicated that linear combinations of the $m_b(M_s)$ data provide more effective discrimination than do straight average values over the two stations. This conclusion should, however, be substantiated by additional case studies.

An example of the improvement obtained by combining short-period discriminants at two stations in a nonlinear way is given in Fig. 10.26, where identification curves for two stations evaluated individually and jointly are compared (Anglin & Israelson, 1973). The data used were the complexity and *TMF* discriminants at Hagfors Observatory and Yellowknife for a set of Eurasian events. Figure 10.26 shows that by combining data from both stations, the deterrence probability rose from about 70 % to 95 % at 1 % false-alarm probability level.

In multistation discrimination it is not necessary to use identically defined discriminants at every station. It seems quite possible to achieve improved discrimination with the multistation method using discriminants different for each station, provided that each discriminant is optimized with regard to the particular receiving conditions at that station.

The applicability of multistation discriminants can increase or decrease as compared to a single-station discriminant, depending on the type of combination formed. For a linear combination of discriminants the applicability becomes equal to that of the least sensitive station. However, for the kind of combination which takes the maximum amplitude over the stations, the applicability becomes equal to that of the station with the highest sensitivity.

Few studies of multistation discrimination have been published so far, and these studies usually involve only two or a few stations. This is mainly due to the difficulties in establishing significant multistation data bases. This is in turn due to the fact that discriminants are not routinely calculated and reported. The promising results of the multistation studies conducted so far should serve to encourage and stimulate to an extended international exchange of discriminant data.

10.6 OPERATIVE ASPECTS OF DISCRIMINATION

The presentation so far has largely been dealing with seismic discriminants tested in case studies which usually comprise only small samples of explosions and earthquakes. The events have mostly been sampled in a rather arbitrary manner and do not represent the result of a systematic search for events in a given region and over a given period of time. Moreover, the events usually cover regions related to different tectonic settings rather than territories of interest in the seismic monitoring of a CTB. Although by now the basic properties of different discriminants are fairly well mapped, the interaction between different discriminants is not yet fully understood. There are events for which some discriminants appear explosion-like and others appear earthquake-like. It is therefore difficult to estimate the joint identification capability of a number of discriminants that could be used in a seismic monitoring system.

To assess the identification capabilities in the context of a CTB it is necessary to know the conditions of the treaty, the technical equipment that is to be used, and the procedures available for monitoring. In a seismic monitoring scheme one has to combine information and discriminants of various kinds, such as estimates of epicenter, depth, the $m_b(M_s)$ discriminant, and short-period discriminants. This can be accomplished by a screening procedure in which the information and the discriminant values are analyzed one at a time in a given order. In each step one could either identify as many explosions and earthquakes as possible, or only identify earthquakes as such, and at the end there would then be a residue of explosions and explosion-like earthquakes. The nature of these residual events could be clarified by a more detailed analysis, such as the earlier mentioned method of signal matching, or by acquiring more discriminant data. It is difficult to estimate the capability of a screening procedure in terms of probabilities for false alarm and for deterrence (Kendall & Stuart, 1963, Weichert et al., 1975). The most straightforward way to get an estimate is to conduct full-scale monitoring experiments.

Six independent, but somewhat similar, experiments are discussed here. Table 10.3 summarizes their characteristics and the principal results obtained. Three of the experiments were dealing with events strictly located within a given territory. Two of them deal with USSR events and one with US events. These studies are based on multistation discriminants obtained from observations principally at the Hagfors Observatory, at Yellowknife, and at other North American stations. The other three experiments cover large parts of the Eurasian continent. These studies are essentially based on data obtained at LASA, at Canadian stations, and at the UK-type arrays. One of the Eurasian studies is based on data of the International Seismic Month (ISM). The discriminants used in the case studies are also indicated in Table 10.3.

Each of the experiments was conducted in two steps. First a catalog of events for the region was compiled, and then the events were identified by seismic discriminants. The seismological bulletins of main agencies, such as the US Geological Survey (USGS) and the Institute of Physics of the Earth in Moscow, were used to compile the event catalogs. This sets the threshold for event definition. In some cases, bulletin data were supplemented by independent event definitions based on observations at other station networks and at the large-aperture arrays. The ISM experiment is supposed to have had its 90 % interval detection probability close to m_b(ISM) = 4. This experiment covered only one month, but the threshold value obtained is comparable to thresholds in the USSR studies by Dahlman et al. (1974) and Austegard (1974) and the Eurasian study by Lacoss (1969). For the US study the threshold is somewhat lower (Israelson et al., 1976).

In all the experiments listed in Table 10.3 the discriminants were applied essentially in the following order: event location, focal depth,

TABLE 10.3. Monitoring experiments.

	Reference						
	Lacoss (1969)	Basham & Anglin (1973)	Marshall & Key (1973)	Dahlman et al. (1974)	Austegard (1974)	Israelson et al. (1976)
Region	Eurasia	Major portion of Eurasia	Sino-Soviet areas	USSR except Kamchatka and Kuriles	USSR except Kamchatka and Kuriles	Continental US
Time period	Not specified	ISM[a]	1966	1971	1972	1972
Source of event definition	LASA	ISM	USGS, Blacknest	USGS, MOS[b], HFS	USGS, MOS, HFS, LASA, NORSAR	USGS, HFS
Total no. of events	191	291	348	199	203	268
Stations used for discriminants	LASA	Canadian Network	ESK, GBA, WRA, YKA, and WWSSN stations	HFS, YKA	HFS, YKA	HFS, ALQ, OGD, and Canadian Network
Discriminants[c]	$m_b(M_s)$, SR, C	$m_b(M_s)$, $\bar{A}$, $\bar{S}$	$m_b(M_s)$	$m_b(M_s)$, C, TMF	$m_b(M_s)$, C, TMF	$m_b(M_s)$, C, TMF
Explosions: no. of	35	0	11	19	18	8
Magnitude range	m_b(LASA) =4.4–6.4			m_b(USGS) =4.5–6.4	m_b(USGS) =4.4–6.3	m_b(USGS) =4.4–5.7
Earthquakes: no. of	156	233	311	134	127	64
Magnitude range	m_b(LASA) =3.7–5.6	m_b(ISM) =3.3–5.9		m_b(USGS) =4.1–6.0	m_b(USGS) =3.7–6.0	m_b(USGS) =3.7–5.8
Unclassified events: no. of	0	58	26	46	58	196
Magnitude range		m_b(ISM) =3.2–4.6	m_b(USGS) =3.7–4.9	M_s(MOS) =3.4–4.5	M_s(MOS) =3.5–4.4	m_b(USGS) =3.4–4.7

[a] International Seismic Month (February 20–March 19, 1972).
[b] Institute of Physics of the Earth, Moscow.
[c] $\bar{A}$ = spectral vector. C = waveform complexity. $\bar{S}$ = spectral vector. SR = spectral ratio. TMF = third moment of frequency.

TABLE 10.4. Events identified by different discriminants.

Study	Screening step[a]			
	Epicenter	Depth	$m_b(M_s)$	Short-period discriminants
Basham & Anglin (1973)	152 (Q)	51 (Q)	18 (Q)	12 (Q)
Marshall & Key (1973)		169 (Q)	142 (Q) 11 (X)	
Dahlman et al. (1974)		39 (Q)	63 (Q) 10 (X)	32 (Q) 9 (X)
Austegard (1974)		50 (Q)	47 (Q) 6 (X)	30 (Q) 12 (X)
Israelson et al. (1976)	31 (Q)		32 (Q) 8 (X)	1 (Q)

[a] Figures indicate the number of earthquakes (Q) and explosions (X) classified in each step.

$m_b(M_s)$, and short-period discriminants. The number of explosions and earthquakes identified in each step in the studies using screening procedures are listed in Table 10.4. First, events with epicenters outside the relevant region were eliminated. From Table 10.4 it can be seen that 152 events were taken as earthquakes by Basham & Anglin (1973) for the reason that their epicenters were at sea. The event catalog was then searched for events having a depth estimated to be below a certain limit (which differed with the experiments). Dahlman et al. (1974) and Austegard (1974) classified those events as earthquakes where the estimated focal depth was greater than 100 km. The remaining events for which $m_b(M_s)$ values were available were then classified as explosions or earthquakes depending on their position in the $m_b(M_s)$ diagram. The exact criterion for classifying the events differed slightly from study to study. Finally the remaining, unclassified events were analyzed by short-period discriminants. This number of remaining events was in most studies quite large, which fact illustrates the importance of having effective short-period discriminants.

The events which remain unclassified in Table 10.3 were weak. They illustrate the gap between thresholds for defining and thresholds for identifying seismic events. Most of the unclassified events have a magnitude less than $m_b(USGS)=4$, and about 95 % of the events unclassified in the US experiment were not detected at distances larger than 10 degrees.

These monitoring experiments did not use the full seismological identification capability that could be achieved at present, nor were any of the nonseismological methods utilized which are discussed in Chapter 14. Moreover, the monitoring experiments discussed here reflect only the present pre-treaty situation. It is difficult to describe in precise terms of probabilities the capability of a monitoring system to identify seismic events if attention has to be paid to all conceivable operative aspects. In a general way, however, the monitoring experiments presented here convincingly demonstrate that present seismic discriminants could indeed provide a high degree of deterrence.

11. YIELD ESTIMATION

The yield of a nuclear explosion can be predicted from the construction of the device and the amount of nuclear material. After the explosion the actual yield can be estimated by *in situ* radiochemical measurement of the radioactive products from the explosion. The precision of the radiochemical measurements on which the officially announced yields usually are based is not reported, but it is generally assumed to be ±10% (Ericsson, 1971c; Springer & Kinnaman, 1971). The post-shot estimated yield may differ from that predicted; it has been reported that for about 25% of the US explosions the radiochemically estimated yields were 50% lower than those predicted (Coffer & Higgins, 1968). Yields are usually given in kilotons (kt), where 1 kt is equal to 4.2×10^{12} J, or the energy released by exploding 1 000 tons of trinitrotoluene (TNT).

There are several reasons for estimating the yields of underground nuclear explosions from seismic signals. One reason is the Threshold Test Ban Treaty, where the threshold has been set to 150 kt. To monitor this treaty it is important to decide whether an observed explosion had a yield above or below this threshold. Detection and identification capabilities of seismic stations and station networks are usually estimated in magnitudes and must when related to explosion-monitoring capabilities be interpreted in terms of explosion yield. Another reason for estimating yields is the desire to follow the ongoing development of nuclear weapons by monitoring the testing activity. For such purposes it is important to obtain good yield estimates of the test explosions to correlate them with data about nuclear-weapon systems.

The amount of energy transmitted as seismic signals from an underground nuclear explosion is only a small fraction of the total explosion energy, as was discussed in Chapter 5. The strength of the seismic signals depends strongly on the medium in which the explosion is carried out.

Explosions in hard rocks, like granite, give considerably stronger signals than do explosions in unconsolidated rocks, like clay or alluvium. Seismic signals from an explosion can also be reduced by firing the explosion in a large, underground cavity. This effect can be used for purposes of evasion (to be further discussed in Chapter 13). The signals recorded at a seismic station depend not only on the explosion yield and the shot medium, but also on the wave-transmission properties of the earth. These transmission properties vary considerably from one region to the other, as was discussed in Chapter 4. The strong dependence of seismic signals on both the shot medium and the signal transmission from the test site to the seismological stations makes it important to use seismic data from a fixed station network for the estimation of explosion yields. Explosion yields cannot be calculated directly from seismic signals, but have to be estimated in relation to a nearby calibration explosion the yield of which is known from radiochemical or other measurements.

The yield of an explosion estimated from seismic signals is usually given as the equivalent yield of an explosion in hard rock. This means that the observed seismic signals correspond to those expected from a fully contained explosion in hard rock in the actual test area of equivalent yield. If detailed information about the testing conditions, especially the explosion medium, is available, absolute yield estimates are possible. For a high-yield explosion the equivalent hard-rock yield will be rather close to the real yield, as such an explosion has to be emplaced at a depth where hard-rock conditions usually exist. For a low-yield explosion, which could be contained at shallow depth in unconsolidated alluvial material, the real yield can be several times as high as the estimated equivalent hard-rock yield. The variation in coupling conditions within such an unconsolidated material could be expected to be larger than in hard rock, which fact will increase the uncertainties in the yield estimates of explosions in such media.

In this chapter we discuss theoretical relations between amplitudes of seismic signals and explosion yields for two explosion-source models. We also present a few observed magnitude–yield relations for short-period and long-period signals. As no comprehensive summary of estimated yields of nuclear explosions has been published, we attempt here to estimate the yields of those explosions which have been conducted since the PTB was signed in August 1963, and for which adequate seismological data are available. The chapter is concluded by a brief

YIELD ESTIMATION

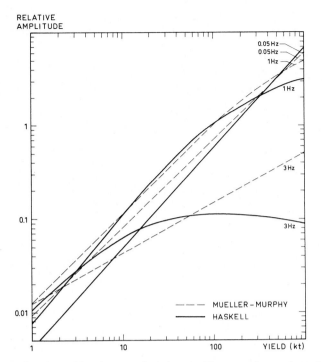

Fig. 11.1. Amplitude-yield relations for the Mueller-Murphy and Haskell source models of explosions in tuff for signal frequencies of 0.05, 1, and 3 Hz.

discussion of the estimated explosion yields in relation to the testing of nuclear weapons in the US and the USSR.

11.1 THEORETICAL AMPLITUDE–YIELD RELATIONS

In Chapter 5 two seismic source models for underground nuclear explosions were briefly presented, one designed by Haskell and the other by Mueller & Murphy. To illustrate the amplitude–yield relation predicted by these two source models, the radiated amplitudes are shown in Fig. 11.1 as functions of the yield for the frequencies 0.05, 1, and 3 Hz. The two latter frequencies define the range in which short-period signals usually fall, and 0.05 Hz corresponds to the 20-s period of the Rayleigh waves. The curves are normalized at 100 kt and 1 Hz. For long-period signals both models predict a linear relation between the logarithm of the amplitude and the logarithm of the yield, with a slope close to 1. At 1 Hz the slopes of the curves for both models decrease with increasing yield. This property is most pronounced for the Haskell model and is due to the very strong amplitude roll-off with

TABLE 11.1. US explosions at NTS with announced yields fired during the period August 1963–December 31, 1976. (From Springer & Kinnaman, 1971, 1975; Nevada Operations Office, 1976.)

Date	Name	Yield (kt)	Medium	Date	Name	Yield (kt)	Medium
630913	Bilby	235	Tuff	670520	Commodore	250	Tuff
631026	Shoal	12	Granite	670523	Scotch	150	Tuff
641009	Par	38	Alluvium	670526	Knickerbocker	71	Rhyolite
641105	Handcar	12	Dolomite	670921	Marvel	2.2	Alluvium
641216	Parrot	1.2	Alluvium	680126	Cabriolet	2.3	Rhyolite [a]
641216	Mudpack	2.7	Tuff	680312	Buggy	5×1.1	Basalt [a]
641218	Sulky	0.092	Basalt	680314	Pommard	1.4	Alluvium
650216	Merlin	10	Alluvium	680426	Boxcar	1200	Rhyolite
650414	Palanquin	4.3	Rhyolite [a]	681208	Schooner	35	Tuff [a]
650514	Cambric	0.75		681219	Benham	1100	Tuff
650611	Petrel	1.2	Alluvium	691029	Cruet	11	Tuff
660224	Rex	16	Tuff	691029	Calabash	110	Tuff
660414	Duryea	65	Rhyolite	700205	Labis	25	Tuff
660505	Cyclamen	13	Alluvium	700306	Cyathus	9.0	Tuff
660506	Chartreuse	70	Rhyolite	700526	Flash	105	Tuff
660527	Discus Thrower	21	Tuff	701217	Carpetbag	220	Tuff
660602	Pile Driver	56	Granite	701218	Baneberry	10	Tuff
660625	Vulcan	25	Alluvium	710708	Miniata	80	Tuff
660630	Halfbeak	300	Rhyolite	720926	Delphinium	15	Alluvium
661220	Greeley	825	Tuff	730426	Starwort	85	Tuff

[a] Crater explosion.

frequency, f, proportional to f^{-4}, compared to the f^{-2} roll-off for the Mueller–Murphy model. The slope of the 3-Hz curve for the Mueller–Murphy model is only 0.55. The Haskell model gives rather strange results for this high frequency, with a maximum amplitude at about 50 kt.

From these two source models and Fig. 11.1 the following predictions can be made: (a) for long-period signals the relation between the logarithm of the explosion yield and the logarithm of the signal amplitude (or magnitude) is linear, with a slope close to 1; (b) for short-period signals at 1 Hz and above, particularly at high yields, the relation is nonlinear, with a first derivative < 1.

11.2 OBSERVED m_b – YIELD RELATIONS

The concept of seismic magnitude originally introduced to structurize earthquake statistics has also been used to estimate the yields of nuclear explosions. General formulas, such as $m_b = k_1{}^{10}\log Y + k_2$, have been widely discussed. A SIPRI report (1968) discussed whether a magnitude-4 explosion has a yield of 1, 2, or 3 kt, and whether the slope of the magnitude–log-yield relation should be 0.7 or 1, or something in between. The influence of the earlier discussed, large variation in the signal-transmission properties is most pronounced for weak explosions recorded at only few stations. Considerable systematic errors can therefore be introduced into yield estimates by such general magnitude–yield relations. In this chapter we therefore limit the discussion to yield estimates by data from a fixed station network.

We noted earlier in this chapter that calibration explosions with known yields are essential for accurate yield estimates. Such calibration explosions are today available only for a few places, almost all located in the US. Yields have been released for a number of underground nuclear explosions at NTS (Zander & Araskog, 1973; Springer & Kinnaman, 1971, 1975). The released yields of early, and in many cases weak, explosions from which few or no seismic data are available have limited value for calibration purposes. The number of explosions with announced yields at NTS that have been recorded by a significant number of stations, and thus can be used for calibration purposes, is about 30. The US has also officially announced the yields of four PNE explosions in the western US and of one weapon test at the test site in the Aleutian Islands. The announced yields of US explosions at NTS since August 1963 are listed in Table 11.1.

TABLE 11.2. Short-period amplitude–yield or magnitude–yield relations for NTS explosions.

Relation [a]	No. of explosions	Explosion medium	No. of stations	Distance (km)	Reference
I a $^{10}\log(A/T) = 0.99\, ^{10}\log Y + C_1$	9	Tuff and rhyolite	7	1900–5000	Springer & Hannon (1973)
b $^{10}\log(A/T) = 0.63\, ^{10}\log Y + C_2$	9	Tuff and rhyolite	4	200–700	Springer & Hannon (1973)
II a $^{10}\log A = 1.09\, ^{10}\log Y + 3.07$	11	Tuff and rhyolite	17	2000–5500	Basham & Horner (1973)
b $^{10}\log A = 1.14\, ^{10}\log Y + 3.46$	3	Granite	17	2000–5500	Basham & Horner (1973)
III $m_b(\text{CAN}) = 0.93\, ^{10}\log Y + 3.49$	6	Tuff and rhyolite	19	1300–5500	Ericsson (1971b)

[a] A = Amplitude (μm); T = Period (s); Y = Yield (kt); C_1, C_2 = constants.

TABLE 11.3. Long-period amplitude–yield or magnitude–yield relations for NTS explosions.

Relation [a]	No. of explosions	Explosion medium	No. of stations	Distance (km)	Reference
I $^{10}\log(A/T) = 1.09\, ^{10}\log Y + C$	9	Tuff and rhyolite	11	200–5000	Springer & Hannon (1973)
II [b] $M_s(\text{CAN}) = 1.23\, ^{10}\log Y + 1.56$	13	Tuff, rhyolite and shale	17	2000–5500	Basham & Horner (1973)
III $M_s(\text{CAN}) = 1.19\, ^{10}\log Y + 2.67$	6	Tuff and rhyolite	19	1300–5500	Ericsson (1971b)
IV [b] $M_s = {^{10}\log Y} + 2.0$	31	Consolidated rock			Marshall et al. (1971)

[a] See notes to Table 11.2. [b] Path correction included in magnitude calculation.

The USSR has not officially announced any yield of nuclear explosions at its main test sites. However, yields, but not places and times, have been reported for several peaceful nuclear explosions in the USSR, as is discussed in Chapter 12. Six explosions with published yields of 8, 8, 30, 47, 125 and 3×15 kt, respectively, are in Table 12.2 tentatively associated with observed seismological data.

Two large chemical explosions, the Medeo explosions, in the Alma-Ata region on October 21, 1966, having yields of 1.7 and 3.6 kt TNT, respectively, have also been described (Aptikayev et al., 1967). Seismological data have been reported from these explosions, which therefore can be used for calibration (Marshall, 1970). However, the limited amount of available calibration data reduces the accuracy of yield estimates for explosions in the USSR.

Most of the published relations between yield and short-period amplitude or magnitude are based on US explosions at the Nevada Test Site. Some of these are summarized in Table 11.2. They are all based on nearly the same calibration explosions, but the seismic data are obtained at different stations and analyzed in different ways.

The two first formulas (Ia and b) are based on data from stations in the US and correspond to signals with different frequency content, around 2 Hz for the short-distance formula, and a little below 1 Hz for the other formula. The constants to estimate the absolute yield level have not been published.

The next two relations (IIa and b) were obtained using data from Canadian stations for 14 explosions at NTS. Station corrections were used to account for the relative differences in transmission properties to the individual stations in the network (Basham & Horner, 1973).

Relation III is based essentially on data from the same stations as relations IIa and b, but magnitudes are used instead of amplitudes. The data were also analyzed in a somewhat different way, and a magnitude–yield relation was estimated for each station. All these relations were given the same slope but different constants. The constant in the table is the average for all stations; the individual station constants may differ by up to 0.8 from this value (Ericsson, 1971c).

The constants in the relations in Table 11.2 cannot be compared, as somewhat different parameters and definitions are used. The yield coefficients, however, can be compared; if we disregard the formula obtained at short distances, the differences are not very pronounced. At teleseismic distances the slopes of the observed log-amplitude–

yield curves are between 0.9 and 1.1 for NTS explosions, in good agreement with theoretical predictions for hard-rock explosions. The slope of 0.63 of the short-distance formula is significantly lower than the slope of the formula for teleseismic distances, but is in fair agreement with the slope of 0.55 predicted by the Mueller–Murphy source model for a frequency of 3 Hz.

Due to lack of calibration explosions in the USSR, it is more difficult to make accurate yield estimates of USSR explosions; this explains why very few such estimates have been published. Ericsson (1971b) and Basham & Horner (1973) propose that the formulas III and IIb, respectively, in Table 11.2 might, in the absence of more adequate calibration data, be used for estimating yields of USSR explosions from Canadian data. This would give the equivalent NTS yield, by which is meant the yield to be exploded at NTS to obtain in the Canadian network the same magnitude estimate as that from the USSR explosions. Ericsson also transferred the relation based on Canadian data into formulas based on the magnitudes of USGS and some Scandinavian stations by estimating a formula relating these magnitudes to Canadian magnitudes for events in the USSR.

Seismic signals recorded at HFS from explosions in the USSR are significantly different from signals recorded at the same station from explosions in the US (cf. Fig. 7.18). We therefore feel that one has to be very careful when applying a magnitude–yield relation estimated for explosions in one region to explosions in quite another part of the world.

11.3 OBSERVED M_s–YIELD RELATIONS

Yield estimates based on long-period surface waves are generally considered to be subject to less regional variation than estimates based on short-period data. The wavelengths of the surface waves are considerably longer than those of the P waves, and are thus less influenced by small-scale heterogeneities at the test site and along the wave path. To take into account the large-scale regional variations which do exist also for these waves, a new path-corrected surface-wave magnitude scale has been introduced. This scale, which was discussed in Chapter 4, is supposed to give surface-wave magnitude estimates which are independent of where the explosions and the stations are located.

A few examples of published relations between explosion yields and observed surface-wave amplitudes or magnitudes are shown in Table

11.3†. Relation I is based on long-period data from the same explosions and the same stations that were used in estimating the short-period-amplitude–yield relations Ia and Ib in Table 11.2. For long-period waves there is just one relation covering the whole distance interval from 200 to 5 000 km. Relation II is obtained from data from 22 Canadian stations for 13 NTS explosions. The magnitudes are estimated taking the path correction into account. Relation III is obtained by the same method as relation III for short-period data (Table 11.2), and is essentially based on the same data as relation II, but the magnitudes are estimated without path corrections. Relation IV is based on long-period data obtained at a variety of stations from 31 explosions at various places in the US, the USSR, and Africa. For some of the explosions the yields were officially announced, whereas for others they are obtained from nonofficial sources, such as newspapers. The magnitudes are calculated using the magnitude formula that includes path correction. The relation is supposed to be valid for explosions in consolidated rocks at any site.

Comparing the coefficients of these relations, we see that they all are equal to or greater than 1, and thus somewhat higher than the corresponding values for the short-period data, and also generally somewhat higher than the theoretically predicted value of about 1. The difference in the constants of relations II and III, which are based on nearly the same data, is due to different magnitude scales. Relations II and IV are based on different data, but both of them utilize path-corrected magnitudes. The relations look rather different, but give almost the same results for yields from 10 to 1000 kt. For 10 kt, relation II gives $M_s = 2.8$ and relation IV $M_s = 3.0$; for 1000 kt relation II gives $M_s = 5.25$ and relation IV $M_s = 5.0$.

When evaluating different methods for yield estimation one has to consider not only their accuracy but also their applicability—the same as is the case when methods of identification are compared. The applicability of M_s–yield relations is limited by the detection capability of the long-period stations. As we have seen earlier, the present detection capability corresponds roughly to $M_s = 3$ in most areas of interest for explosion monitoring, which is equivalent to an explosion yield of about 10 kt in consolidated rocks. The relations based on short-period data, however, can be applied to explosions with yields down to about 1 kt.

---

†Table 11.3 is found on page 268.

11.4 YIELD ESTIMATES OF UNDERGROUND NUCLEAR EXPLOSIONS 1963–1976

Explosion yields have been officially announced only for a small portion of all nuclear explosions so far carried out, and no comprehensive summary of estimated explosion yields has been published. We are here making an attempt to estimate the yields of all underground explosions carried out during the period August 5, 1963–December 31, 1976, for which adequate seismological data are available. These estimates are based on short-period amplitude and period data most of which were published by USGS and ISC. Estimates have been made for those explosions for which data are reported from at least three stations. The data were analyzed by a method using simultaneously estimated relative source and transmission functions. The estimated explosion yields are given in Appendix 2. We have attempted to update this list as far as possible with data available at the time of publication of this book. Figures containing yield information, however, refer to the time period 1964–1975.

11.4.1 Computational procedure

The general principle of this method of analyzing the relative source and transmission functions is close to that of the method of joint relative epicenter location discussed in Chapter 8 (Dahlman, 1974). In both cases a fixed station network is used to analyze closely spaced seismic events. The relative source and transmission functions show the relative differences in seismic signal parameters between closely spaced explosions and globally distributed recording stations, respectively. The signal parameters may be either amplitude and period, or energy, amplitude, and phase spectra. For the computation, reported ratios of the short-period amplitude to the period are used. The method is based on the following assumptions.

> The source functions are spherically symmetric, i. e., an explosion radiates the same seismic signal in all directions.
>
> It is indifferent where an explosion is located within a limited test area, i. e., the transmission function to each station is the same from all parts of the test area.
>
> The transmission functions are time invariant, which mainly means that the characteristics of the seismological stations are constant in time.

YIELD ESTIMATION

If these assumptions are fulfilled, the signal parameter a_{ij}, in our case the short-period amplitude/period ratio, observed at station i from explosion j can be written

$$a_{ij} = t_i s_j,$$

where t_i is the transmission function for station i, and s_j is the source function for explosion j. This can be rewritten as

$$^{10}\log a_{ij} = {}^{10}\log t_i + {}^{10}\log s_j.$$

For such relations to hold for m stations recording n explosions within a specified area, a correction ϵ_{ij} has to be introduced to take care of stochastic errors. The expected mean of ϵ_{ij} is zero. A system of equations can then be written as

$$^{10}\log a_{ij} = {}^{10}\log t_i + {}^{10}\log s_j + \epsilon_{ij},$$

where $i = 1, \ldots, m$, and $j = 1, \ldots, n$. This system thus has $n + m$ unknowns and nm equations, provided that data from all explosions are obtained at all stations, which however happens very seldom in practice. As the equations are linearly dependent, we introduce the additional condition that the average of the transmission functions should be zero. The way the normalization is carried out does not affect the relative source or transmission functions, but only the error estimates of these functions.

11.4.2 US explosions

For the US explosions yield estimates are made only for those carried out at NTS. The yields of explosions for peaceful purposes carried out in western US have been officially announced, as has the yield of one of the explosions in the Aleutian Islands. Unofficial yields given in press reports for the other two explosions in the Aleutian Islands agree well with available seismological data. We have therefore no reason to believe that these reported yields should not be fairly correct.

The relative source functions of 179 explosions at NTS have been computed using 3 500 amplitude and period data from 120 seismological stations. For 30 of these explosions the yields and explosion media have been officially announced. The estimated source functions of the announced explosions are shown in Fig. 11.2. The least-squares estimates of the relation between yield and estimated source function for explosions in tuff and rhyolite are based on yields above 20 kt. The

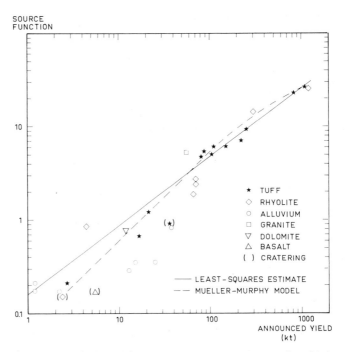

Fig. 11.2. The estimated source function versus the announced yield for NTS explosions in various media. The least-squares relation between source function and yield for explosions in tuff and rhyolite has a slope of 0.74. The Mueller–Murphy model for explosions in tuff has been fitted to the data.

yield coefficient of the relation, 0.74, is somewhat lower than the value, 0.90, obtained by similar calculations on a smaller data base (Dahlman, 1974) and also lower than the coefficients given in Table 11.2, except that for short distances. The amplitude–yield relation predicted by the Mueller–Murphy source model for tuff and rhyolite for a frequency of 1 Hz is also shown in the figure and seems to give a somewhat better fit to the data. The figure also shows that the explosions in alluvium give considerably lower values for the source function, and that the explosion in granite gives a value somewhat higher than that expected for an explosion of the same yield in tuff.

In this preliminary evaluation of the data the yields are estimated in a straightforward way. The equivalent yields for explosions in tuff and rhyolite are estimated from the individual source functions using the least-squares relation in Fig. 11.2. If the explosion is reported to have been carried out in tuff or rhyolite, the equivalent yield is also an estimate of the actual yield. If the explosion is carried out in alluvium, the difference in coupling between tuff and alluvium is taken into account by multiplying the equivalent tuff yield by a factor 3 to get an estimate of the actual yield. We do not try to make any correction for the depth of the explosion, for the moisture content of the shot media, or for the depth of the static water table.

11.4.3 USSR explosions

For the explosions in the USSR the computations are in principle made in the same way. Two circumstances do, however, complicate the estimation of the USSR explosion yields. One problem is the lack of adequate calibration explosions with announced yields, and another is that the USSR explosions are carried out at many places, spread over a large area. The nuclear weapon tests are supposed to be concentrated to two test areas, one in Eastern Kazakh, where most of the tests are conducted, and one at Novaya Zemlya, where most of the high-yield explosions are carried out. In our computations we use as calibration explosions the two peaceful explosions in Uzbekistan in 1966 and 1968, see Chapter 12, and the two chemical explosions at Medeo near Alma-Ata. The two chemical explosions had yields of 1.7 and 3.6 kt TNT and were detonated in granite. A chemical explosion is reported to be 3–4 times as efficient in generating seismic waves as a nuclear explosion of the same yield (Aptikayev et al., 1976). The Alma-Ata explosions were used for dam construction and were not fully contained, and we therefore tentatively suggest that they had equivalent nuclear yields of 3.4 and 7.2 kt for fully contained explosions in granite. The two explosions in Uzbekistan had yields of 30 and 47 kt, respectively, and were fired in an oil field. Data from explosions under the same conditions in the US show that such an environment has a coupling corresponding to that of tuff (Basham & Horner, 1973). The seismic coupling for explosions in tuff is about a factor 2 lower than that for explosions in granite. We thus get equivalent granite yields of 15 and 24 kt, respectively, for the two Uzbekistan explosions. We further assume that the slope of the source-function–yield curve is the same, 0.74, for the USSR as for the US explosions.

The USSR explosion sites are distributed over a very large area, which fact must be taken into account when calculating the source and transmission functions for USSR explosions. Most of the explosions have been concentrated to the Eastern Kazakh test area, and therefore a special computation is made for these explosions and the four calibration explosions. The calibration explosions were made at some dis-

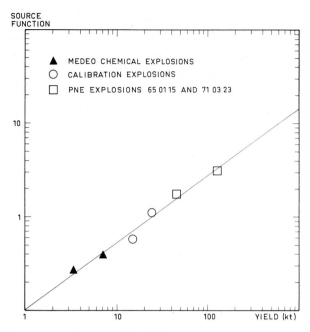

Fig. 11.3. The estimated source function versus the equivalent hard-rock nuclear-explosion yield for two chemical explosions in Alma-Ata (Medeo) and two PNE explosions (on 660930 and 680521) in Uzbekistan, used as calibration explosions. A line with a slope of 0.74 has been fitted to these four explosions. Source functions for two other PNE explosions (on 650115 and 710323) are shown for comparison.

tance from the Eastern Kazakh test site, around 700 km for Alma-Ata and almost 2 000 km for Uzbekistan. In the computation we have to assume that the individual transmission functions do not vary over this area. The source functions of the calibration explosions are shown in Fig. 11.3 as a function of their equivalent hard-rock yield. A line with a slope of 0.74, as estimated for the NTS explosions, is fitted through the four calibration explosions. The source functions of two other presumed PNE explosions, one on 650115, supposed to have had a yield of 125 kt, and one on 710323, supposed to have consisted of three simultaneously detonated 15-kt devices, are also shown in the figure. These two explosions are supposed to have been fired in water-saturated material. The seismic coupling for the two PNE explosions would therefore probably be similar to fully contained explosions in hard rock. Aptikayev et al. (1967) report that an explosion in water-saturated material is 2—5 times as efficient in generating seismic signals as an explosion of the same yield in dry material. It is also interesting

YIELD ESTIMATION

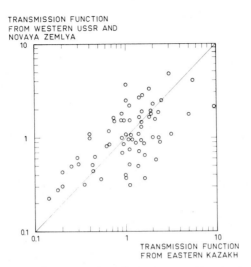

Fig. 11.4. Comparison of transmission functions to 69 seismic stations from explosions in Eastern Kazakh, and in western USSR and Novaya Zemlya.

to note that the source functions of these two explosions agree with those expected for explosions in granite.

The yields of the explosions are estimated from the calculated source functions using the linear relation in Fig. 11.3. The yields estimated in this way are equivalent yields of explosions in granite or comparable hard rock. As described in Chapter 8, the Eastern Kazakh test area consists of three different test sites located within about 100 km from each other. The central and western test sites contain granite bedrock. The eastern test site, which is the one most seldom used, has a carboniferous bedrock. Available geological maps do not indicate large deposits of unconsolidated ground material at any of the test sites. It is therefore reasonable to assume that most of the explosions in the Eastern Kazakh test area are fired in granite or comparable material.

To estimate the yields of explosions at other places in the USSR, we assume that the transmission functions do not change over those parts of the USSR where the explosions are located. There are obviously differences between the various regions, and estimates of the relative transmission functions to the different stations from explosions in the western USSR and Novaya Zemlya show that they differ from the transmission functions to the same stations from the Eastern Kazakh explosions, Fig. 11.4. Most of the transmission functions are, however, within a factor of 4, which in a way might justify our assumption. We

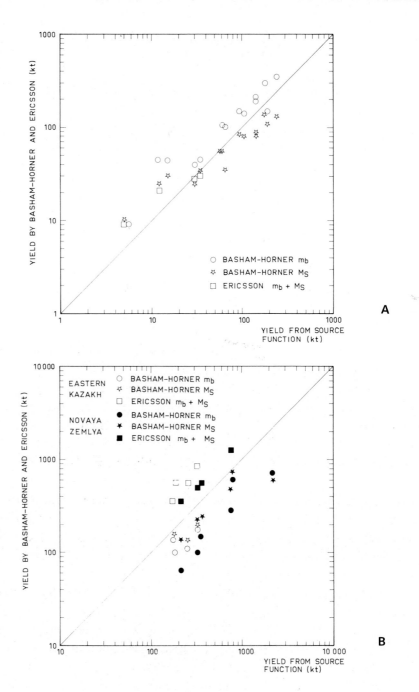

Fig. 11.5. Comparison of yields estimated from source functions and yields estimated by Ericsson (1971b) and by Basham & Horner (1973) from m_b and M_s data, for (A)–NTS explosions in tuff, (B)–USSR explosions in Eastern Kazakh and Novaya Zemlya.

thus take all the observations from all the USSR explosions, a total of more than 3 700 amplitude and period data from 116 seismic stations, and estimate the source function of each of the 159 explosions. The yields of the explosions outside the Eastern Kazakh test area are then estimated from these source functions in the same way as described above for the Eastern Kazakh explosions.

11.4.4 Comparison of yield estimates

Yield estimates published by Ericsson (1971b) and by Basham & Horner (1973) for some explosions at NTS, in the Eastern Kazakh, and at Novaya Zemlya are in Fig. 11.5 compared with the yield estimates based on the source functions discussed above. This comparison gives fairly concordant results for the NTS explosions. A systematic difference between the estimates can, however, be observed. The yields estimated by Basham & Horner from M_s data are generally lower than their estimates based on m_b data. The yield estimates based on source functions and those of Ericsson, which are based both on short-period and long-period data, are somewhere in between.

For the USSR explosions, both in the Eastern Kazakh and at Novaya Zemlya, the systematic difference between the yield estimates is, however, considerably larger than that observed for the NTS explosions. The maximum ratio of the estimates amounts to about 6. The yields reported by Ericsson are systematically higher, and those published by Basham & Horner are systematically lower than those estimated from our source functions. This larger systematic difference between the various estimates is probably due to inadequate calibration data. The yield estimates from source functions are based on the few yield data available on USSR explosions, and are made under the assumption that the explosions are carried out in granite. The estimates by Basham & Horner are based on their relation obtained for explosions in granite at NTS, whereas the relation used by Ericsson is based on data for NTS explosions in tuff. This different assumption about explosion media and the use of different calibration explosions could explain part of the observed differences.

11.4.5 Accuracy of yield estimates

When discussing the accuracy of yield estimates, we must consider both the accuracy in estimating the equivalent yields for explosions

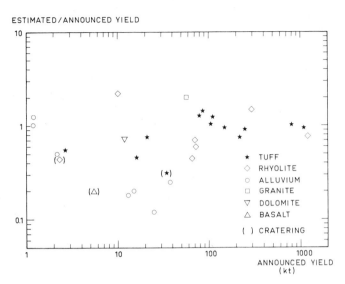

Fig. 11.6. Ratio of estimated yield to announced yield for 30 explosions in different media at NTS.

in tuff or hard rock and the accuracy in converting such equivalent-yield estimates into real explosion yields. The estimates of equivalent tuff yields for NTS explosions are probably the most accurate ones that can be made with calibration data available today. Ericsson (1971b) and Basham & Horner (1973) suggest that the average error of such estimates is of the order of 10–50 %. A comparison of yield estimates for NTS explosions by different methods in Fig. 11.5A does, however, indicate a somewhat larger divergence.

The accuracy of the yields estimated from source functions is illustrated in Fig. 11.6, where the ratio of estimated yield to announced yield is shown for 30 explosions. The figure shows that the explosions in tuff and rhyolite have the same average value, although the scatter may be somewhat larger for the rhyolite explosions. An error distribution for the tuff and rhyolite explosions, Fig. 11.7, shows that for 90 % of these explosions the error in estimated yield is less than 50 % of the announced yield. Figure 11.6 also shows that for explosions in alluvium, the estimated yields are considerably less than those announced. For the larger explosions, recorded by 50 or more stations, the standard deviation of the source-function estimates is only about 10 %, which is less than the total errors shown in Fig. 11.7. For weaker explosions the uncertainty in the source-function estimates is greater and may correspond to a difference by a factor 2.

YIELD ESTIMATION

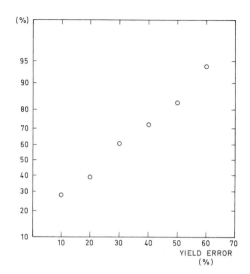

Fig. 11.7. Distribution of the difference between announced and estimated yields for NTS explosions in tuff and rhyolite, in percent of the announced yield.

Due to the fair number of available NTS calibration explosions within a wide range of yields, there seems to be little risk that systematic errors are built into the yield estimates of these explosions. Most of the yields estimated for explosions in tuff and rhyolite should therefore be within 50 % of the true value. The uncertainty for explosions in alluvium is larger, even when a correction for the average difference in seismic coupling is applied. This uncertainty is larger the weaker the explosion.

For USSR explosions it is considerably more difficult to give a measure of the accuracy of the yield estimates. The relative estimates of equivalent hard-rock yields of explosions in the Eastern Kazakh test area are probably good, as these explosions are located close to each other, which fact reduces the systematic errors. The signals are also recorded by a large number of stations. The standard deviations of the source-function estimates for these explosions are somewhat less than those for the US explosions, and range between 10 and 40 %. For estimates of absolute yield we are critically dependent on the calibration data. For the explosions scattered over the western USSR, the distances to the calibration explosions are quite large, and the uncertainty in the estimates might therefore increase. Also for these explosions the estimates of relative equivalent hard-rock yields are more accurate than estimates of absolute yields.

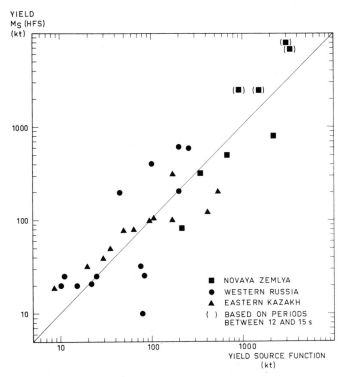

Fig. 11.8. Comparison of yields estimated from source functions and from M_s(HFS) for 12 explosions in Eastern Kazakh and 21 explosions in Western USSR and Novaya Zemlya.

Another way of assessing the consistency of yield estimates from source functions is to compare them with estimates based on long-period surface waves. The latter are estimated from the long-period surface-wave magnitude, M_s(HFS), obtained at the Hagfors Observatory, using the formula $^{10}\log Y = M_s(\text{HFS}) - 2.0$. Data on M_s(HFS) are available for 12 explosions at the Eastern Kazakh test site and for 21 explosions in the western USSR including Novaya Zemlya. Yield estimates from M_s(HFS) and from source functions are compared in Fig. 11.8. The agreement is fairly good for Eastern Kazakh explosions. The scatter is larger for the explosions at Novaya Zemlya and in western USSR; for some of these the surface-wave magnitudes are based on periods in the range of 12–15 s. No significant systematic bias is observed, however; this is an important result, as the yields are estimated by uncorrelated methods and with different calibration data. This general agreement between the two methods and the concordant yield estimates for the six reference explosions in Fig. 11.3 enhance the

YIELD ESTIMATION

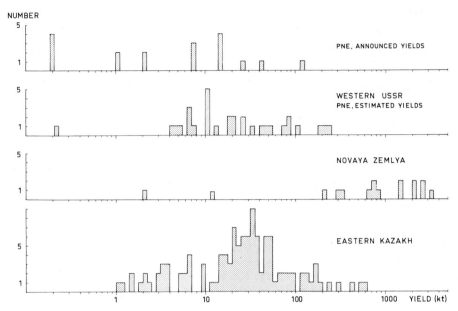

Fig. 11.9. Distribution of estimated yields for explosions in Western USSR, Novaya Zemlya, and Eastern Kazakh in 1964–1975, and of announced yields for PNE explosions in W. USSR.

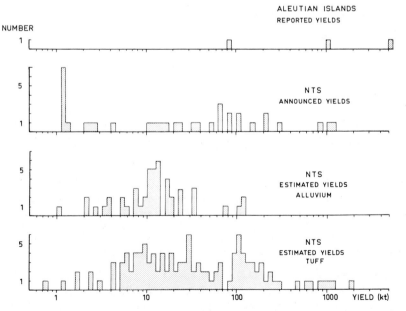

Fig. 11.10. Distribution of estimated and announced yields for explosions fired at the US test sites in Nevada and the Aleutian Islands. Explosions in tuff and in alluvium are shown separately.

credibility for our yield estimates also for the USSR explosions. Despite this, the systematic uncertainty of yield estimates for USSR explosions is significantly greater than that for explosions at the NTS. The bilateral exchange of calibration data between the US and the USSR foreseen in the Threshold Test-Ban Treaty would mean a significantly increased accuracy of the yield estimates for USSR explosions. We therefore hope these data will be made generally available.

11.4.6 Interpretation of yield estimates

The distribution of estimated yields of USSR and US explosions are shown in Figs. 11.9 and 11.10, respectively. The USSR explosions have been divided into three groups according to location: Western USSR, Novaya Zemlya, and Eastern Kazakh. The first-mentioned group is supposed to be the PNE explosions, and the announced PNE yields discussed in Table 12.2 are also shown for comparison, although we do not know whether these yields refer to explosions fired in the time period, 1963–1976, discussed here. In Fig. 11.10 US explosions at NTS have been separated into two groups, those in alluvium and those in tuff. In the tuff group are included also those explosions for which no information on the medium is available. The distribution of the announced yields of the NTS explosions since 1964 is also given in Fig. 11.10, together with the reported yields of the three explosions in the Aleutian Islands. PNE explosions in the US outside NTS are not discussed here, as their yields have been announced, see Table 12.1.

The nuclear-weapon tests in both the US and the USSR seem to cover the same yield interval, that is, from about 1 kt up to 2 or 3 Mt. Most of the USSR explosions have estimated yields between 10 and 100 kt. In the US, also many explosions with yields below 10 kt and around and above 100 kt have been fired. Figure 11.9 also shows that the Novaya Zemlya test site is almost entirely used for high-yield tests.

When interpreting the test activity in terms of yields of tested nuclear weapons, we have not considered each single test separately, but rather made the assumption that throughout the development of a nuclear weapon several tests with the same intended explosion yield are carried out. Thus, we try to find series of tests over a limited time interval having roughly the same estimated yields. The time and yield distributions of the USSR and the US explosions are illustrated in Figs. 11.11 and 11.12, respectively. These figures do not contain any data on announced or presumed PNE explosions. We also do not try to explain

YIELD ESTIMATION

all observed yields, but rather to find out the large-scale pattern. Most of the interpretation is done visually. For some observed test series statistical variance analysis is used to tell whether they differ in average yield. A more detailed discussion of this interpretation is given by Dahlman & Elvers (1977). It might also well be that all weapon tests were not carried out at the full, intended explosion yield. Some tests may, for example, include only the fission explosion intended as an initiator to a fusion device of higher yield, whereas other tests may be related to more fundamental research in nuclear-weapon technology.

USSR explosions

When interpreting the yields of the USSR explosions, we must be aware of the possible risk of a bias in our estimates. The relative yield estimates at each of the two test sites are probably fairly concordant, but the absolute yield level, and thus the comparison of yield estimates at the two test sites, is more uncertain.

Since Jan. 1, 1973, six explosions have been carried out at Novaya Zemlya, with estimated yields between 1.5 and 3 Mt. In late 1972 and in 1973, three explosions with estimated yields around 500 kt were conducted at the Eastern Kazakh test site, and in 1975 one more explosion of roughly the same estimated yield was observed. In 1970–1974 four explosions with somewhat higher yields, around 1 Mt, have also been observed at Novaya Zemlya. In 1966–1969, a few explosions having yields between 100 and 300 kt were observed both in Novaya Zemlya and Eastern Kazakh. Explosions having estimated yields around 100 kt were also conducted in 1970–1972 and in 1974. It is difficult to tell whether these explosions had different nominal yields. From 1964 to 1972, a large number of tests were conducted with estimated yields in the range of from 20 to 50 kt. In 1973–1974, seven tests with estimated yields around 20 kt were observed. The yields of these explosions differ significantly from those of seven later tests having estimated yields of around 50 kt. A few tests in the yield range of 4–7 kt were observed in 1965–1969. Four explosions with estimated yields of 2 to 3 kt were conducted in 1974.

US explosions

The interpretation of the yield estimates of US explosions is facilitated by the official announcement of yields of a number of explosions. The

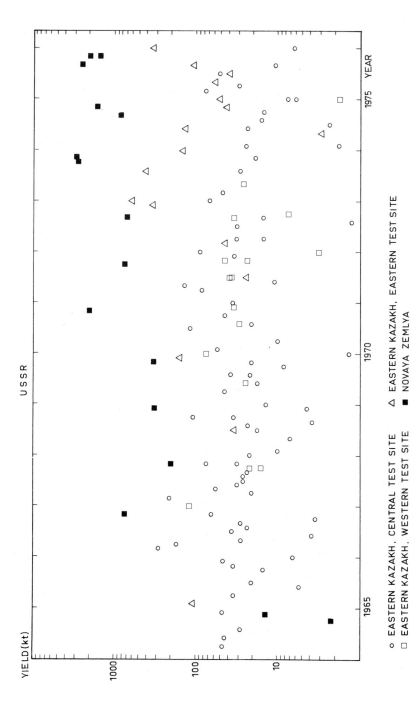

Fig. 11.11. Distribution in time of estimated yields of USSR explosions in Eastern Kazakh and Novaya Zemlya.

YIELD ESTIMATION 287

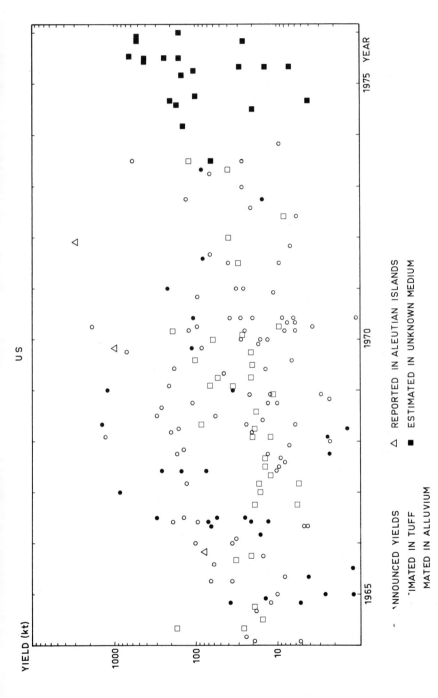

Fig. 11.12. Distribution in time of announced and estimated yields of explosions at NTS and of reported yields of explosions in the Aleutian Islands.

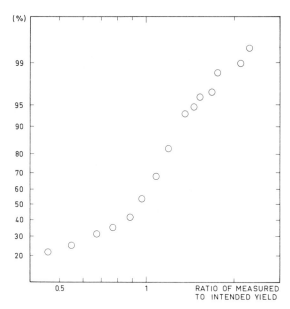

Fig. 11.13. Distribution of the ratio between radiochemically measured and intended yields of NTS explosions. (From Coffer & Higgins, 1968).

announced explosion yields, ranging from 1 to 1 000 kt, reduce the risk of introducing any systematic bias into our interpretation. The announced yields are estimated from close-in, probably radiochemical, measurements. According to Coffer & Higgins (1968), the measured (announced) yield of an explosion may differ from the intended or nominal yield, as can be seen in Fig. 11.13. This diagram suggests that the probability of the actual yield being considerably below the intended is fairly high.

The explosion "Cannikin" in 1971, for the purpose of testing the "Spartan" warhead of the US antiballistic missile systems, is officially reported to have had a yield of less than 5 Mt. In 1967–1970, several explosions with announced and estimated yields around 1 Mt were carried out at NTS. Several explosions with estimated yields around or somewhat below 500 kt were fired during the latter part of 1975 and the beginning of 1976. In 1966–1969 several explosions were carried out with announced and estimated yields around or somewhat below 200 kt. Explosions with estimated yields in the same range were also conducted in 1974–1976. Five explosions with announced yields between 56 and 80 kt were conducted in 1965–1967. Explosions with announced and estimated yields from 60 to 120 kt have been

YIELD ESTIMATION

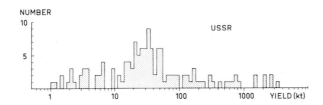

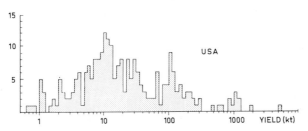

Fig. 11.14. *Distributions of estimated yields of explosions fired in 1964–1975 at the weapon-test sites Eastern Kazakh and Novaya Zemlya, USSR, and NTS (all media) and the Aleutian Islands, US.*

observed in 1969–1975. It is not possible to establish whether the nominal yields of these explosions differ from those with somewhat lower announced yields in 1965–1967. A large number of tests in the yield range of 10–40 kt were conducted from 1966 to 1972. It is difficult to split these explosions into different groups according to yield. However, many of both the announced and the estimated yields in 1966–1968 are around or somewhat below 20 kt, whereas several explosions in 1970 and 1971 have estimated yields around 30 kt. Several explosions in 1967 and late 1968 have yields that cluster around 10 kt. Seven explosions conducted during three months in 1970 had estimated yields between 6 and 10 kt. Five explosions, two of which with announced and three with estimated yields, are in the range of 2–3 kt.

Comparison of US and USSR explosion yields

Comparing the nuclear-test activity of the two superpowers during the period of 1964–1975, the most striking difference that we find is that the US conducted almost twice as many tests at their test sites as did the USSR at the Eastern Kazakh and Novaya Zemlya test sites. The distributions of the estimated yields of the tests, compared in Fig. 11.14, show the same yield range and the same general pattern, with a large number of tests in the 5–50 kt range, and less tests at higher and lower yields. Most of the US explosions seem to have been between 5 and 20 kt, whereas most of the USSR tests have somewhat higher yields, 10–50 kt. The maximum yields tested by the two countries seem to have been of the same order, the "Cannikin" test, with a yield below 5 Mt, and some tests around 1 Mt in the US, and some tests in the 1.5–3 Mt range in the USSR. To interpret the test activity in the subkiloton range is difficult, and we have refrained from any attempt to do so. Subkiloton explosions, not included in our yield estimates, have, however, been observed at NTS. No explosion with an estimated yield of less than 1 kt has been observed at the USSR test sites in Eastern Kazakh and Novaya Zemlya during 1964–1976. One of the PNE explosions in the USSR had, however, an estimated yield of 0.2 kt, and reported PNE projects in the USSR involved nuclear explosions in that yield range, see Table 12.2.

The most important observation to be made when comparing the test activities of the US and the USSR is, in our opinion, the great similarity in the explosion-yield distribution. The observed test explosions do not give any support to the often expressed opinion that the USSR nuclear-weapon systems, contrary to the American, should contain mainly megaton charges. Both countries seem to develop nuclear-weapon systems with yields ranging from one or a few kilotons to one or a few megatons. Interpretation of estimated explosion yields in terms of nuclear weapon systems is, however, outside the scope of this book.

12. PEACEFUL NUCLEAR EXPLOSIONS

Peaceful uses of nuclear explosions have been discussed for more than twenty years. Large-scale projects involving nuclear explosions have been suggested, and a number of test explosions have been carried out in the US and the USSR. Environmental and economical concern has decreased the interest in peaceful nuclear explosions, *PNE,* in the US, whereas in the USSR the interest in this technique seems to remain strong. India has also conducted a nuclear explosion, officially said to be for peaceful purposes (Gandhi, 1974). The development of nuclear devices for peaceful purposes will give a country technical experience and knowledge of considerable value for acquiring nuclear warheads. The nuclear proliferation through PNE is therefore a problem of considerable international concern (CCD/454, 1974; CCD/456, 1974).

Our intention here is to discuss possible applications of PNE, to review some of the PNE projects carried out by the US and the USSR, and to discuss the future prospects of PNE. The most comprehensive technical discussions on the use of PNE have taken place at the IAEA International Panels for Peaceful Nuclear Explosions held in Vienna in 1970, 1971, 1972, and 1975, and most of the data presented here are taken from the Proceedings of these meetings (IAEA, 1970, 1971, 1974, and 1975; CCD/455, 1975).

12.1 POSSIBLE APPLICATIONS OF PNE

According to their purpose, PNE can be divided into three main categories: *crater explosions,* for the excavation of large volumes of ground material, *fully-contained underground explosions,* for creating large underground cavities or for crushing large volumes of rock at great depths, and *explosions for scientific purposes.* In the first two types

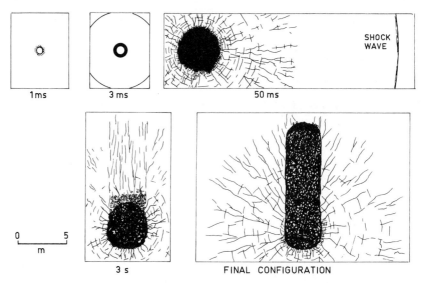

Fig. 12.1. Underground cavity and chimney formed by an explosion. Cavity-chimney formation history for a 5-kt explosion in granite (Nordyke, 1970).

of application the nuclear explosive is used as an alternative to a conventional explosive or to other excavation methods, whereas in scientific applications the properties specific to a *nuclear* explosion are utilized.

The formation of craters by nuclear explosions has been thoroughly investigated, and a large number of cratering explosions have been conducted. Craters of considerable dimensions can be formed by such explosions. The 100-kt explosion "Sedan" at the Nevada Test Site (Fig. 14.4), for example, formed a crater having a radius of 185 meters and a depth of 98 meters. Discussed excavation projects using crater explosions are, e.g., the building of large canals, water reservoirs, and harbors.

If an explosion is carried out at such a depth that it is fully contained, it will form an underground cavity. The originally formed cavity is nearly spherical, and the size of the cavity depends on the explosion yield, the shot depth, and the properties of the surrounding medium. In most media the cavity collapses after some time, and a rubble-filled cylindrical volume, a chimney, is formed, the height of which will be 2–6 times the cavity radius, Fig. 12.1 (Nordyke, 1970). Cavity radii and chimney heights for some explosions for peaceful purposes in the USSR are given in Table 12.2.

Various civil applications of the fully-contained underground nuclear explosion have been suggested and in some cases tested. The use of nuclear explosions for stimulating gas and oil production by fracturing rock to increase its permeability has been widely discussed, and such gas- and oil-stimulation projects have also been conducted, both in the US and in the USSR. Underground cavities have been suggested as storage for gas and oil and for radioactive and other highly poisonous material. To produce noncollapsing cavities, the explosions have to be carried out in a plastic material, such as salt, and test explosions in salt have been conducted both in the USSR and in the US. At least one of the USSR cavities is being used on an industrial basis for storage of liquefied gas (M. Nordyke, personal communication, 1976). An underground nuclear explosion can also be used for mining purposes. The explosion crushes large volumes of rock, which then can be mined either by conventional mining techniques or by *in situ* leaching—the latter method has been considered for mining of low-grade copper-ore deposits. In this last technique an acid or oxygen-rich solution is pumped down into the explosion chimney to leach out copper from the crushed ore. The solution is then pumped back to the surface, where copper is removed by conventional separation methods. Another application of the disintegrating effect of deep, underground explosions is for recovering geothermal energy. One or more charges are detonated in an area and at a depth where the geothermal conditions are appropriate. Water is pumped down and circulated through the broken rock in the cavity, or between the cavities of multiple explosions, where it is heated, and the heat is then recovered at the surface. Fracturing experiments in hard rock have been conducted both in the US and in the USSR.

In the USSR, PNE has also been used to extinguish gas-well fires by exploding a nuclear device close to the runaway well in a plastic layer and thus blocking the gas flow at a considerable depth below ground.

The third type of application of PNE utilizes the unique characteristics of the nuclear explosion. Scientific applications of nuclear explosions have included the production of some ten new transplutonium elements, and the measurement of neutron cross-sections. Nuclear explosions have also been used as known and strong energy sources for seismic measurements over global distances, and much information about the earth's interior has been revealed through such explosions.

Fig. 12.2. Cavity of the Indian explosion in the Rajasthan desert on May 18, 1974.

12.2 NUCLEAR EXPLOSIONS CONDUCTED FOR PEACEFUL PURPOSES

Nuclear explosions for peaceful purposes have been conducted by the US, the USSR, and India. France and the United Kingdom have conducted nuclear explosions and have programs for studying the applications of PNE, but no explosions specifically designed for peaceful purposes have been announced. No information is available as to whether China has a PNE program. Within the framework of projects for the peaceful utilization of nuclear explosions, considerable efforts have been made to investigate the technical and economical aspects of PNE, and several supporting, small-scale experiments using high explosives have also been conducted. As the topic of this review is the PNE that have been carried out, these latter efforts will not be discussed here.

In the US, the so-called Plowshare project was established by the USAEC in 1957 to study and develop industrial uses of nuclear explosions. This project is unclassified, and the data presented here on PNE explosions in the US are based on official announcements. Technical descriptions of several nuclear explosions for peaceful purposes in the USSR have been reported in connection with the Vienna panel discussions of PNE. Places and times of these explosions were, however, omitted in the reports. Seismological data indicate that a large number of explosions, in fact far more than those for which technical descriptions have been made available, have been carried out by the USSR outside the ordinary test sites, and these explosions may be part of the Soviet PNE program. The Indian underground nuclear explosion on May 18, 1974, was carried out in the Rajasthan desert in Northwestern India. The cavity of the Indian explosion is shown in Fig. 12.2.

The US has announced 38 nuclear explosions as being directly associated with the Plowshare project. These explosions are listed in Table 12.1. Of the scientific experiments, all of which concerned studies of heavy elements or neutrons, ten were conducted in conjunction with weapon tests. The US PNE tests can be summarized according to purpose as follows:

Scientific studies (heavy-element or neutron physics)	16
Crater studies	6
Underground effects	3
Device development	10
Gas stimulation	3
Total	38

In the crater experiments, not only the cratering mechanism but also the dispersion of radioactive material were studied. During the development tests of explosion devices, an explosive of very low fission yield as well as special emplacement techniques were developed to reduce the amount of radioactivity released to the atmosphere (M. Nordyke, personal communication, 1976). The three gas-stimulation experiments were carried out by USAEC in cooperation with petroleum companies. The Rio Blanco experiment, where the cavities of three explosions, one on top of the other, were planned to join and form one large cavity, failed in the way that three separate cavities were formed (AEC, 1974; Woodruff & Guido, 1975; Toman, 1975). This illustrates that problems still exist in the prediction of close-in explosion effects. In states where gas-stimulation explosions were conducted, public concern about the release of radioactivity was expressed.

During the period of 1962–1966, the US conducted 24 nuclear explo-

TABLE 12.1. Announced US explosions for peaceful purposes.

Date	Name	Yield[a] (kt)	Medium	Place	Purpose
611210	Gnome	3.1	Salt	N. Mexico	Underground effects experiment
620706	Sedan	100	Alluvium	NTS	Crater study
621127	Anacostia	L	Tuff	NTS	Heavy-element experiment
630221	Kaweah	L	Alluvium	NTS	Heavy-element experiment
630625	Kennebec[b]	L	Alluvium	NTS	Heavy-element experiment
631011	Tornillo	L	Alluvium	NTS	Device-development test
631122	Greys[b]	L	Alluvium	NTS	Heavy-element experiment
640123	Oconto[b]	L	Tuff	NTS	Heavy-element experiment
640220	Klickitat	L	Tuff	NTS	Device-development test
640611	Ace	L–I	Alluvium	NTS	Device-development test
640716	Bye[b]	L	Tuff	NTS	Heavy-element experiment
641009	Par	L–I	Alluvium	NTS	Heavy-element experiment
641105	Handcar	38	Dolomite	NTS	Underground-effects experiment
641216	Parrot[b]	12	Alluvium	NTS	Neutron-physics experiment
641218	Sulky	1.2	Basalt	NTS	Crater experiment
650414	Palanquin	0.09	Rhyolite	NTS	Crater experiment
650521	Tweed[b]	4.3	Tuff	NTS	Heavy-element experiment
650611	Petrel[b]	L	Alluvium	NTS	Heavy-element experiment
660324	Templar	1.2	Tuff	NTS	Device-development test
660414	Duryea[b]	L	Rhyolite	NTS	Heavy-element experiment
660505	Cyclamen[b]	65	Alluvium	NTS	Heavy-element experiment
660615	Kankakee[b]	13	Dolomite	NTS	Heavy-element experiment
660625	Vulcan	L–I	Alluvium	NTS	Heavy-element & device-development experiment
660728	Saxon	25	Tuff	NTS	Device-development test
661105	Simms	L	Alluvium	NTS	Device-development test
670223	Persimmon[b]	L	Alluvium	NTS	Neutron-physics experiment

a, b See notes at bottom of table (next page).

TABLE 12.1 (continued).

Date	Name	Yield[a] (kt)	Medium	Place	Purpose
670622	Switch	L	Tuff	NTS	Device-development test
670921	Marvel	2.2	Alluvium	NTS	Underground-effects experiment
671210	Gasbuggy	29	Shale	N. Mexico	Gas stimulation
680126	Cabriolet	2.3	Rhyolite	NTS	Crater experiment
680312	Buggy I	5×1.1	Basalt	NTS	Row-charge cratering experiment
680917	Stoddard	L–I	Tuff	NTS	Device-development test
681208	Schooner	35	Tuff	NTS	Cratering experiment
690716	Hutch[b]	I	Alluvium	NTS	Heavy-element experiment
690910	Rulison	40	Sandstone	Colorado	Gas stimulation
700526	Flask	105	Tuff	NTS	Device-development test
710708	Miniata	80	Tuff	NTS	Device development for gas stimulation
730517	Rio Blanco	3×30	Sandstone	Colorado	Gas stimulation

[a] L=Low (<20 kt), L–I=Low to intermediate (20–200 kt), I=Intermediate (200–1000 kt).
[b] Conducted in conjunction with weapon test.

TABLE 12.2. Reported USSR nuclear explosions for peaceful purposes.

Yield (kt)	Medium	Description	Reference
		Cratering experiments	
1.1	Siltstone	Project "1003". To study crater formation, radioactive contamination, and seismic close-in effects. Shot depth 46 m. Crater diameter 141 m, depth 29 m, volume 10^5 m^3.	Kedrovski (1970)
125	Sandstone	Project "1004". Explosion immediately adjacent to river bed to block the river and form a water reservoir. Shot depth 178 m. Crater diameter 408 m, depth 100 m, volume 6×10^6 m^3. From seismological data the explosion is estimated to have been conducted on 650115 at 0600 GMT at lat. 49.9°N, long. 79.0°E.	Nordyke (1974)
			Marshall (1972)
0.2	Sandstone	Project "T1". Shot depth 31 m. Crater diameter 70–80 m, depth 21 m.	Nordyke (1974)
3×0.2	Sandstone	Project "T2". In same area as Project "T1". Spacing between devices 40 m. Crater length 150 m, width 61–69 m, depth 16 m.	Nordyke (1974)
3×15	Alluvium (water saturated)	Test for the Pechora–Kama canal. Fired near the southern end of the section proposed for nuclear construction. From seismological data the explosion is estimated to have been conducted on 710323 at 0700 GMT at lat. 61.3°N, long. 56.5°E. Radioactive particles from the explosion were observed in Sweden.	CCD/455 (1975)
			Eriksen (1972)
		Underground cavities	
1.1	Salt	Cavity for underground storage. Shot depth 161 m. Cavity height 37 m, radius 12–14 m, volume 1.4×10^3 m^3,	Kedrovski (1970)

TABLE 12.2 (continued).

Yield (kt)	Medium	Description	Reference
25	Salt	Same purpose as the foregoing test, and in a nearby location. Shot depth 600 m. Spherical cavity with radius 32 m, volume 1.5×10^4 m^3.	Kedrovski (1970)
15	Salt	Liquefied-gas reservoir. Shot depth 1140 m. Cavity volume 10^4 m^3.	Kedrovski et al. (1975)
		Oil stimulation	
2×2.3	Limestone	200 m between charges at depth of 1378 m. Not detonated simultaneously.	Kedrovski (1970)
8	Limestone	Same oil field as in foregoing experiment. Detonated 3 1/2 months later, 350 m from one of the two 2.3-kt charges at depth of 1350 m.	
8 + 8	Limestone	About 1200 m between the charges, one explosion at depth of 1212 m, the other at 1208 m. Not detonated simultaneously. From seismological data the charges are estimated to have been detonated on 690902 and 690908 at 0500 GMT at lat. 57.4°N, long. 55.0°E.	
		Control of runaway wells	
30	Clay	At Urtabulak oil field at a depth of 1520 m. From seismological data the explosion is estimated to have occurred on 660930 at 0559 GMT at lat. 38.8°N, long. 64.5°E.	Nordyke (1974) Marshall (1972)
47	Salt	Nearby the above oil field. Explosion depth 2440 m. From seismological data the explosion is estimated to have occurred on 680521 at 0359 GMT at lat. 38.9°N, long. 65.2°E.	Nordyke (1974) Marshall (1972)

sions for peaceful purposes, compared to 12 during the following five-year period, 1967–1971, and only two since 1971. This illustrates a decreased interest in PNE in the US, partly due to a growing uncertainty about the economical feasibility of PNE. The increased environmental concern in the US also makes it more difficult to carry out projects involving potential environmental hazards.

The USSR has not released enough information to permit a complete description of its PNE program to be made. More or less incomplete descriptions, all excluding date and place, have however been published or reported for 13 PNE projects in the USSR. This information is summarized in Table 12.2 (Nordyke, 1974; CCD/455, 1975).

A large number of seismic events have been located to areas in Western USSR outside active seismic regions and known test areas for nuclear weapons. By means of the seismological identification methods discussed in Chapter 10, most of these events have been identified as underground explosions. We presume that these explosions were made for civil purposes. Most of the explosions had also such high yields that it is reasonable to assume that they were nuclear explosions. We have estimated the equivalent hard-rock yields of those explosions for which we have adequate seismological data. The locations of the presumably peaceful explosions, 40 in all, are shown on the map in Fig. 12.3, and source parameters and yield estimates are given in Table 12.3. We and others have made efforts to associate the explosions identified from seismological data with the PNE explosions described by USSR scientists. It has been suggested that the explosions observed from seismological data on 660930 and 680521 are the two nuclear explosions used to extinguish gas-well fires in the Buchara region (Marshall, 1972). It has also been suggested that explosion 650115 close to, but outside, the ordinary test area in Eastern Kazakh was a 125-kt PNE cratering experiment, as a significant amount of atmospheric radioactivity was reported to be associated with this event (Koyama et al., 1966). We suggest that the reported cratering explosion consisting of three 15-kt charges, associated with the preparation work for the Pechora–Kama canal, is the explosion identified from seismological data 710323 at latitude 61.3°N and longitude 56.5°E (Dahlman et al., 1974). We further believe that the two 8-kt charges detonated close to each other for the purpose of oil stimulation might be the explosions seismologically observed on 690902 and 690908 and located in an oil-producing region at latitude 57.4°N and longitude 54.9°E. The yields of these two ex-

PEACEFUL NUCLEAR EXPLOSIONS

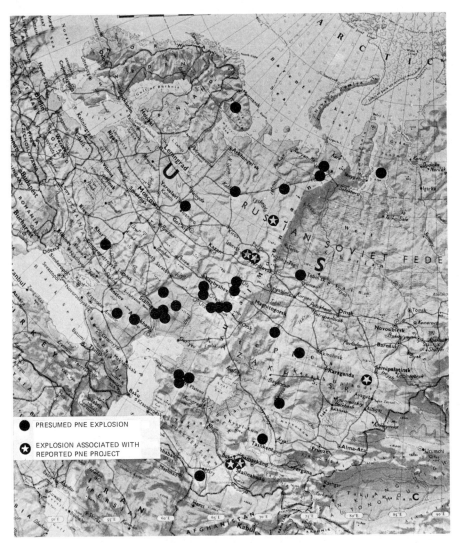

Fig. 12.3. Map showing the location of presumed PNE in Western USSR.

plosions are from seismological data estimated to have been 11 kt, which is in close agreement with the reported yields of 8 kt. The almost identical seismic signals from these two explosions that were obtained at the Hagfors Observatory are shown in Fig. 4.14.

Comparing the presumed PNE activities in the USSR and the US for the period 1970–1974, we find 27 presumed PNE projects in the USSR against three in the US. There can be several reasons for this difference in activity between the two countries. The USSR might have

TABLE 12.3. Presumed peaceful nuclear explosions in the USSR.

Date	Time	Lat. (°N)	Long. (°E)	Yield Estd. (kt)	Yield Ann. (kt)	Date	Time	Lat. (°N)	Long. (°E)	Yield Estd. (kt)	Yield Ann. (kt)
650115	055959	49.89	78.97	110	125	720411	060005	37.37	62.00	7	
660422	025804	47.90	47.70	-		720709	065958	49.78	35.40	6	
660930	055953	38.80	64.50	[b]	30	720714	145949	50.00	46.40[c]	0.2	
671006	070002	57.69	65.27	-		720820	025958	49.46	48.18	87	
680521	035912	38.92	65.16	[b]	47	720904	070004	67.69	33.44	7	
680701	040202	47.92	47.95	46		720921	090001	52.13	51.99	21	
690902	045957	57.41	54.86	11	8	721003	085958	46.85	45.01	88	
690908	045956	57.36	55.11	11	8	721124	090008	52.78	51.07	11	
690926	065956	45.89	42.47	78		721124	095958	51.84	64.15	20	
691206	070257	43.83	54.78	100		730815	015958	42.71	67.41	28	
700625	045952	52.20	55.69	5		730828	025958	50.55	68.39	14	
701212	070057	43.85	54.77	190		730919	025957	45.63	67.85	11	
701223	070057	43.83	54.85	240		730930	045957	51.61	54.58	22	
710323	065956	61.29	56.47	51	3×15	731026	055958	53.66	55.38	7	
710702	170002	67.66	62.00[c]	7		740708	060002	53.80	55.20	-	
710710	165959	64.17	55.18	27		740814	145958	68.91	75.90	45	
710919	110007	57.78	41.10	4		740829	150000	67.23	62.12	20	
711004	100002	61.61	47.12	11		750425	045957	47.50	47.50	-	
711022	050000	51.57	54.54	34		750929	105958	69.59	90.40	6	
711222	065956	47.87	48.22	210		760729	045958	47.78	48.12	-	

[a] Estd. = estimated; Ann. = announced.
[b] Calibration explosion.
[c] ISC location.

had more large-scale projects suitable for PNE in fairly remote areas with low population density and with valuable mineral and petroleum deposits. The two countries may also have different criteria for evaluating the profitability of PNE projects and for assessing the environmental impact of such explosions.

12.3 FUTURE PROSPECTS FOR PNE

The future usefulness of PNE depends to a large extent on nontechnical considerations, among which the current public and political reactions to the environmental impact of such explosions are among the more important ones at least in the western countries. The use of PNE is also limited by some international treaties. The Partial Test Ban Treaty prohibits a nuclear explosion to be carried out "if such explosion causes radioactive debris to be present outside the territorial limits of the states under whose jurisdiction or control such explosion is conducted". This is a severe limitation to large-scale excavation projects. On the one hand, the Non-Proliferation Treaty prohibits non-nuclear-weapon states parties to the treaty from manufacturing or otherwise acquiring nuclear weapons or devices, including devices for peaceful use, and, on the other hand, it obligates nuclear-weapon states to make available nuclear explosives for peaceful purposes to non-nuclear-weapon states under appropriate international observation and control. The Antarctic Treaty prohibits the introduction and use of any kind of nuclear explosives on the Antarctic continent.

The bilateral Threshold Test Ban Treaty, banning underground nuclear-weapon tests with yields above 150 kt, as of March 31, 1976, has a companion bilateral treaty, signed May 28, 1976, containing provisions for underground nuclear explosions for peaceful purposes. This PNE Treaty, the full text of which is reproduced in Appendix 1, prohibits:

"(a) any individual explosion having a yield exceeding 150 kilotons;
 (b) any group explosion:
    (1) having an aggregate yield exceeding 150 kilotons except in ways that will permit identification of each individual explosion and determination of the yield of each individual explosion in the group in accordance with the provisions of Article IV of and the Protocol to this Treaty;
    (2) having an aggregate yield exceeding one and one-half megatons;
 (c) any explosion which does not carry out a peaceful application;
 (d) any explosion except in compliance with the provisions of the Treaty Bann-

ing Nuclear Weapon Tests in the Atmosphere, in Outer Space and Under Water, the Treaty on Non-Proliferation of Nuclear Weapons, and other international agreements entered into by that Party."

The treaty also includes provisions for carrying out individual explosions with yields exceeding 150 kt:

"The question of carrying out any individual explosion having a yield exceeding the yield specified in paragraph 2(a) of this article will be considered by Parties at an appropriate time to be agreed."

The PNE Treaty has an extensive control procedure, set forth in a protocol to the treaty, to ensure that an explosion was conducted for peaceful purposes and also that the yields of individual explosions in a group explosion do not exceed 150 kt. The control procedure includes *inter alia* provisions for on-site inspection at the explosion site, including the conduct of certain measurements to verify explosion yields.

A more extensive discussion of legal aspects of nuclear explosions for peaceful purposes has been conducted by the IAEA Ad Hoc Advisory Group on Nuclear Explosions for Peaceful Purposes (IAEA, 1976).

The question how PNE should be handled under a CTB treaty has been raised many times. From a visual inspection at the site of an explosion it might be difficult to ascertain whether an explosion was carried out entirely for peaceful purposes, and whether military advantages were gained or not. To get an adequate verification, nuclear explosives used for peaceful projects under a CTB treaty probably have to be well specified in advance with regard to yield, type and amount of fissile material, and perhaps also with regard to some other quantities (CCD/454, 1975). By analyzing the radioactive products obtained from the explosion cavity, it seems possible to confirm whether the explosion was conducted in accordance with given specifications or not. Such radiochemical analyses can be carried out by an international agency or by national laboratories on radioactive samples obtained under appropriate international control (De Geer & Persson, 1975). It might thus be possible, using fairly intrusive inspection techniques, including comprehensive measurements at the explosion site, to technically verify that a PNE is not used also for weapon development. The conduct of nuclear explosions for peaceful purposes, also in accordance with agreed procedures, may, however, preserve the competence necessary to resume weapon testing at a later date, if deemed necessary. It is therefore still uncertain whether nuclear explosions for peaceful purposes could be allowed under a CTB treaty, and it is also uncertain whether a higher threshold could be used for PNE than for weapon tests in a possible future threshold treaty having a threshold considerably lower than 150 kilotons.

If future test-ban treaties will have provisions for PNE, what PNE applications can then be foreseen in the future? PNE projects can probably not be carried out if a significant amount of radioactivity will be released in an uncontrolled way. This will severely limit large-scale excavation projects. Deep underground nuclear explosions for oil or gas stimulation, for mineral mining, or for creating underground storages may be of future interest, if such explosions can be economically competitive, and if the societies concerned can be convinced that the associated risks are acceptable. Very deep underground explosions for recovery of geothermal energy and for building caves for storage of radioactive and poisonous material could be of considerable interest, as in these applications the radioactivity released by the explosions will be contained in a closed system. The same reasoning may apply to oil- and gas-recovery explosions, if oil or gas is converted to some other kind of energy at the well site. PNE might also be the veil or pretext under which a non-nuclear country would develop nuclear devices and thereby demonstrate its nuclear capability. The existence of such a capability is of considerable military importance as a main technical step towards the production of nuclear weapons.

13. EVASION

The so-called *evasion methods* have been discussed in connection with the monitoring of a CTB. By such methods the seismic signals from underground nuclear explosions are supposed to be, either reduced below the detection threshold of the monitoring network, or manipulated in such a way that the explosion will be identified as an earthquake.

Different evasion methods have been suggested. Some are based on published experimental data, whereas others are based on theoretical calculations (CCD/404, 1973). The suggested evasion methods can be divided into three groups: decoupling, multiple explosions, and hide-in-earthquake. In *decoupling,* the strength of the seismic signals is reduced by firing the explosion in a low-coupling medium or in a cavity. *Multiple explosions* means that several explosions placed in a defined geometrical pattern are fired in a certain time sequence to produce seismic signals intended to be identified as coming from an earthquake. In the proposed *hide-in-earthquake* method, an explosion is carried out simultaneously with a large earthquake, so that the seismic signals from the explosion are hidden in the signals from the earthquake.

13.1 DECOUPLING

Any monitoring station network will have a threshold below which seismic signals cannot be detected. The explosion yield corresponding to such a threshold depends on the medium and the dimension of the cavity in which the explosion is conducted. We first consider decoupling by firing explosions in a low-coupling medium, such as dry alluvium. Many low-yield explosions have been fired in alluvium at the Nevada Test Site (Springer & Kinnaman, 1971). The seismic signals from explosions in alluvium are much weaker than those from explo-

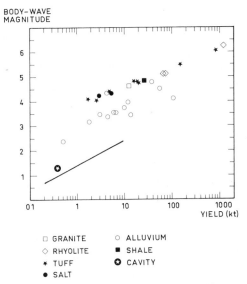

Fig. 13.1. Magnitude-yield data for NTS explosions in different media showing the decoupling effect of alluvium. Data from the "Sterling" cavity-decoupling project are also shown. The straight line shows the calculated magnitude of fully decoupled explosions in mined cavities (from Rodean, 1971a).

sions of an equal yield in hard rock. This is illustrated in Fig. 13.1, where the observed body-wave magnitude is given as a function of the explosion yield for some NTS explosions in different media (Rodean, 1971a). The decoupling factor is usually defined as the ratio of the maximum amplitude of the seismic signals from a fully contained explosion in hard rock to the maximum amplitude of those from a decoupled explosion of the same yield. The decoupling factor for dry alluvium, estimated from a substantial amount of data, is up to 10 for low-yield explosions.

Theoretical calculations have been carried out to study the behavior of different materials in the source region of an underground nuclear explosion. On the basis of such calculations it has been proposed that partially saturated, glacial material will have a decoupling factor comparable to that of dry alluvium (Butkovich, 1968). However, no decoupling experiment in such material has been reported.

To carry out clandestinely a decoupling test under a CTB, one has not only to prevent the seismic signals from being detected, but also to avoid leakage of radioactive material and permanent surface deformation. The maximum yield that can be fully contained without creating any detectable surface deformation in a decoupling material is thus dependent on the thickness of the deposit in which the explosion is carried out. The alluvial deposit at the Nevada Test Site has a thickness of up to 600 m, which might fully contain explosions with yields up to 10 kt without creating subsidence craters (Holzer, 1971).

Unsaturated, alluvial deposits with a thickness of up to about 300 m also exist in the USSR, south and southwest of Lake Balkhash (Lukasik, 1971). It might be possible also in these deposits to detonate explosions with yields of a few kilotons without getting subsidence craters, although no such explosions have been reported. There seem to be no places in the US or the USSR where deposits of glacial or similar material have the thickness, of the order of hundreds of meters, necessary to contain even low-yield explosions.

We thus conclude that it seems possible to reduce the amplitude of the seismic signals from an underground nuclear explosion by a factor of up to 10 compared to that obtained from a hard-rock explosion, by firing the explosion in a low-coupling material, such as alluvium. Provided that a permanent surface deformation must not be created, the applicability in the US and the USSR of this evasion method is limited to explosions with yields of a few kilotons. Such explosions can only be carried out at a limited number of places, where thick deposits of such low-coupling media exist.

Another way of reducing the seismic signals from an underground explosion is to fire the explosion in a large cavity. The degree of decoupling is determined by the yield of the explosion, the size and form of the cavity, the physical properties of the ground material, and by the static pressure on the cavity walls. An explosion is said to be fully decoupled if the shock-wave deformation of the cavity walls is elastic. Theoretical calculations have shown that a fully decoupled explosion should have a decoupling factor of about 120 (Lamb, 1960; Patterson, 1966; Rodean, 1971b). If the size of the cavity is less than that corresponding to full decoupling, the decoupling factor is also less, and we have what is usually called *partial decoupling*. Results from a few decoupling experiments have been reported. In 1960 the USAEC conducted a decoupling experiment, known as the "Cowboy" project, where the decoupling of seismic signals recorded at distances of up to 15 km was studied from about ten chemical explosions with yields between 100 and 1 000 kg (Carder & Mickey, 1960; Herbst et al., 1961). Decoupling values of up to 100 were observed, and the decoupling was found to be frequency dependent, with higher values for higher frequencies.

In 1966 another decoupling experiment, the "Sterling" project, was carried out in which a nuclear device with a yield of 0.38 kt was detonated in the cavity formed by a 5.3-kt nuclear explosion previously detonated in a salt dome (Springer et al., 1968). The cavity had a volume of 20 000 m³, corresponding to a sphere with a radius of about 17 m, which is just enough to fully decouple a 0.38-kt explosion. The seismic signals from the decoupled explosion were recorded at 14 seismic stations at distances of up to 110 km. The observed average amplitude ratio of the two explosions was around 1000 at 1 Hz. The decoupling factor was about 70, when the differences in explosion yields had been taken into account. This decoupling factor is of the same order as that theoretically predicted for a mined cavity (ca. 120). Calculated magnitudes of fully decoupled explosions of yields below 10 kt in mined cavities are shown in Fig. 13.1.

It has thus been shown by theoretical calculations and by a few, rather small-scale, experiments, that a reduction of the seismic signals by a factor of about 100 may be obtained by firing the explosion in a cavity of appropriate dimensions. The main difficulty in applying this decoupling technique is to obtain large, standing cavities. Such cavities must be constructed in hard rock. Salt domes have been especially considered for such constructions, as salt can be conveniently mined by the solution-mining technique. In the US, a project, "Payette", was conducted to study the feasibility of building a cavity to fully decouple a 5-kt explosion (Fenix & Scission, 1970). To fully decouple such an explosion, a spherical cavity with a diameter of 100 m at a depth of 800 m is needed. The construction of such a cavity was considered possible, although no cavity of this size with the necessary requirements as to form has so far been built. Such a project will involve the excavation of more than 400 000 m³ of salt and is estimated to take four years. As was mentioned above, in the case of the "Sterling" explosion, cavities can also be formed by large nuclear explosions. The cavity generated by a 10-kt explosion can be used to decouple a 1-kt explosion, that of a 100-kt explosion can be used to decouple a 10-kt explosion, and so on. Fig. 13.2 illustrates the size of cavities needed to fully decouple explosions of yields between 1 and 100 kt at different explosion depths. We see that the cavities needed to fully decouple also rather weak explosions have considerable dimensions. A cavity needed to fully decouple a 1-kt explosion at a depth of 1 km must have a diameter of about 40 meters, or a volume around 30 000

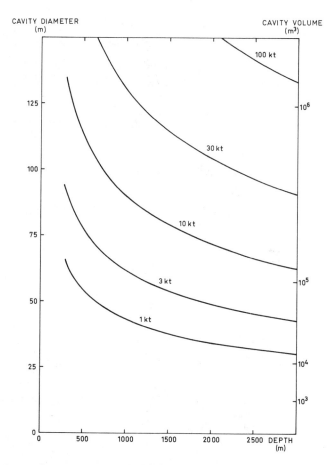

Fig. 13.2. Cavity dimension needed to fully decouple explosions with yields between 1 kt and 100 kt, as functions of the explosion depth. (From Rodean, 1971b.)

m^3. Cavities of that size have been mined, but the feasibility of constructing considerably larger cavities is still unproved.

As the size of cavities needed to fully decouple also low-yield explosions is very large, it might be interesting to see what seismic amplitude reduction might be obtained by partial decoupling in smaller cavities. The decrease in decoupling factor with increasing explosion yield for explosions fired in a cavity designed to fully decouple a 1-kt explosion is illustrated in Fig. 13.3. For a yield of 5 kt, the decoupling factor is about 50, but at 10 kt the factor has decreased to 10. This means that a fully decoupled 1-kt explosion gives signals equivalent to those of a 0.01-kt tamped explosion. The corresponding values for 5-kt and 10-kt explosions are respectively 0.1 kt and 1 kt.

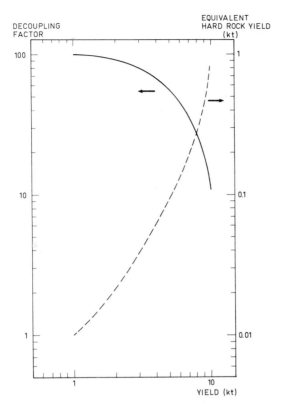

Fig. 13.3. Decoupling factor and equivalent hard-rock yields for explosions with yields between 1 kt and 10 kt fired in a cavity designed to fully decouple a 1-kt explosion.

13.2 MULTIPLE EXPLOSIONS

This proposed evasion technique utilizes several nuclear explosions, placed in a specific geometrical pattern and fired in a certain time sequence, to produce seismic signals similar to those generated by earthquakes. The explosions must be detonated in a seismic region, and the signals generated by the explosions are intended to be detected but to be misinterpreted as coming from an earthquake. It has been suggested that yields of 50—100 kt could be accommodated in such test series with reasonable probability of being identified as earthquakes (Holzer, 1971; Lukasik, 1971). These figures probably refer to the possibilities to evade the $m_b(M_s)$ criterion.

The multiple-explosions technique is based on the superposition of seismic signals from several explosions, a technique that has been used in seismic prospecting, where several small charges are detonated in

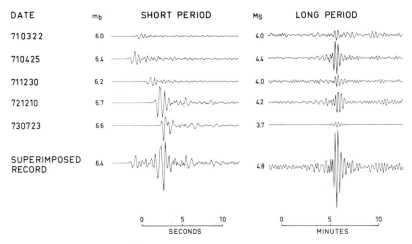

Fig. 13.4. Superposition of short-period and long-period signals recorded at the Hagfors Observatory from five nuclear explosions at the Eastern Kazakh test site in the USSR, to illustrate the multiple-explosions technique.

such a way that the seismic signals are enhanced in a certain direction. A similar explosion technique is also used in mining, where intervaled detonations are used to reduce the seismic vibrations from the explosions. Few studies have so far been published on the generation of seismic waves by the detonation of multiple, underground explosions and on the use of this technique for simulating earthquake signals (Power, 1972; Kolar & Pruvost, 1975). No experimental study of the use of multiple explosions to simulate earthquake signals has been published. Kolar & Pruvost (1975) present a theoretical scenario in which short-period signals from an explosion in the Eastern Kazakh test area in the USSR and long-period signals from an earthquake in the Kermadec Islands, recorded at three seismological stations, are used. The short-period signals were shifted in time and added to simulate a sequence of eight explosions with yields in the range of from 3 to 100 kt fired within a time interval of about 6 seconds. By this theoretical superposition, the body-wave magnitude could be reduced by half a magnitude unit. By a similar superposition, the surface-wave magnitude could be increased by half a magnitude unit compared to the original earthquake magnitude. From these superposition calculations, based on short-period signals from one explosion and long-period signals from one earthquake, it was considered possible to get $m_b(M_s)$ data from multiple explosions that would appear earthquake-like on an $m_b(M_s)$ diagram.

To illustrate this evasion technique, we have in Fig. 13.4 displayed short-period and long-period records obtained at the Hagfors Observatory from five explosions at the Eastern Kazakh test site in the USSR. The m_b values range from 6.0 to 6.7, and the M_s values from 3.7 to 4.4. It is worth noting that an explosion giving a large P-wave signal does not necessarily give a large Rayleigh-wave signal, which is usually assumed in scenarios where signals are scaled. The signals shown at the bottom in Fig. 13.4 are the sum of the short-period and long-period signals added with the time delays shown in the figure. Various time delays of up to 4 s have been tested, and the result given in Fig. 13.4 is a typical example of what can be achieved in suppressing m_b and enhancing M_s. Magnitudes estimated from these superimposed signals are m_b = 6.4 and M_s = 4.8. The superimposed long-period signal has an amplitude which is equal to the sum of the amplitudes of the five individual signals. The minimum short-period signal amplitude expected from multiple explosions should be equal to or larger than that of the weakest explosion. This gives in our example a minimum short-period magnitude value of 6.0. The $m_b(M_s)$ and short-period identification data for the synthetic signals are in Fig. 13.5 compared with data from the Eurasian explosions and earthquakes discussed in Chapter 10. In the $m_b(M_s)$ diagram, also the set of values m_b = 6.0, M_s = 4.8 is shown. From the two diagrams in Fig. 13.5 we can see that the synthetic signals are neither typical explosion signals nor typical earthquake signals. This scenario, which is just one of the many that can be constructed, illustrates the difficulties involved in an attempt to evade various identification methods, also by theoretical computations. The evader not only has to produce nonexplosion-like signals, he further has to produce signals that, at a majority of monitoring stations, look like those coming from an earthquake in the actual region.

In a UK working paper on safeguards against the employment of multiple explosions to simulate earthquakes (CCD/459, 1975) it is argued that broad-band recording would be an efficient way to counteract evasion by multiple explosions. It is estimated that, with equipment today operating in the UK, events with short-period body-wave magnitudes down to 5.5 can be observed on broad-band records. The deployment of an array of such instruments at a continental site might increase the applicability down to m_b = 4.5.

So far we have discussed the possibilities of constructing theoretical scenarios, using recorded or theoretically computed signals, whereby

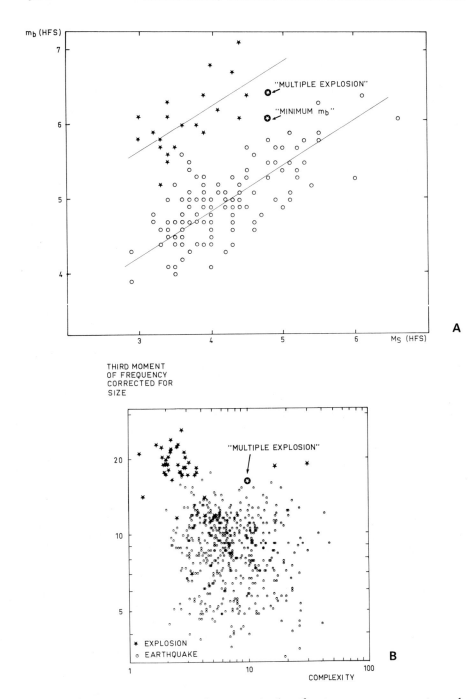

Fig. 13.5. (A)—$m_b(M_s)$, and (B)—short-period identification parameters estimated from the theoretically obtained record of five superimposed explosion signals, shown together with data obtained at the Hagfors Observatory from explosions and earthquakes in Eurasia.

different identification methods might be evaded by multiple explosions. The practical applicability of the method is, however, quite another and still more difficult problem. Although two or more underground nuclear explosions have been exploded simultaneously both in the US and in the USSR, no experimental study of the multiple-explosions technique of evasion has been reported. The difficulties of predicting even large-scale effects of underground nuclear explosions can be illustrated by two recent US explosions. The explosion "Baneberry", detonated in December 1970 at the Nevada Test Site, a place where a most comprehensive experience in nuclear testing has been collected, gave an unpredicted and strong venting of radioactive debris into the atmosphere, and a considerable area had to be evacuated. The other example is the "Rio Blanco" explosion discussed in Chapter 12. Three nuclear devices, each with a yield of 30 kt, were placed one on top of the other in a 2 km deep borehole, with a distance of 130 meters between the charges. The idea was to form one large cavity from which oil could be extracted. Postshot drilling revealed that three small cavities instead of one large had been formed (Chedd, 1973; AEC 1973, 1974).

This illustrates the difficulties involved in the prediction of the effects of nuclear explosions in areas where considerable experience of nuclear testing has been acquired and where unlimited preshot investigations can be conducted. We therefore consider it to be difficult to predict and control, with reasonable confidence, the seismic signals at a number of globally distributed stations from a number of explosions detonated in a seismic region. Predictions in such areas should be even more difficult than at ordinary test sites.

13.3 HIDE-IN-EARTHQUAKE METHOD

In this proposed evasion method a nuclear explosion is supposed to be fired so that seismic signals from the explosion are hidden in signals from a large earthquake (Lukasik, 1971; CCD/404, 1973). It has been suggested that either large nearby earthquakes or very large earthquakes in any part of the world be used to hide the explosion signals. No experiment where signals from an explosion have purposely been hidden in signals from an earthquake has been reported. Surface waves from explosions have, however, as discussed in Chapter 7, on several occasions been disturbed by interfering surface waves from large earthquakes.

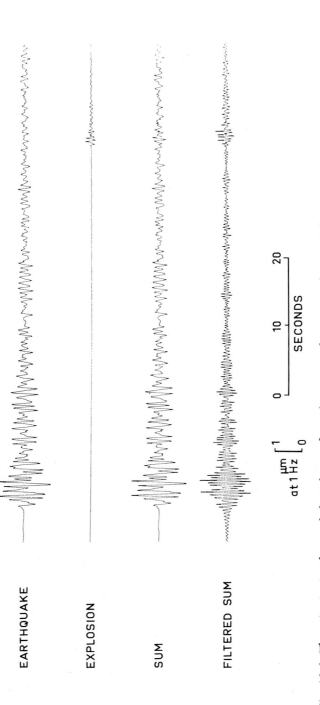

Fig. 13.6. *The seismic signal recorded at the Hagfors Observatory from an explosion at the Kazakh test site in the USSR has been superimposed on the signal from a Kashmir earthquake, to illustrate the hide-in-earthquake technique. The time difference between the two signals is 52 seconds. The events have the following source data (USGS):*

Event	Date	Origin time	Epicenter	m_b
Earthquake	720903	164828	36.0N 73.4E	6.3
Explosion	720816	031657	49.8N 78.1E	5.2

The summed signal is also filtered between 2 Hz and 4 Hz.

It has been discussed to hide either both the short-period and the long-period signals, or just the long-period signals and let the P waves be detected, so-called *partial hiding*. We will consider here the hiding of the short-period explosion signals. The long-period signals, which for explosions are about one tenth as strong as those for earthquakes of equal short-period amplitude, will in such a case be completely hidden.

Comprehensive studies of the possibilities of hiding explosion signals in signals from nearby earthquakes in the US and the USSR have been conducted by Jeppsson (1975a, b), and most of the data presented here are taken from his studies. In a simulation experiment short-period seismic signals recorded at the Hagfors Observatory from explosions in the US and the USSR were added, with different scaling and different time delays, to signals recorded from earthquakes in nearby regions. In Fig. 13.6 is shown an example of such mixed signals, where signals from an explosion at the Eastern Kazakh test site were added with a time delay of 52 s to signals from an earthquake in Kashmir. The body-wave magnitude of the earthquake is 1.1 unit higher than that of the explosion. A suite of such signals was analyzed by a number of seismologists to disclose the explosion signals. It was found that when the explosion signal arrived between half a minute and a few minutes after the onset of the earthquake signal, the former signal could have an amplitude of at most 1/30 of that of the earthquake signal to pass undetected.

By nearby earthquakes we here mean earthquakes occurring within 500 km from the point where the nuclear device is supposed to be placed. The decision to fire the explosion must be made as soon as the earthquake parameters have been estimated by a local network of seismological stations and found to be appropriate. The explosion must be timed so that the signals from the explosion will, at all monitoring stations, arrive after those from the earthquake. The time difference between the earthquake signals and the explosion signals at a seismic station depends on the difference in origin time between the explosion and the earthquake and on the relative location of the earthquake, the explosion, and the stations. If, for example, the distance between the explosion and the earthquake is 300 km, and if the explosion is fired when the earthquake signal arrives at the explosion point, the time difference for stations at 60 degrees distance from the test site will vary from 25 to 60 s. The amplitude of an earthquake signal decreases

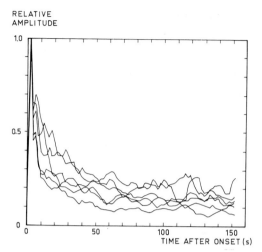

Fig. 13.7. Typical examples of P-wave coda amplitudes for some earthquakes recorded at the Hagfors Observatory.

rapidly during the first half of a minute and then remains fairly constant, as is illustrated in Fig. 13.7.

We have so far discussed the possibilities to detect the explosion signals in the presence of earthquake signals. The next problem is to identify the signals as coming from an explosion. The signal-to-noise ratio needed to detect an event is always less than that needed to identify the event. The signal-to-noise ratio has to be two to three times as high for identification as for detection. This means that for a hidden explosion to evade identification by short-period identification methods, its signals must have amplitudes at most about one tenth of those of the earthquake (equal to a difference in strength of one magnitude unit).

This discussion on magnitude differences needed for hiding an explosion is based on data obtained at one single station. The amplitudes of both the earthquake and explosion signals vary significantly from one station to the other. To confidently hide the explosion signals at all monitoring stations, the evader must probably increase the difference in strength by another half or whole magnitude unit.

In the following discussion of the number of opportunities that might exist to hide explosions in earthquakes, we will use a magnitude difference of 1.5 unit.

The number of opportunities to hide a clandestine explosion in an earthquake depends on the size of the explosion to be hidden and on

EVASION

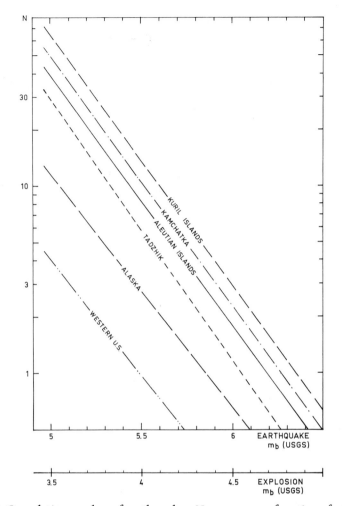

Fig. 13.8. Cumulative number of earthquakes, N, per year as a function of m_b (USGS) for the area with a 500-km radius that gives the highest seismicity in each of the six regions: Alaska, Western US, Aleutian Islands, Tadzhik, Kamchatka, and Kuril Islands. The number of expected opportunities to hide explosions of different magnitudes in the six regions can also be obtained from the explosion m_b scale in the figure.

the seismicity in the actual region. Based on estimates of the seismicity in six regions of interest in a CTB situation, Alaska, Western US, the Aleutian Islands, Tadzhik, Kamchatka, and the Kuril Islands, the number of earthquakes at different magnitudes expected to occur per year within 500 km of a predetermined shot point has been estimated and is shown in Fig. 13.8. The tentative explosion points were chosen to give the maximum number of earthquakes within 500 km in each

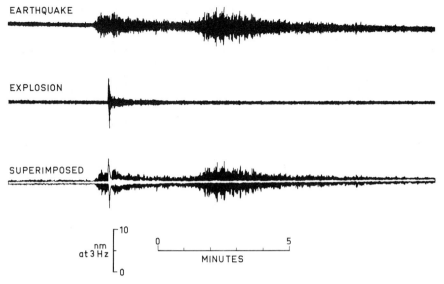

Fig. 13.9. *Short-period signals filtered between 2 Hz and 4 Hz obtained at Hagfors from an earthquake in New Guinea at 100 degrees distance and from an Eastern Kazakh explosion at a distance of 37 degrees. The events have the following source data (USGS):*

Event	Date	Origin time	Epicenter		Depth (km)	m_b
Earthquake	710110	071704	3.1S	139.7E	33	7.3
Explosion	720706	010258	49.7N	77.9E	0	4.2

region. Estimates of the expected number of opportunities per year to fire explosions with different expected magnitudes can also be obtained from Fig. 13.8. The number of opportunities to hide large explosions is small, although the seismicity is rather high, especially in the Aleutian Islands, Kamchatka, and the Kuril Islands. The expected number of opportunities per year to hide a magnitude 4.5 explosion, corresponding to a fully tamped explosion in hard rock with a yield of 3–5 kt, varies from 0.3 in the Western US to 3 in the Kuril Islands. Corresponding figures for a magnitude 5 explosion having a hard-rock yield of the order of 10 kt varies between 0.05 and 0.5. Filson (1973) has studied the time interval available in which an explosion could be fired in Central Asia following an earthquake in the Kuril–Kamchatka region without LASA being able to detect it. This study shows that, from the evader's point of view, there would have been approximately one interval of two minutes per year during which a clandestine magnitude 5 explosion could be hidden in earthquakes from the Kuril Islands with a minimum risk of detection at LASA.

It has also been suggested that explosions could be hidden in very large earthquakes occurring at large distances from the explosion point (Davies, 1973). Due to the shadow of the earth core, short-period signals recorded in the distance ranges of 100—135 degrees and 150—180 degrees from the source have low amplitudes. Seismic stations in either of these two distance intervals from the earthquake and within 90 degrees from the explosion might thus have a high probability of detecting also weak explosion signals. To illustrate this situation, we have in Fig. 13.9 reproduced short-period signals recorded at Hagfors from an earthquake in New Guinea with m_b (USGS) = 7.3 at a distance of 110 degrees and signals from an explosion in Eastern Kazakh with m_b (USGS) = 4.7 at a distance of 37 degrees. These time-compressed records show that the signal from the rather weak explosion, with an estimated yield of a few kilotons, could with high probability have been discerned from the earthquake signals at Hagfors.

With a well-distributed station network there is a high probability of having a number of stations in a similar situation. To illustrate this we show in Fig. 13.10 the areas which are within 90 degrees from the Eastern Kazakh test area and in the distance intervals 100—135 and 150—180 degrees, respectively, from the New Guinea earthquake. There are a considerable number of seismological stations in these areas.

Partial hiding, the case where only surface waves are hidden, will yield many test opportunities. One estimate (CCD/404, 1973) gives between five and ten opportunities per year to test an explosion with a surface magnitude of 4, which corresponds to about 100 kt, without getting the long-period signals detected at the present long-period array stations. Masking the long-period signals means that the $m_b(M_s)$ identification method cannot be applied. The short-period signals are, however, not affected, and, as we have seen in Chapter 10, the short-period identification methods now available could identify the event with a high degree of confidence.

13.4 THE FEASIBILITY OF EVASION

Contrary to other evasion methods, the multiple-explosions method is supposed to give discernible seismic signals at a number of stations. It is critically dependent on the possibility to predict and control the seismic signals from a number of explosions fired in a seismic region. As the explosions will become detected and located, the rather ex-

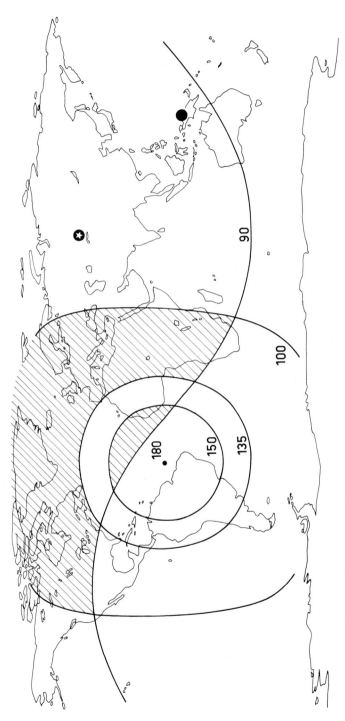

Fig. 13.10. Map showing the areas (hatched) that are within 90 degrees of the Eastern Kazakh test area and in the distance intervals 100–135 degrees and 150–180 degrees from New Guniea.

tensive operation at the test site also has to be concealed from satellite reconnaissance. Based on past failures to predict and control even the large-scale effects of underground explosions at established test sites, we doubt that this method could with any credibility be suggested as a way to carry out clandestine testing under a CTB.

Considering the hide-in-earthquake technique, we think that it might be technically feasible to have the device emplaced in advance in a seismic area, waiting for an earthquake to occur. The risk of making a wrong estimate of the signatures of the earthquake signals at globally distributed stations seems, however, too great to make the method credible. To our mind, hiding an explosion in a distant, very large, earthquake involves also a very large risk that explosion signals will be clearly recorded at some stations being in the shadow zone of the earthquake, provided that an adequate, global monitoring network exists.

In our opinion the hiding of the surface waves of an explosion will not prevent the explosion from being confidently identified by short-period identification methods.

It has been demonstrated by many experiments at the Nevada Test Site that the seismic signals from low-yield explosions fired in a low-coupling medium, such as alluvium, could be reduced by a factor of up to 10 compared to those from hard-rock explosions. There are also strong reasons to believe that a similar decoupling can be achieved for low-yield explosions in large cavities, although this decoupling technique has been demonstrated so far only for explosions with a yield below 1 kt. Decoupling techniques might therefore be used to conceal a low-yield test. They further reduce the signals below the detection threshold of the monitoring network and thus do not leave any discernible signals. By this technique it seems possible to decouple the seismic signals from explosions with yields up to a few kilotons, so that the signals will be weaker than those from a fully tamped 1-kt explosion in hard rock.

Our conclusion is thus that the decoupling technique, either in a low-coupling medium or in larger cavities, is the only evasion technique so far presented that could with reasonable credibility be used for clandestine testing of explosions with yields of a few kilotons, provided that the monitoring system consists of teleseismic stations only. Decoupled explosions can, however, be carried out only in limited areas, where either large deposits of unconsolidated material exist, or where large cavities exist or can conveniently be mined. These limited areas can be monitored by local seismic networks, which make it possible to discern also explosions having equivalent hard-rock yields far below one kiloton. In the PNE Treaty, provisions already exist for the establishment of local seismic networks to estimate explosion yields under certain conditions.

14. NONSEISMOLOGICAL IDENTIFICATION

Throughout the long discussions about a Test Ban, seismological methods have been most thoroughly discussed among the methods of verification, but other verification methods have also been proposed as supplements. Since the CTB discussion started, the US has maintained that on-site inspection is necessary for adequate verification of a CTB. The USSR has opposed such obligatory inspections, and has repeatedly declared that national means would provide adequate verification of a CTB. These opposing positions of the two superpowers have been the main technical reason why CTB negotiations have not made any substantial progress over the years.

The USSR has never defined what should be understood by "national means of verification". A similar wording, "national technical means of verification", is also used to describe verification methods accepted to assure compliance with the provisions of the so-called Threshold Test Ban Treaty and the SALT agreements. Although there is no definition of "national means of verification" in these treaties, it is clear that seismological methods and satellite methods, respectively, are methods of main importance in that respect.

The intention with this chapter is to shortly review some nonseismological verification methods that have been discussed in the context of a CTB, and to discuss their potential usefulness for monitoring underground nuclear explosions. Two nonseismological methods, on-site inspection and reconnaissance by satellite, must be regarded as complements to seismological methods for monitoring such explosions. These methods can hardly be used to detect any new explosions, but rather to increase the capability to identify events already detected and located by seismological means. On-site inspection, where foreign observers search a certain area to find an illicit test, is an intrusive method, whereas the use of reconnaissance satellites, whereby detailed pictures can be obtained of any part of a country, is regarded as a nonintrusive method.

Technical and nontechnical intelligence methods used to acquire information in other countries can be independent of seismological data. The usefulness of such methods for the verification of a CTB is difficult to evaluate and quantify. Generally, such intelligence methods, and at present also the satellite methods, are more applicable to the verification of bilateral agreements between the two superpowers than to the verification of multilateral agreements. In a multilateral treaty all parties should have access to the information needed to monitor the treaty or to the information on which a certain country may base its withdrawal from the treaty.

14.1 ON-SITE INSPECTION

The US has repeatedly stated that an adequate verification of a CTB must include provisions for on-site inspection. In the early sixties there was some bargaining between the US and the USSR on the number of inspections needed and considered acceptable for verification of a CTB, but since then the USSR has refused to accept on-site inspection. Provisions for on-site inspection for verifying PNE explosions are, however, included in the recent US–USSR PNE Treaty. According to US proposals, on-site inspection in connection with a CTB should be carried out when a detected and located seismic event cannot confidently be identified by seismological methods alone. An inspection team should then be sent to the area where the suspect event has been located in order to search for evidence of an illicit nuclear explosion.

Over the years, there have been a lot of general and political arguments about on-site inspection, but few technical discussions of on-site inspection techniques, their expected capability, etc. In the early sixties, the US presented rather detailed plans of how fairly extensive on-site inspection operations, involving inspection teams of about 20 persons and a considerable amount of equipment, should be carried out (Long, 1963). Among methods of on-site inspection suggested in those days were: seismic aftershock studies, magnetic-field surveys to find ferromagnetic material, studies of solid-state effects of the shockwave in ground material close to the explosion, radiochemical sampling of fission products released by venting from a subsurface cavity, visual inspection and aerial photography to find a subsidence crater or other signs of human activities that could be associated with an illicit test, and, finally, seismic profile shooting and drilling to find the explosion cavity (George, 1963). In a "Draft Treaty Banning Nuclear Weapon Tests in All Environments", proposed by the US on August 27, 1962, the establishment of an International Scientific Commission was foreseen, with the task of coordinating the monitoring system and carrying out on-site inspections (Joint Committee on Atomic Energy, 1963).

Since then the US, although still insisting on on-site inspections, has been less specific on what inspection methods should be used, how an inspection should be carried out, and from where the inspectors should come. In 1971, the Director of the US Advanced Research Projects Agency stated that "of all the on-site inspection techniques studied only two appear to be useful: Visual inspection and radiochemical analysis. Sufficiently deep burial will preclude surface effects and seepage of radioactive gas to the surface. While careful planning and execution of a clandestine test would defeat on-site inspection, any nation must contend with human fallibility." (Lukasik, 1971). This statement suggests that only few on-site inspection methods can be utilized, and also that the overall efficiency of these methods is low. The applicability of on-site inspection is limited to those events which have been detected and located by seismological means. On-site inspections can therefore not increase the detection capability of the verification system. Also, such inspections cannot counteract the evasion techniques of cavity decoupling and hiding-in-earthquake, as the basic idea behind these techniques is that the seismic signals from an illicit explosion should go undetected.

On-site inspection of an illicit explosion would allow inspection in an area where such an explosion has been conducted. Another way of using on-site inspections might therefore be to identify those earthquakes that look explosion-like according to certain seismological identification criteria. On-site inspection of earthquakes has so far not been discussed, and no special methods for such an inspection have been proposed. Of the two methods, visual inspection and radiochemical analysis, suggested for on-site inspection of explosions, only visual inspection is applicable to earthquakes. Earthquakes could be identified as such, either by observing the earthquake effect on the environment, or indirectly by the lack of human activity in the actual area. However, only very few earthquakes cause discernible seismic effects on the environment.

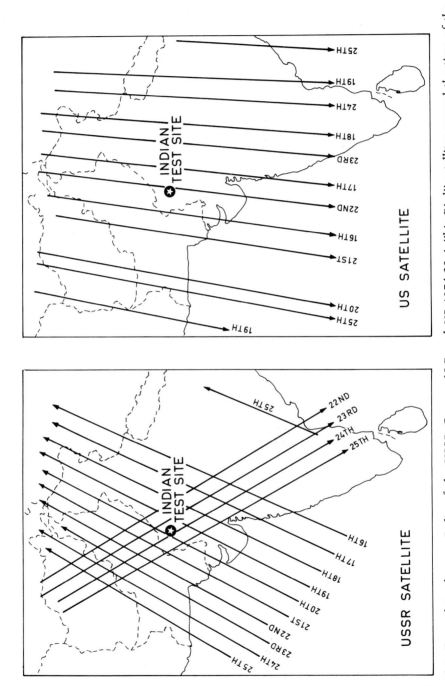

Fig. 14.1. Ground tracks over India of the USSR Cosmos-65 B and US 1974-20 A "big Bird" satellites around the time of the Indian explosion on May 18, 1974 (modified after SIPRI, 1975).

The lack of human activity in an area can be observed by on-site inspection, but such visual inspection might, however, be more conveniently carried out by satellite methods.

Analyzing the technical arguments, it is difficult to see why on-site inspection, in the way it has been proposed, is regarded as a necessary verification method to achieve an adequate verification of a CTB.

Another way of using on-site inspections or observations is to verify that explosions for peaceful purposes are not used for the development of nuclear weapons. The PNE Treaty contains provisions for fairly extensive on-site inspections, including measurements at the site, to verify the yields of group explosions having an aggregate yield exceeding 150 kilotons. Such on-site observations, to verify either PNE or large chemical explosions that could be misinterpreted as clandestine nuclear-weapon tests, seem to be an important element in a verification scheme for a CTB. Observations might for chemical explosions be confined to visual inspections. For PNE, if ever allowed under a CTB, such observations have to include comprehensive measurements to ensure that the explosion was carried out according to the specifications given in advance, as is further discussed in Chapter 12.

14.2 RECONNAISSANCE SATELLITES

Reconnaissance satellites equipped with various sensor systems have dramatically increased the amount of information available to the two superpowers about conditions all over the world. The availability to the superpowers of this detailed and rapidly accessible information on military and civilian facilities and activities in other countries may have severe economical and political implications. A discussion of such matters is outside the scope of this book. Such satellite data are, however, one main provision for decreasing the tension between the two superpowers by reducing the risk of their misunderstanding or misinterpreting each other's activities. Reconnaissance satellites have also been used to monitor military crises, e. g. in the Middle East in 1973 and in Cyprus in 1974 (SIPRI, 1974, 1975). Both a US and a USSR satellite also passed over, or close to, the area in northwestern India where a nuclear explosion was conducted on May 18, 1974, around the time of the explosion, see Fig. 14.1 (SIPRI, 1975).

Satellites are also playing an important role in monitoring arms-limitation treaties, such as SALT, although satellite verification is not explicitly mentioned in the treaty texts.

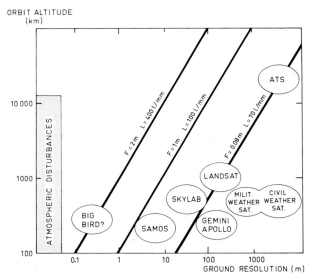

Fig. 14.2. Ground resolution and orbit altitude of some US satellite systems. The lines show the ground resolution as functions of orbit altitude and camera-film system. F denotes the focal length of the camera system. L gives the maximum line density of the film (from Orhaug & Dyring, 1973).

Fig. 14.3. Comparison of (A)–LANDSAT and (B,C)–Apollo photos of the eastern part of Sicily. The enlarged Apollo photo (C) shows the town of Catania.

B

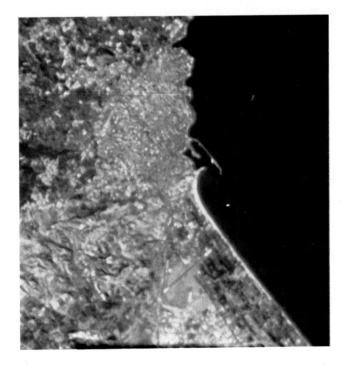

C

Surveys of reconnaissance satellites launched for different purposes have been regularly published by SIPRI (1973, 1974, 1975), and we will here limit our discussion to the possible use of reconnaissance satellites for monitoring a CTB. Data on the equipment and capability of the most sophisticated reconnaissance satellites are classified. Some conclusions on the possible capability of at least the photographic reconnaissance systems can, however, be drawn from the satellite systems used for civil purposes, such as the US LANDSAT and Skylab. The ground resolution and orbit altitude for some US satellite systems are shown in Fig. 14.2 (Orhaug & Dyring, 1973). We see from this figure that LANDSAT, operating at an altitude of about 900 km, has a ground resolution of the order of 100 m, and that Apollo, operating at a somewhat lower altitude, has a somewhat higher resolution. Photographic pictures by LANDSAT and Apollo of the western part of Sicily with Mount Etna are shown in Fig. 14.3. On the enlargement of the Apollo photo, showing the town of Catania, the piers in the harbor can be clearly discerned. Figure 14.4 shows a LANDSAT photo of the northern part of Nevada Test Site. The crater of the "Sedan" explosion, with a diameter of 390 m, is clearly seen. Smaller craters from other nearby explosions can also be discerned.

The reconnaissance satellites are operating at lower altitudes and use cameras with larger focal lengths, which will increase the ground resolution, as can be seen from Fig. 14.2.

Photo-reconnaissance satellites are used either for surveying large areas with rather low resolution, or for closeup inspection of specific targets, where high-resolution techniques are used. In areal surveying a resolution of the order of 10 m might be adequate, and about 14 000 photos will then be needed to cover the USSR. If the ground resolution is increased to 10 cm, a resolution that might be possible to achieve with the most modern equipment, about 140 million pictures will be needed to cover the same area (Orhaug & Dyring, 1973).

These figures illustrate that the amount of information needed to cover large areas with high resolution is so large that continuous monitoring of, e. g., the superpowers seems unrealistic. A reasonable way of utilizing photo-reconnaissance satellite data for verification purposes is to use such data as supplement to seismic data. In a CTB monitoring operation it is important to have ways to verify the rare earthquakes that according to seismic discrimination criteria might appear explosion-like. Such events are detected and located by seismological data, and satellite data are then used to disclose whether human activities that could be associated with an explosion can be observed in the area or not. The resolution needed to monitor human activities is probably of the order of a few meters, a resolution which is significantly lower than what can be technically achieved today. Satellite data should thus only be applied to events already detected and located from seismological data, and the area and time interval to be searched would be limited, which would reduce the amount of satellite data that has to be analyzed. The time interval between successive surveys of an area should be related to the time it takes to complete the field operation for a nuclear explosion and could probably be of the order of months.

The precautions needed to fully hide from satellite reconnaissance the logistical operation associated with the firing of a nuclear device at a remote place would considerably increase the complexity and difficulty of the operation. These difficulties will increase with the size of the operation, and the use of any evasion technique would undoubtly increase the size and complexity of the field operation. This method of indirect identification of earthquakes is applicable only to earthquakes located in areas where there normally is no human activity of the kind that can be associated with the field-work for a nuclear explosion, e. g. the southeastern USSR. If the event is located in an area with extensive human activity, and especially in a mining district, satellite data would be of less value.

We have so far been discussing only photographic reconnaissance data, because such data could be of value in certain situations, and also because the capability of photographic satellite systems can be estimated with reasonable accuracy. Other methods of satellite reconnaissance might also prove to be of value for monitoring a CTB, but adequate information to allow an evaluation of these potential methods is not generally available today.

Reconnaissance-satellite data are today available only to the US and the USSR. Information on the operative usefulness of such satellite data for CTB monitoring has not been published. If satellite data should be included for monitoring a CTB to which a large number, hopefully all, of the states on the globe are parties, then such satellite data must be generally and easily available. The launching of satellites and the operation of an adequate data-collection system are complicated and expensive operations, and few states will in a foreseeable future have the ability to deploy their own systems. The two superpowers, with the

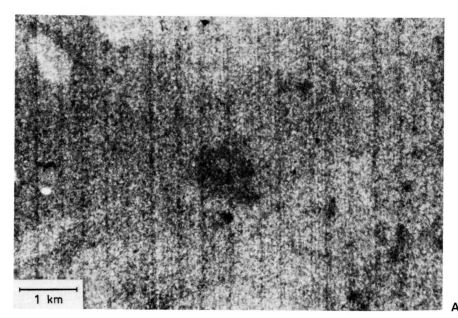

Fig. 14.4. The "Sedan" crater in the northern part of Nevada Test Site. (A)–LANDSAT photo. (B)–Aerial photo.

necessary equipment operating today, must here undertake to distribute pertinent parts of the collected data to countries wanting to monitor the CTB. Satellite data could be distributed either directly to a requesting state or to an international data center established primarily for collecting and evaluating seismological data, but which also may, if necessary, use satellite data to verify the occurrence of earthquakes. Such an international data center is further discussed in Chapter 15.

14.3 INTELLIGENCE METHODS

Little is published on technical and nontechnical intelligence methods used by states to acquire information on other countries. To maintain secrecy about what can be achieved by such methods is a matter of prime interest to the states and organizations involved. It is therefore not possible to estimate the kind or amount of information on underground nuclear testing that can be achieved by such methods. It has been suggested that it could be possible to follow radio communications in connection with nuclear test experiments either from satellite-based or ground-based communication stations (SIPRI, 1972). Radio communication during a test experiment can be replaced by telephone or highly directional, electromagnetic or optic communication links,

NONSEISMOLOGICAL IDENTIFICATION 335

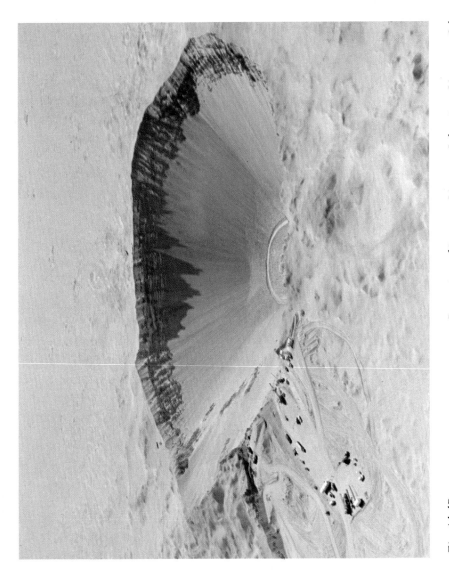

Fig. 14.4B. By courtesy of Lawrence Livermore Laboratory, Livermore, Cal.

which will prevent this type of eavesdropping. Such precautions will, however, increase the complexity of the operation. Generally it could be said that the efficiency of the various intelligence methods does not depend on the yield of the tested explosion, but rather on the overall size and structure of the operation.

Although such intelligence information might be of considerable importance for monitoring bilateral agreements, it seems difficult to include such information, obtained through unspecified sources, in any organized international data exchange or to present such information as evidence of a clandestine test. The US presentation of aerial photographs in the United Nations as proof of the Soviet missile installations in Cuba in October 1962 is one of the few occasions when intelligence data have been used to support a political action (Kennedy, 1969). Even if an intelligence source might give certain countries additional information, such information is in our opinion not needed to achieve an adequate verification of a CTB.

One source of information that should not be overlooked is reports in the press and other mass media, and the public debate in the various countries. Today the amount of publicly available information differs considerably from one country to the other. We think, however, that such information could be of value not only for the assessment of a specific event, e. g. an earthquake in a populated area, but also as a measure of the more general public reactions to certain proposals, e. g. further PNE. A growing public concern for environmental and health risks will today probably make it most difficult for a number of countries to detonate nuclear explosives for any purpose within their territories, regardless of any international treaties.

15. MONITORING A COMPREHENSIVE TEST-BAN TREATY

In this chapter we will briefly discuss the available technical capabilities to monitor a CTB in relation to political requirements. The political monitoring requirements put forward so far have only been stated in general terms, like "adequate verification", which are difficult to translate into technical requirements. In an attempt to formulate such general political requirements in technical terms, we summarize and interpret the few statements that have been made on this issue. We also shortly discuss how the achievable capabilities to detect, locate, and identify explosions and earthquakes discussed in this book meet the political requirements. There is always a significant difference between what can be achieved in principle, or in a special case study, and what is routinely achievable. We therefore finally present an outline of an operative monitoring system that, in our opinion, would provide an adequate verification of a CTB. This system consists of a global network of seismological stations, operated by various host countries, and an international data collection and evaluation center.

15.1 POLITICAL REQUIREMENTS OF THE VERIFICATION OF A CTB

The ultimate purpose of a CTB is to ban all nuclear weapon tests in all environments and by all countries. A CTB can be regarded as a complement to the NPT, as it primarily will limit the test activity of the nuclear powers, whereas the nonnuclear weapon states by signing the NPT have already renounced their nuclear option. Today, it seems unlikely that a CTB can be signed without the final adherence of all present nuclear-weapon states. A CTB would make it most difficult for the nonnuclear-weapon states to develop nuclear weapons. For those countries that today have nuclear weapons, a CTB would limit their possibilities to further develop their nuclear devices—this statement is based on the assumption that it would probably be very difficult even for a superpower, although most experienced in the designing of nuclear weapons, to develop a completely new nuclear device without testing it. The most important component to be developed further to obtain a more efficient nuclear-weapon system is in fact not the nuclear device itself, but rather the navigation and guidance system. It has also been reported that up to July 1971, the US had never conducted nuclear tests with stockpiled weapons solely for the purpose of checking their reliability (Scoville, 1973). According to Panofsky (1973), the necessary stockpile control of existing devices can be confidently carried out without test explosions.

However, at an informal CCD Expert Meeting, Mr. Fakley stated (CCD/492, 1976):

> "The maintenance of stockpiled weapons in a safe yet serviceable condition is of vital importance whilst such weapons continue to exist. Like all other types of hardware, nuclear weapons age; in fact they tend to age more rapidly than most other weapons because of the characteristics of the special materials used in the construction. From time to time they have to be repaired and this frequently requires some of the components to be replaced. These replacements are likely to be somewhat different from the originals if only because available materials will have changed and because safety and security requirements have been made more stringent. The changes involved are usually small but it cannot be ascertained whether or not they are significant without nuclear testing. In order to be sure that the refurbished weapons remain safe and serviceable, it is necessary to carry out a full scale test but a reduced-scale test can offer the required assurances in some circumstances."

Nuclear weapons have so far been developed under the condition that unlimited testing can be carried out. We have, however, the strong feeling that nuclear weapons could be designed to maintain their long-term reliability without testing.

It would be impossible to create a verification system that would secure the timely detection of any violation of a treaty at any time. The level of control considered necessary would depend on the consequences of clandestine violation. The expected internal pressure on parties to violate a treaty would also have to be considered. The pressure to violate a treaty of course depends on what the treaty forbids.

In any worthwhile agreement, one basic assumption would have to be that the parties to the treaty generally have the intention to adhere to the provisions of the treaty and not to carry out any illicit actions. The purpose of a system designed to monitor a test-ban treaty would be to counteract unfounded suspicions arising out of lack of information and to monitor with such a high credibility that it deters parties from carrying out clandestine tests. For the latter purpose the system should have enough capability to disclose clandestine test programs significant to the military and political balance. On the other hand, the monitoring system should not give rise to a substantial number of false alarms about earthquakes. It is striking to see how the verification issue has dominated the CTB discussion, much more than in the discussions of other disarmament treaties. The proposed verification requirements for the CTB are indeed considerably more far-reaching than for existing treaties on nuclear-weapon matters.

The Partial Test Ban Treaty does not, as we have seen in Chapter 3, limit the test activity nor the development of nuclear devices—it only forces the countries that are parties to the treaty to make their testing underground. Only little pressure to violate the treaty could therefore be expected, and verification procedures were not explicitly included in the treaty. This simplicity was, however, facilitated by the ease with which the radioactive radiation from such explosions could be monitored, and by the nonexistence of obvious, natural phenomena which could cause false alarms.

The Treaty of Tlatelolco and the Sea-Bed Treaty involve various degrees of inspection, either on the basis of free access, or on the basis of consultation, cooperation, and invitation. The Tlatelolco Treaty also includes the establishment of an international exchange of data and reports and of a special agency for the prohibition of nuclear weapons in Latin America, OPANAL. The verification of the Non-Proliferation Treaty is carried out through an extensive book-keeping control by IAEA of nuclear material and by the inspection of reactors and other nuclear facilities in the nonnuclear-weapon states. The Threshold Test Ban Treaty limits weapon tests in the US and the USSR to yields below 150 kt. In 1969–1974, however, only 10 % of the tests in these two countries had yields above this threshold. The increased accuracy of nuclear-weapon delivery systems has also decreased the need for high-yield devices. The TTBT can therefore not be regarded as a nuclear disarmament measure. In the TTBT an exchange of data on test sites and earlier explosions is foreseen for calibration of the seismological methods considered to be the most important national means of verification. Nuclear explosions under the Threshold Test Ban Treaty shall further be carried out only at specified test sites, and when conducted tests should be reported. This treaty could thus in fact be considered as a Comprehensive Test Ban Treaty outside the test sites.

It is also of interest to note that elaborate evasion techniques have been extensively analyzed in the CTB context, whereas in the SALT agreement of 1972 the use of evasion methods was simply ruled out in the treaty by the following stipulation: "Each party undertakes not to use deliberate concealment measures which impede verification by national technical means of compliance with the provisions of this Interim Agreement."

In designing a verification procedure for a Comprehensive Test Ban Treaty, it is necessary to know the requirements of the parties involved. These verification requirements have so far been stated only in most general terms. The Soviet Union considers national means of verification, assisted by an international data exchange, as adequate, and is against obligatory on-site inspections as required by the US. The US has repeatedly stated that it wants to achieve a CTB adequately verified, and that adequate verification includes on-site inspection. The US definition of adequate verification was also stated as follows: "We consider adequate verification as that which would reduce to an acceptable level the risk that clandestine test programs of military significance could be conducted under a CTB." (CCD/PV.553, 1972). What are then "acceptable risk level" and "test programs of military significance"? Little or no guidance has been given in interpreting the phrase "acceptable risk level". Perhaps one can assume that acceptable risk levels would have to be lower the more significant the tests are.

The "military significance" of a test would generally increase with the yield of the explosion. Low-yield tests could therefore more easily be dispensed of than could higher-yield ones. In 1973, Dr. John Foster, Director of Research and Development in the United States Department of Defense touched on this question in a statement before a United States Senate Committee: "A kiloton is a very significant yield. Now I will admit there are some cases where a half kiloton can be very significant or a quarter of a kiloton. But certainly I know from my experience that a kiloton is a very significant yield from a military point of view." This statement indicates that the military significance of sub-kiloton tests would be essentially less than that of larger tests. The discussion of the verification of a CTB has also so far mainly been limited to events above magnitude 4, corresponding to a hard-rock yield of about 1 kt, which is generally considered to be the lower limit of teleseismic detection. On-site inspection, suggested by the US to improve the verification of events detected by seismological means, cannot improve on this detection capability.

Such a threshold of 1 kt would be of interest primarily for the monitoring of the present nuclear-weapon states. As all the first nuclear explosions of the present nuclear-weapon powers had yields around or above 10–20 kt, that yield range could be a sufficient lower limit for monitoring those nonnuclear states that might have the ambition of going nuclear.

The political requirements of a monitoring system for a CTB, as presented so far, can thus be summarized as follows: The monitoring system shall have the ability to disclose test programs involving explosions having yields of the order of 1 kt and above, which does not mean that all explosions in a test program necessarily have to be disclosed. The monitoring system must further not give rise to false alarms to such an extent that it will deteriorate the credibility of the treaty.

The monitoring of such a multilateral treaty must further be an international undertaking, so that all parties to the treaty can utilize the verification system.

15.2 A MONITORING SYSTEM

With the preceding discussion on political requirements as a background, we outline a possible seismic system for global monitoring of a Comprehensive Test Ban Treaty.

The system is designed to enable also states having limited resources in detection seismology to make an independent assessment of globally collected and preanalyzed data. The design of this system is largely based on the assessment of the seismological methods and capabilities of detection, location, depth estimation, and identification presented in Chapters 7–10. The system consists of a network of selected seismological stations, a communication network, and an international data center. This system should be regarded as a tentative proposal, to promote and facilitate further discussion on this matter.

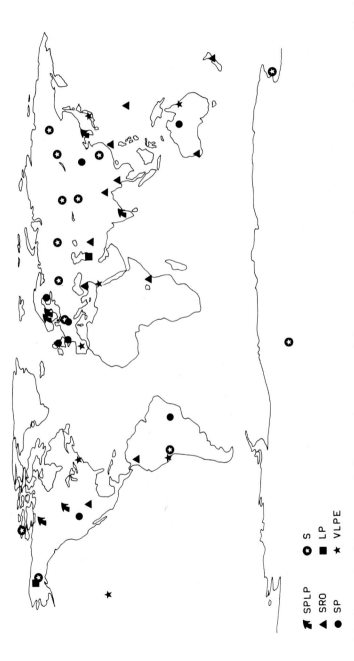

Fig. 15.1. Geographical locations of the seismic stations of the monitoring network. SPLP = Array with short-period and long-period systems. SRO = Seismological Research Observatory. SP = Array with short-period system. S = Single station with short-period system. LP = Array with long-period system. VLPE = Very-Long-Period Experimental Station.

15.2.1 Seismic station network

One basic idea behind the design of the seismic network presented here is to utilize, to the greatest possible extent, already existing or planned seismological installations. By using existing facilities the necessary investments are kept at a low level, and the monitoring system can be put into operation with short delay.

A network should consist of highly sensitive, single stations and arrays. There is no requirement of uniformity in equipment or detection procedures at the stations. Each station could be optimized for the local receiving conditions at its site. By leaving to the individual countries to decide on the design and operation of their stations, many standardization difficulties would be avoided.

The operational and economic responsibilities for the individual stations should stay with the owner countries. They could also be discharged in bilateral cooperation between countries. In this connection the notion of "sister observatories" and other organizational matters have been taken up by the Swedish delegation at CCD (CCD/405, 1973).

It is important that the individual stations of a network have comparable, if not equal, capabilities. A station with an exceptional sensitivity could yield detections which could not be verified by other stations. Similarly, a relatively insensitive station would make little contribution to the joint capability of a network.

The number of stations in a network should be kept fairly small. It has been demonstrated, for example during the International Seismic Month in 1972, that carefully analyzed data from appropriately selected stations can provide detection and location capabilities comparable with and even superior to those routinely achieved at present from the analysis of data from all seismological stations on the globe. A limited number of stations will also keep the data flow at a manageable level.

A global seismic station network might be supplemented by networks of local stations, to monitor regions of special interest, e. g. areas containing thick deposits of low-coupling media or areas of high seismicity.

It is important that careful analysis be carried out at each individual station (Needham, 1975). Besides arrival times, signal amplitudes, and period data, which are generally reported at present, the analysis should extract special identification parameters, based on short-period as well as long-period recording, and also information on depth phases and

TABLE 15.1. Seismic stations selected for the monitoring-network example.

State	Station Name	Code	Type[a]	China	France[c]	India	UK	US	USSR
Australia	Narrogin	WRA	SRO	x	x	x	-	-	-
	Warramunga	CTA	SP	x	x	x	-	-	x
	Charters Towers		VLPE	x	x	x	-	-	-
Bolivia	Zongo	ZLP	S, VLPE	-	x	-	x	x	-
Brazil	Brazilia	BDF	SP	-	x	-	x	x	-
Canada	Yellowknife	YKA	SPLP	x	x	-	x	x	x
	Mould Bay	MBC	S	x	x	x	x	x	x
China	Peking	PEK	SP	(x)	-	x	x	-	x
	Wuhan	WHN	S	(x)	-	x	x	-	x
	Wulumuchi	WMQ	S	(x)	-	x	x	-	x
Colombia	Bogota	BOG	SRO	-	x	-	x	x	-
Finland	Jyväskylä	JYSA	SP	x	-	x	x	x	x
France	Domestic CEA network		SP	x	-	x	x	x	x
FRG	Graefenberg	GRF	SP	x	x	x	x	x	x
GDR	Moxa	MOX	S	x	x	x	x	x	x
India	Gauribidanur	GBA	SPLP	x	-	(x)	x	-	x
	Shillong	SHL	SRO	x	-	(x)	x	-	x
Iran	Teheran	ILPA	LP	x	-	x	x	-	x
	Mashad	MSH	SRO	x	-	x	x	-	x
Israel	Eilat	EIL	VLPE	x	-	x	x	-	x
Japan	Matsushiro	MAT	VLPE	x	-	x	x	x	x
Kenya	Nairobi	NAI	SRO	x	-	x	x	-	x
New Zealand	Wellington	WEL	SRO	-	x	-	-	-	-
	Scott Base	SBA	S	-	x	-	-	-	-

a,b,c See notes at bottom of table (next page).

TABLE 15.1 (continued).

State	Station			Station located within 90 degrees from[b]					
	Name	Code	Type[a]	China	France[c]	India	UK	US	USSR
Norway	Norsar	NAO	SPLP	x	-	x	x	x	x
South Korea	Seoul	KSRS	SPLP	x	-	x	x	x	x
Spain	Toledo	TOL	VLPE	x	-	x	x	x	x
Sweden	Hagfors	HFS	SPLP	x	-	x	x	x	x
Taiwan	Taipei	TAP	SRO	x	-	x	-	-	x
Thailand	Chiengmai	CHG	SRO	x	-	x	-	-	x
Turkey	Ankara	ANK	SRO	x	-	x	x	x	x
UK	Eskdalemuir	ESK	SP	x	-	x	(x)	x	x
US	Lasa	LAO	SPLP	-	x	-	x	x	x
	Albuquerque	ALQ	SRO	-	x	-	x	(x)	x
	Guam	GNA	SRO	x	x	x	-	-	x
	Unita Basin	UBO	SP	-	x	-	x	(x)	x
	College	COL	S	x	x	x	x	(x)	x
	South Pole	SPA	S	-	x	-	-	-	-
	Alpa	FBK	LP	x	x	x	x	(x)	x
	Honolulu	HNL	VLPE	x	x	-	-	(x)	x
	Ogdensburg	OGD	VLPE	-	x	-	x	(x)	x
USSR	Bodaybo	BOD	S	x	-	x	x	x	(x)
	Eltsouka	ELT	S	x	-	x	x	x	(x)
	Obinsk	OBN	S	x	-	x	x	x	(x)
	Sverdlovsk	SVE	S	x	-	x	x	x	(x)
	Yakutsk	YAK	S	x	-	x	x	x	(x)

[a] SPLP= Array with short-period and long-period systems. SRO=Seismological Research Observatory. SP=Array with short-period system. S=Single station with short-period system. LP=Array with long-period system. VLPE=Very-long-period experiment station.
[b] (x) denotes that the station is located inside the country to be monitored.
[c] Pacific test site.

other secondary phases. The identification parameters could well differ from station to station depending on the facilities available and on the receiving conditions for seismic signals. The analysis at the stations would materially reduce the amount of data to be transmitted for joint analysis at the data center described below. Today, only few stations routinely conduct the type of analysis outlined here.

The operation of several of the arrays is at present in a more or less experimental stage. In a monitoring situation one would have to put strong demands on maintenance and calibration, to obtain high-quality data and to enhance the operation reliability of the stations.

The map in Fig. 15.1 shows the geographical distribution of the seismic stations selected here to constitute an example of a highly sensitive seismic station network having global coverage. Networks of local stations, which might be desirable, are not discussed here. The stations, which are listed in Table 15.1, can be grouped into the following categories according to their equipment:

Category	No. of stations
Array with short-period and long-period systems	6
Seismological Research Observatory, SRO	11
Array with short-period system	8
Single station with short-period system	13
Array with long-period system	2
Very-Long-Period Experiment station, VLPE	7

This makes 47 stations in all, 38 of which have short-period and 26 long-period instruments. The stations are distributed over 26 countries, as can be seen from Table 15.1. In order not to make the network critically dependent on certain stations, the total number of stations is somewhat larger than the minimum necessary. The capabilities of this network would not be significantly reduced even if some ten stations became inoperative for certain events or ceased to operate in the network.

The station network in Fig. 15.1 has been designed to provide global coverage, utilizing as many stations as possible that already exist or are under installation. The stations have been selected among those used during the International Seismic Month (ISM) presented in Chapter 7. Stations in China and in the USSR as well as the SRO stations have been added. The final installations for some of the SRO stations have not been completed at the time of this writing. Moreover, little has

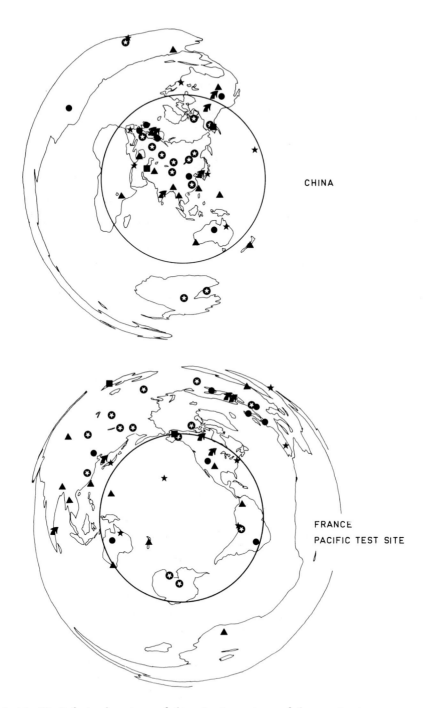

Fig. 15.2 (A–F). Relative locations of the seismic stations of the monitoring network and the territories of the nuclear countries. The circles enclose areas within 90 degrees from the center of the respective territories. Symbols as in Fig. 15.1.

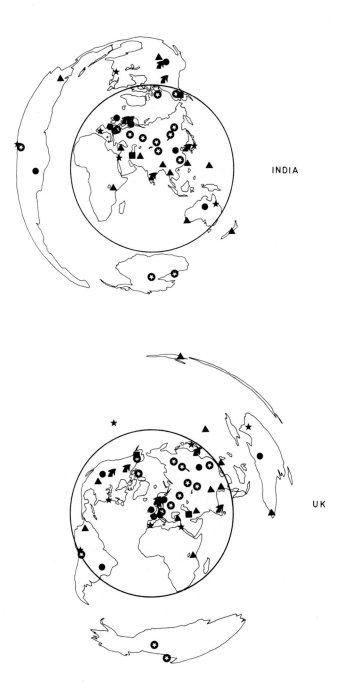

Fig. 15.2 (C, D). For legend, see page 347.

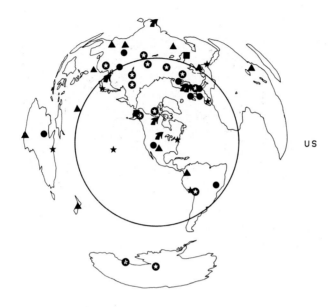
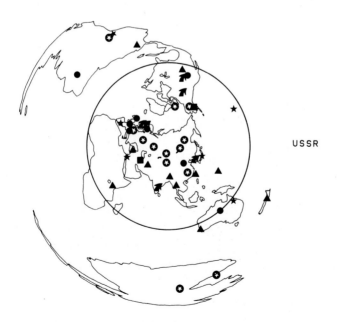

Fig. 15.2 (E, F). For legend, see page 347.

TABLE 15.2. Number of stations within 90 degrees from the nuclear countries.

Station type[a]	Country					
	China	France[b]	India	UK	US	USSR
SPLP	5	2	3 (1)	6	4 (1)	6
SRO	8	5	7 (1)	7	2 (1)	8
SP	5 (1)	3	6	6 (1)	5 (1)	7
S	8 (2)	5	10	11	8 (1)	5 (5)
LP	2	1	2	2	- (1)	2
VLPE	5	4	4	5	3 (2)	5
Total number of stations with short-period recording	26 (3)	15	26 (2)	30 (1)	19 (4)	26 (5)
Total number of stations with long-period recording	20	12	16 (2)	20	9 (5)	21

Numbers without (within) parentheses represent stations outside (inside) the respective country.

[a] SPLP = Array with short-period and long-period systems. SRO = Seismological Research Observatory. SP = Array with short-period system. S = Single station with short-period system. LP = Array with long-period system. VLPE = Very-long-period experiment station.
[b] Pacific test site.

been published on observed or expected capabilities of the SRO. Their modern technical design and installations in deep bore holes make them, however, superior to standard single stations at corresponding sites. Only one African station has been selected. The extremely favorable recording condition in Africa would, however, make it worthwhile to consider the establishment of a few more, well-equipped stations, especially in the Central and Northern parts of the continent.

The five USSR stations listed in Table 15.1 are standard stations belonging to the USSR network of conventional stations which is used for acquiring data on world seismicity. It has been claimed that several arrays, so-called *KOD* seismograph networks, have been used for detection and identification of nuclear explosions (Shishkevish, 1975). These arrays, which are reported to operate in various regions of the USSR since 1966, consist of three to five stations each, equipped with 3-component digital-recording systems. The *KOD* arrays would be most valuable in a global monitoring network, but cannot be included here since their actual locations are not known.

To get some idea of the geographical coverage of the monitoring network the distribution of the stations of the network relative to the territories of the six nuclear powers is shown in Fig. 15.2. The maps in this figure represent projections with the central parts of the various countries as projection points. The UK has carried out its recent testing at NTS in the US, but the map for UK is representative for any European country. The 90-degree distance circles from the projection points have also been drawn in the figures.

It is indicated in Table 15.1 whether a station falls within these 90-degree distance circles. It is clear that if a station falls within a 90-degree circle of a country, this does not mean that it covers all parts of that country. Similarly, if a station falls outside 90 degrees, it may well cover some parts of the country. Even if most of the stations are within 90 degrees to three or more of the nuclear countries, only the stations in Alaska, COL and FBK, cover all six nuclear countries. The total numbers of stations of the different categories within 90 degrees of the different nuclear countries are summarized in Table 15.2. These numbers indicate that the network provides a fairly uniform coverage of the different nuclear countries. The number of stations covering the French Pacific test area and the US is only slightly below the number of those covering the other four nuclear countries.

15.2.2 *Data exchange*

Parameters extracted from the records at the individual stations should be transmitted regularly in the form of bulletins to the data center. The total number of parameters in the station bulletins is estimated to be not more than 20 per event. A highly sensitive array with both short-period and long-period recording systems can be expected to detect roughly 30 events per day. This would mean that the average number of parameters to be transmitted from a station is not greater than 600 per day. It is important that every station transmit the data in a consistent format, but it is not necessary that all stations use the same format.

Ordinary telex channels provide a convenient way of transmitting seismic bulletin data. The global telecommunication system of the World Meteorological Organization has been discussed for transmission of seismological data for this and for other purposes. These systems are simple and easily accessible to most countries, and are also compatible with computers. Exchange of full recordings, which is supposed to be conducted for events of special interest only, could probably be handled by ordinary mail. If recordings are to be sent as digital magnetic tapes, the problems with achieving compatibility between computers must be considered. Large and technically advanced computer-communication networks like the ARPA NET would be helpful but not necessary for handling the limited amount of data discussed here.

15.2.3 *Data center*

The task of the international data center would be to assist states wishing to monitor the CTB, by collecting and processing data about seismic events. The data and the results arrived at would be made available to the parties. In this way the data center would be particularly valuable to countries with only limited expertise and facilities. The political assessment of the events analyzed by the data center should be left to the parties of the treaty.

In particular, the data center should have the following main functions:

— To collect data from the network of selected seismological stations.

— To define seismic events and estimate their epicenters and depths.

— To apply various seismic identification criteria to the events defined.

— To distribute the results to interested parties.

— To conduct consultations and make inquiries with designated institutes to obtain additional data and information about events insufficiently described by the data routinely available.

— To provide experts to observe, by invitation, nuclear explosions for peaceful purposes and large chemical explosions.

The procedure of analyzing seismic events, which could be undertaken in various ways, would have to be agreed upon among the parties concerned. Even if the current seismic event-identification techniques are highly effective, there will probably be a small residue of events defined by the data center which could not be sufficiently analyzed by routinely applied teleseismic methods. Further analysis would then be made by acquiring additional seismological data, e. g. from local seismic station networks, or by applying more refined analysis methods than would be applied on a routine basis. Satellite photographs of relevant areas could also be requested and analyzed in order to facilitate the clarification of residual events.

If the nature of a seismic event cannot be confidently clarified, the center could consult designated agencies in the country to which the event had been located for further clarification. If doubts still remain about the nature of the event, the countries concerned should be notified, and they could then make their own interpretation of such events and take the steps they may find appropriate. Obviously, the data center should confine itself to scientific tasks. Political evaluation of the events should be left to the Governments parties to the treaty.

The monitoring system should operate with a time delay that would be sufficient in order to take appropriate political action after a violation of the treaty. The length of this time delay would have to be agreed upon.

The observation, by invitation, of nuclear explosions for peaceful purposes and of large chemical explosions would be an important element in the monitoring of a CTB. The participation in such activities would be an important task for the center.

To conduct the tasks outlined above, a staff of about 40 professionals and technicians and appropriate technical equipment and computers would be required for the data center. The estimated cost of such a data center is of the order of two million dollars per year.

The data center could be set up as an independent international institute or be included in an already existing international agency, such as IAEA, or part of a disarmament monitoring agency of the kind earlier proposed by Myrdahl (1974).

15.3 GENERAL CONCLUSIONS

In this book we have tried to give a comprehensive discussion of the possibilities to detect, locate, and identify earthquakes and explosions by seismological means, and we will here only briefly summarize our conclusions. It is demonstrated that seismic events with magnitudes down to about 4 can be detected by seismological stations at teleseismic distances. To obtain an operative capability to achieve this detection threshold, an adequate station network has to be established, and such a network is presented above. To achieve a significantly lower detection threshold, stations at short distances from the event must be utilized. It has further been demonstrated that in most cases seismic events can be located with an accuracy of 10–20 km if data from about ten well distributed stations are available. A further increase in the number of stations will not significantly increase the location accuracy. If, however, calibration data from earlier events in the actual region are available, e. g. the data on nuclear explosions foreseen to be exchanged between the US and the USSR in connection with the Threshold Test Ban Treaty, then the location error would be less than 5 km.

The identification method generally referred to, the $m_b(M_s)$ method, can be applied at least down to the detection thresholds of the long-period stations, which for the most sensitive stations range between $M_s = 2.5$ and $M_s = 3.5$. This means that the $m_b(M_s)$ method can be applied to most earthquakes with a magnitude, m_b, of about 4.0 and above. Identification criteria based solely on short-period data, which are gradually gaining acceptance, can be applied to both earthquakes and explosions close to the short-period detection threshold. The applicability of various depth-estimation methods is still to be evaluated, and will probably prove to be of great operative value. The $m_b(M_s)$ method and the short-period identification methods can, with a high degree of confidence, identify most of the events to which they are applied. The confidence increases when identification data from many stations are utilized. There might, however, be a few, mostly low-magnitude, earthquakes which cannot be confidently identified using seismological data alone. The expected number of earthquakes in various regions that cannot be confidently identified using seismological data only is not known, as no identification study has so far been conducted utilizing data from a considerable number of globally distributed stations. There does not exist today any routine exchange of seismological identification parameters, and such an exchange, as suggested above, must be instituted if an operative capability to routinely identify occurring events shall be achieved. Some earthquakes occur in remote and uninhabited areas, which means that satellite photos might support the identification of these earthquakes by some sort of "negative evidence" through the absence of human activities. Various elaborate evasion techniques have been discussed, by which clandestine explosions of high yields might be conducted. In our opinion, the decoupling technique is the only one that might be used in practice, and then only for explosions with a yield of one or a few kilotons. Such explosions can, however, be carried out only in limited areas where appropriate geological conditions exist, and such areas could be monitored by special, local seismic networks.

Summarizing this technical discussion, we conclude that it is possible to implement a monitoring system, consisting of (1) a global network of seismological stations, (2) access to photographic satellite data, and (3) an international data collection and evaluation center, by which magnitude-4 earthquakes and explosions, corresponding to hard-rock explosion yields of about 1 kt, can be detected, located, and identified with a high degree of confidence.

APPENDIX 1
TREATIES RELATED TO NUCLEAR EXPLOSIONS

CONTENTS

The Partial Test-Ban Treaty (*Treaty banning nuclear weapon tests in the atmosphere, in outer space and under water*) 358

The Non-Proliferation Treaty (*Treaty on the non-proliferation of nuclear weapons*) 360

The Threshold Test-Ban Treaty (*Treaty between the United States of America and the Union of Soviet Socialist Republics on the limitation of underground nuclear weapon tests*) 365

    Protocol to the Threshold Test-Ban Treaty 366

The PNE Treaty (*Treaty between the United States of America and the Union of Soviet Socialist Republics on underground nuclear explosions for peaceful purposes*) 368

    Protocol to the PNE Treaty 371

    Agreed Statement 383

TREATY BANNING NUCLEAR WEAPON TESTS IN THE ATMOSPHERE, IN OUTER SPACE AND UNDER WATER

The Governments of the United States of America, the United Kingdom of Great Britain and Northern Ireland, and the Union of Soviet Socialist Republics, hereinafter referred to as the "Original Parties",

Proclaiming as their principal aim the speediest possible achievement of an agreement on general and complete disarmament under strict international control in accordance with the objectives of the United Nations which would put an end to the armaments race and eliminate the incentive to the production and testing of all kinds of weapons, including nuclear weapons,

Seeking to achieve the discontinuance of all test explosions of nuclear weapons for all time, determined to continue negotiations to this end, and desiring to put an end to the contamination of man's environment by radioactive substances,

Have agreed as follows:

ARTICLE I

1. Each of the Parties to this Treaty undertakes to prohibit, to prevent, and not to carry out any nuclear weapon test explosion, or any other nuclear explosion, at any place under its jurisdiction or control:

(a) in the atmosphere; beyond its limits, including outer space; or under water, including territorial waters or high seas; or

(b) in any other environment if such explosion causes radioactive debris to be present outside the territorial limits of the State under whose jurisdiction or control such explosion is conducted. It is understood in this connection that the provisions of this subparagraph are without prejudice to the conclusion of a treaty resulting in the permanent banning of all nuclear test explosions, including all such explosions underground, the conclusion of which, as the Parties have stated in the Preamble to this Treaty, they seek to achieve.

2. Each of the Parties to this Treaty undertakes furthermore to refrain from causing, encouraging, or in any way participating in, the carrying out of any nuclear weapon test explosion, or any other nuclear explosion, anywhere which would take place in any of the environments described, or have the effect referred to, in paragraph 1 of this Article.

ARTICLE II

1. Any Party may propose amendments to this Treaty. The text of any proposed amendment shall be submitted to the Depositary Governments which shall circulate it to all Parties to this Treaty. Thereafter, if requested to do so by one-third or more of the Parties, the Depositary Governments shall convene a conference, to which they shall invite all the Parties, to consider such amendment.

2. Any amendment to this Treaty must be approved by a majority of the votes of all the Parties to this Treaty, including the votes of all of the Original Parties. The amendment shall enter into force for all Parties upon the deposit of instruments of ratification by a majority of all the Parties, including the instruments of ratification of all of the Original Parties.

ARTICLE III

1. This Treaty shall be open to all States for signature. Any State which does not sign this Treaty before its entry into force in accordance with paragraph 3 of this Article may accede to it at any time.

2. This Treaty shall be subject to ratification by signatory States. Instruments of ratification and instruments of accession shall be deposited with the Governments of the Original Parties – the United States of America, the United Kingdom of Great Britain and Northern Ireland, and the Union of Soviet Socialist Republics – which are hereby designated the Depositary Governments.

3. This Treaty shall enter into force after its ratification by all the Original Parties and the deposit of their instruments of ratification.

4. For States whose instruments of ratification or accession are deposited subsequent to the entry into force of this Treaty, it shall enter into force on the date of the deposit of their instruments of ratification or accession.

5. The Depositary Governments shall promptly inform all signatory and acceding States of the date of each signature, the date of deposit of each instrument of ratification of and accession to this Treaty, the date of its entry into force, and the date of receipt of any requests for conferences or other notices.

6. This Treaty shall be registered by the Depositary Governments pursuant to Article 102 of the Charter of the United Nations.

ARTICLE IV

This Treaty shall be of unlimited duration.

Each Party shall in exercising its national sovereignty have the right to withdraw from the Treaty if it decides that extraordinary events, related to the subject matter of this Treaty, have jeopartized the supreme interests of its country. It shall give notice of such withdrawal to all other Parties to the Treaty three months in advance.

ARTICLE V

This Treaty, of which the English and Russian texts are equally authentic, shall be deposited in the archives of the Depositary Governments. Duly certified copies of this Treaty shall be transmitted by the Depositary Governments to the Governments of the signatory and acceding States.

IN WITNESS WHEREOF the undersigned, duly authorized, have signed this Treaty.

DONE in triplicate at the city of Moscow the fifth day of August, one thousand nine hundred and sixty-three.

TREATY ON THE NON-PROLIFERATION OF NUCLEAR WEAPONS

1. The States concluding this Treaty, hereinafter referred to as the "Parties to the Treaty",
2. Considering the devastation that would be visited upon all mankind by a nuclear war and the consequent need to make every effort to avert the danger of such a war and to take measures to safeguard the security of peoples,
3. Believing that the proliferation of nuclear weapons would seriously enhance the danger of nuclear war,
4. In conformity with resolutions of the United Nations General Assembly calling for the conclusion of an agreement on the prevention of wider dissemination of nuclear weapons,
5. Undertaking to co-operate in facilitating the application of International Atomic Energy Agency safeguards on peaceful nuclear activities,
6. Expressing their support for research, development and other efforts to further the application, within the framework of the International Atomic Energy Agency safeguards system, of the principle of safeguarding effectively the flow of source and special fissionable materials by use of instruments and other techniques at certain strategic points,
7. Affirming the principle that the benefits of peaceful applications of nuclear technology, including any technological by-products which may be derived by nuclear-weapon States from the development of nuclear explosive devices, should be available for peaceful purposes to all Parties to the Treaty, whether nuclear-weapon or non-nuclear-weapon States,
8. Convinced that, in furtherance of this principle, all Parties to the Treaty are entitled to participate in the fullest possible exchange of scientific information for, and to contribute alone or in co-operation with other States to, the further development of the applications of atomic energy for peaceful purposes,
9. Declaring their intention to achieve at the earliest possible date the cessation of the nuclear arms race and to undertake effective measures in the direction of nuclear disarmament,
10. Urging the co-operation of all States in the attainment of this objective,
11. Recalling the determination expressed by the Parties to the 1963 Treaty banning nuclear weapon tests in the atmosphere, in outer space and under water in its Preamble to seek to achieve the discontinuance of all test explosions of nuclear weapons for all time and to continue negotiations to this end,
12. Desiring to further the easing of international tension and the strengthening of trust between States in order to facilitate the cessation of the manufacture of nuclear weapons, the liquidation of all their existing stockpiles, and the elimination from national arsenals of nuclear weapons and means of their delivery pursuant to a treaty on general and complete disarmament under strict and effective international control,

13. Recalling that, in accordance with the Charter of the United Nations, States must refrain in their international relations from the threat or use of force against the territorial integrity or political independence of any State, or in any other manner inconsistent with the Purposes of the United Nations, and that the establishment and maintenance of international peace and security are to be promoted with the least diversion for armaments of the world's human and economic resources,

Have agreed as follows:

ARTICLE I

Each nuclear-weapon State Party to the Treaty undertakes not to transfer to any recipient whatsoever nuclear weapons or other nuclear explosive devices or control over such weapons or explosive devices directly, or indirectly; and not in any way to assist, encourage, or induce any non-nuclear-weapon State to manufacture or otherwise acquire nuclear weapons or other nuclear explosive devices, or control over such weapons or explosive devices.

ARTICLE II

Each non-nuclear-weapon State Party to the Treaty undertakes not to receive the transfer from any transferor whatsoever of nuclear weapons or other nuclear explosive devices or of control over such weapons or explosive devices directly, or indirectly; not to manufacture or otherwise acquire nuclear weapons or other nuclear explosive devices; and not to seek or receive any assistance in the manufacture of nuclear weapons or other nuclear explosive devices.

ARTICLE III

1. Each non-nuclear-weapon State Party to the Treaty undertakes to accept safeguards, as set forth in an agreement to be negotiated and concluded with the International Atomic Energy Agency in accordance with the Statute of the International Atomic Energy Agency and the Agency's safeguards system, for the exclusive purpose of verification of the fulfillment of its obligations assumed under this Treaty with a view to preventing diversion of nuclear energy from peaceful uses to nuclear weapons or other nuclear explosive devices. Procedures for the safeguards required by this article shall be followed with respect to source or special fissionable material whether it is being produced, processed or used in any principal nuclear facility or is outside any such facility. The safeguards required by this article shall be applied on all source or special fissionable material in all peaceful nuclear activities within the territory of such State, under its jurisdiction, or carried out under its control anywhere.

2. Each State Party to the Treaty undertakes not to provide: (a) source or special fissionable material, or (b) equipment or material especially designed or prepared for the processing, use or production of special fissionable material, to any non-nuclear-weapon State for peaceful purposes, unless the source or special fissionable material shall be subject to the safeguards required by this article.

3. The safeguards required by this article shall be implemented in a manner designed to comply with article IV of this Treaty, and to avoid hampering the economic or technological development of the Parties or international co-operation in the field of peaceful nuclear activities, including the international exchange of nuclear material and equipment for the processing, use or production of nuclear material for peaceful puposes in accordance with the provisions of this article and the principle of safeguarding set forth in the Preamble of the Treaty.

4. Non-nuclear-weapon States Party to the Treaty shall conclude agreements with the International Atomic Energy Agency to meet the requirements of this article either individually or together with other States in accordance with the Statute of the International Atomic Energy Agency. Negotiation of such agreements shall commence within 180 days from the original entry into force of this Treaty. For States depositing their instruments of ratification or accession after the 180-day period, negotiation of such agreements shall commence not later than the date of such deposit. Such agreements shall enter into force not later than eighteen months after the date of initiation of negotiations.

ARTICLE IV

1. Nothing in this Treaty shall be interpreted as affecting the inalienable right of all the Parties to the Treaty to develop research, production and use of nuclear energy for peaceful purposes without discrimination and in conformity with articles I and II of this Treaty.

2. All the Parties to the Treaty undertake to facilitate, and have the right to participate in, the fullest possible exchange of equipment, materials and scientific and technological information for the peaceful uses of nuclear energy. Parties to the Treaty in a position to do so shall also co-operate in contributing alone or together with other States or international organizations to the further development of the applications of nuclear energy for peaceful purposes, especially in the territories of non-nuclear-weapon States Party to the Treaty, with due consideration for the needs of the developing areas of the world.

ARTICLE V

Each Party to the Treaty undertakes to take appropriate measures to ensure that, in accordance with this Treaty, under appropriate international observation and through appropriate international procedures, potential benefits from any peaceful applications of nuclear explosions will be made available to non-nuclear-weapon States Party to the Treaty on a non-discriminatory basis and that the charge to such Parties for the explosive devices used be as low as possible and exclude any charge for research and development. Non-nuclear-weapon States Party to the Treaty shall be able to obtain such benefits, pursuant to a special international agreement or agreements, through an appropriate international body with adequate representation of non-nuclear-weapon States. Negotiations on this subject shall commence as soon as possible after the Treaty enters into force. Non-nuclear-weapon States Party to the Treaty so desiring may also obtain such benefits pursuant to bilateral agreements.

ARTICLE VI

Each of the Parties to the Treaty undertakes to pursue negotiations in good faith on effective measures relating to cessation of the nuclear arms race at an early date and to nuclear disarmament, and on a treaty on general and complete disarmament under strict and effective international control.

ARTICLE VII

Nothing in this Treaty affects the right of any group of States to conclude regional treaties in order to assure the total absence of nuclear weapons in their respective territories.

APPENDIX 1

ARTICLE VIII

1. Any Party to the Treaty may propose amendments to this Treaty. The text of any proposed amendment shall be submitted to the Depositary Governments which shall circulate it to all Parties to the Treaty. Thereupon, if requested to do so by one third or more of the Parties to the Treaty, the Depositary Governments shall convene a conference, to which they shall invite all the Parties to the Treaty, to consider such an amendment.

2. Any amendment to this Treaty must be approved by a majority of the votes of all the Parties to the Treaty, including the votes of all nuclear-weapon States Party to the Treaty and all other Parties which, on the date the amendments is circulated, are members of the Board of Governors of the International Atomic Energy Agency. The amendment shall enter into force for each Party that deposits its instrument of ratification of the amendment upon the deposit of such instruments of ratification by a majority of all the Parties, including the instruments of ratification of all nuclear-weapon States Party to the Treaty and all other Parties which, on the date the amendment is circulated, are members of the Board of Governors of the International Atomic Energy Agency. Thereafter, it shall enter into force for any other Party upon the deposit of its instrument of ratification of the amendment.

3. Five years after the entry into force of this Treaty, a conference of Parties to the Treaty shall be held in Geneva, Switzerland, in order to review the operations of this Treaty with a view to assuring that the purposes of the Preamble and the provisions of the Treaty are being realized. At intervals of five years thereafter, a majority of the Parties to the Treaty may obtain, by submitting a proposal to this effect to the Depositary Government, the convening of further conferences with the same objective of reviewing the operation of the Treaty.

ARTICLE IX

1. This Treaty shall be open to all States for signature. Any State which does not sign the Treaty before its entry into force in accordance with paragraph 3 of this article may accede to it at any time.

2. This Treaty shall be subject to ratification by signatory States. Instruments of ratification and instruments of accession shall be deposited with the Governments of the Union of Soviet Socialist Republics, the United Kingdom of Great Britain and Northern Ireland and the United States of America, which are hereby designated the Depositary Governments.

3. This Treaty shall enter into force after its ratification by the States, the Governments of which are designated Depositaries of the Treaty, and forty other States signatory to this Treaty and the deposit of their instrument of ratification. For the purposes of this Treaty, a nuclear-weapon State is one which has manufactured and exploded a nuclear weapon or other nuclear explosive device prior to 1 January 1967.

4. For States whose instruments of ratification or accession are deposited subsequent to the entry into force of this Treaty, it shall enter into force on the date of the deposit of their instruments of ratification or accession.

5. The Depositary Governments shall promptly inform all signatory and acceding States of the date of each signature, the date of deposit of each instrument of ratification or of accession, the date of the entry into force of this Treaty, and the date of receipt of any requests for convening a conference or other notices.

6. This Treaty shall be registered by the Depositary Governments pursuant to article 102 of the Charter of the United Nations.

ARTICLE X

1. Each Party shall in exercising its national sovereignty have the right to withdraw from the Treaty if it decides that extraordinary events, related to the subject matter of this Treaty, have jeopardized the supreme interests of its country. It shall give notice of such withdrawal to all other Parties to the Treaty and to the United Nations Security Council three months in advance. Such notice shall include a statement of the extraordinary events it regards as having jeopardized its supreme interests.

2. Twenty-five years after the entry into force of the Treaty, a conference shall be convened to decide whether the Treaty shall continue in force indefinitely, or shall be extended for an additional fixed period or periods. This decision shall be taken by a majority of the Parties to the Treaty.

ARTICLE XI

This Treaty, the Chinese, English, French, Russian and Spanish texts of which are equally authentic, shall be deposited in the archives of the Depositary Governments. Duly certified copies of this Treaty shall be transmitted by the Depositary Governments to the Governments of the signatory and acceding States.

In witness whereof the undersigned, duly authorized, have signed this Treaty.

TREATY BETWEEN THE UNITED STATES OF AMERICA AND THE UNION OF SOVIET SOCIALIST REPUBLICS ON THE LIMITATION OF UNDERGROUND NUCLEAR WEAPON TESTS

The United States of America and the Union of Soviet Socialist Republics, hereinafter referred to as the Parties,

Declaring their intention to achieve at the earliest possible date the cessation of the nuclear arms race and to take effective measures toward reductions in strategic arms, nuclear disarmament, and general and complete disarmament under strict and effective international control,

Recalling the determination expressed by the Parties to the 1963 Treaty Banning Nuclear Weapon Tests in the Atmosphere, in Outer Space and Under Water in its preamble to seek to achieve the discontinuance of all test explosions of nuclear weapons for all time, and to continue negotiations to this end,

Noting that the adoption of measures for the further limitation of underground nuclear weapon tests would contribute to the achievement of these objectives and would meet the interests of strengthening peace and the further relaxation of international tension,

Reaffirming their adherence to the objectives and principles of the Treaty Banning Nuclear Weapon Tests in the Atmosphere, in Outer Space and Under Water and of the Treaty on the Non-Proliferation of Nuclear Weapons,

Have agreed as follows:

ARTICLE I

1. Each Party undertakes to prohibit, to prevent, and not to carry out any underground nuclear weapon test having a yield exceeding 150 kilotons at any place under its jurisdiction or control, beginning 31 March 1976.

2. Each Party shall limit the number of its underground nuclear weapon tests to a minimum.

3. The Parties shall continue their negotiations with a view toward achieving a solution to the problem of the cessation of all underground nuclear weapon tests.

ARTICLE II

1. For the purpose of providing assurance of compliance with the provisions of the Treaty, each Party shall use national technical means of verification at its disposal in a manner consistent with the generally recognized principles of international law.

2. Each Party undertakes not to interfere with the national technical means of verification of the other Party operating in accordance with paragraph 1 of this Article.

3. To promote the objectives and implementation of the provisions of the Treaty the Parties shall, as necessary, consult with each other, make inquiries and furnish information in response to such inquiries.

ARTICLE III

The provisions of this Treaty do not extend to underground nuclear explosions carried out by the Parties for peaceful purposes. Underground nuclear explosions for peaceful purposes shall be governed by an agreement which is to be negotiated and concluded by the Parties at the earliest possible time.

ARTICLE IV

This Treaty shall be subject to ratification in accordance with the constitutional procedures of each Party. This Treaty shall enter into force on the day of the exchange of instruments of ratification.

ARTICLE V

1. This Treaty shall remain in force for a period of five years. Unless replaced earlier by an agreement in implementation of the objectives specified in paragraph 3 of Article 1 of this Treaty, it shall be extended for successive five-year periods unless either Party notifies the other of its termination no later than six months prior to the expiration of the Treaty. Before the expiration of this period the Parties may, as necessary, hold consultations to consider the situation relevant to the substance of this Treaty and to introduce possible amendments to the text of the Treaty.

2. Each Party shall, in exercising its national sovereignty, have the right to withdraw from this Treaty if it decides that extraordinary events related to the subject matter of this Treaty have jeopardized its supreme interests. It shall give notice of its decision to the other Party six months prior to withdrawal from this Treaty. Such notice shall include a statement of the extraordinary events the notifying Party regards as having jeopardized its supreme interests.

3. This Treaty shall be registered pursuant to Article 102 of the Charter of the United Nations.

Done at Moscow on 3 July, 1974, in duplicate, in the English and Russian languages, both texts being equally authentic.

PROTOCOL TO THE TREATY BETWEEN THE UNITED STATES OF AMERICA AND THE UNION OF SOVIET SOCIALIST REPUBLICS ON THE LIMITATION OF UNDERGROUND NUCLEAR WEAPON TESTS

The United States of America and the Union of Soviet Socialist Republics, hereinafter referred to as the Parties,

Having agreed to limit underground nuclear weapon tests,

Have agreed as follows:

1. For the purpose of ensuring verification of compliance with the obligations of the Parties under the Treaty by national technical means, the Parties shall, on the basis of reciprocity, exchange the following data:

(a) The geographic co-ordinates of the boundaries of each test site and of the boundaries of the geophysically distinct testing areas therein.

(b) Information on the geology of the testing areas of the sites (the rock characteristics of geological formations and the basic physical properties of the rock, i.e., density, seismic velocity, water saturation, porosity and depth of water table).

(c) The geographic co-ordinates of underground nuclear weapon tests, after they have been conducted.

(d) Yield, date, time, depth and co-ordinates for two nuclear weapon tests for calibration purposes from each geophysically distinct testing area where underground nuclear weapon tests have been and are to be conducted. In this connexion the yield of such explosions for calibration purposes should be as near as possible to the limit defined in Article I of the Treaty and not less than one-tenth of that limit. In the case of testing areas where data are not available on two tests for calibration purposes, the data pertaining to one such test shall be exchanged, if available, and the data pertaining to the second test shall be exchanged as soon as possible after a second test having a yield in the above-mentioned range. The provisions of the Protocol shall not require the Parties to conduct tests solely for calibration purposes.

2. The Parties agree that the exchange of data pursuant to subparagraphs a, b, and d of paragraph 1 shall be carried out simultaneously with the exchange of instruments of ratification of the Treaty, as provided in Article IV of the Treaty, having in mind that the Parties shall, on the basis of reciprocity, afford each other the opportunity to familiarize themselves with these data before the exchange of instruments of ratification.

3. Should a Party specify a new test site or testing area after the entry into force of the Treaty, the data called for by subparagraphs a and b of paragraph 1 shall be transmitted to the other Party in advance of use of that site or area. The data called for by subparagraph d of paragraph 1 shall also be transmitted in advance of use of that site or area if they are available; if they are not available, they shall be transmitted as soon as possible after they have been obtained by the transmitting Party.

4. The Parties agree that the test sites of each Party shall be located at places under its jurisdiction or control and that all nuclear weapon tests shall be conducted solely within the testing areas specified in accordance with paragraph 1.

5. For the purposes of the Treaty, all underground nuclear explosions at the specified test sites shall be considered nuclear weapon tests and shall be subject to all the provisions of the Treaty relating to nuclear weapon tests. The provisions of Article III of the Treaty apply to all underground nuclear explosions conducted outside of the specified test sites, and only to such explosions.

This Protocol shall be considered an integral part of the Treaty.

Done at Moscow on 3 July, 1974.

TREATY BETWEEN THE UNITED STATES OF AMERICA AND THE UNION OF SOVIET SOCIALIST REPUBLICS ON UNDERGROUND NUCLEAR EXPLOSIONS FOR PEACEFUL PURPOSES

The United States of America and the Union of Soviet Socialist Republics, hereinafter referred to as the Parties,

Proceeding from a desire to implement Article III of the Treaty between the United States of America and the Union of Soviet Socialist Republics on the Limitation of Underground Nuclear Weapon Tests, which calls for the earliest possible conclusion of an agreement on underground nuclear explosions for peaceful purposes,

Reaffirming their adherence to the objectives and principles of the Treaty Banning Nuclear Weapon Tests in the Atmosphere, in Outer Space and Under Water, the Treaty on the Non-Proliferation of Nuclear Weapons, and the Treaty on the Limitation of Underground Nuclear Weapon Tests, and their determination to observe strictly the provisions of these international agreements,

Desiring to assure that underground nuclear explosions for peaceful purposes shall not be used for purposes related to nuclear weapons,

Desiring that utilization of nuclear energy be directed only toward peaceful purposes,

Desiring to develop appropriately co-operation in the field of underground nuclear explosions for peaceful purposes,

Have agreed as follows:

ARTICLE I

1. The Parties enter into this Treaty to satisfy the obligations in Article III of the Treaty on the Limitation of Underground Nuclear Weapon Tests, and assume additional obligations in accordance with the provisions of this Treaty.

2. This Treaty shall govern all underground nuclear explosions for peaceful purposes conducted by the Parties after 31 March 1976.

ARTICLE II

For the purposes of this Treaty:

(a) "explosion" means any individual or group underground nuclear explosion for peaceful purposes;

(b) "explosive" means any device, mechanism or system for producing an individual explosion;

(c) "group explosion" means two or more individual explosions for which the time interval between successive individual explosions does not exceed five seconds and for which the emplacement points of all explosives can be interconnected by straight line segments, each of which joins two emplacement points and each of which does not exceed 40 kilometres.

ARTICLE III

1. Each Party, subject to the obligations assumed under this Treaty and other international agreements, reserves the right to:

   (a) carry out explosions at any place under its jurisdiction or control outside the geographical boundaries of test sites specified under the provisions of the Treaty on the Limitation of Underground Nuclear Weapon Tests; and

   (b) carry out, participate or assist in carrying out explosions in the territory of another State at the request of such other State.

2. Each Party undertakes to prohibit, to prevent and not to carry out at any place under its jurisdiction or control, and further undertakes not to carry out, participate or assist in carrying out anywhere:

   (a) any individual explosion having a yield exceeding 150 kilotons;

   (b) any group explosion:

   (1) having an aggregate yield exceeding 150 kilotons except in ways that will permit identification of each individual explosion and determination of the yield of each individual explosion in the group in accordance with the provisions of Article IV of and the Protocol to this Treaty;

   (2) having an aggregate yield exceeding one and one-half megatons;

   (c) any explosion which does not carry out a peaceful application;

   (d) any explosion except in compliance with the provisions of the Treaty Banning Nuclear Weapon Tests in the Atmosphere, in Outer Space and Under Water, the Treaty on Non-Proliferation of Nuclear Weapons, and other international agreements entered into by that Party.

3. The question of carrying out any individual explosion having a yield exceeding the yield specified in paragraph 2(a) of this article will be considered by Parties at an appropriate time to be agreed.

ARTICLE IV

1. For the purpose of providing assurance of compliance with the provisions of this Treaty, each Party shall:

   (a) use national technical means of verification at its disposal in a manner consistent with generally recognized principles of international law; and

   (b) provide to the other Party information and access to sites of explosions and furnish assistance in accordance with the provisions set forth in the Protocol to this Treaty.

2. Each Party undertakes not to interfere with the national technical means of verification of the other Party operating in accordance with paragraph 1(a) of this article, or with the implementation of the provisions of paragraph 1(b) of this article.

ARTICLE V

1. To promote the objectives and implementation of the provisions of this Treaty, the Parties shall establish promptly a Joint Consultative Commission within the framework of which they will:

   (a) consult with each other, make inquiries and furnish information in response to such inquiries, to assure confidence in compliance with the obligations assumed;

   (b) consider questions concerning compliance with the obligations assumed and related situations which may be considered ambiguous;

(c) consider questions involving unintended interference with the means for assuring compliance with the provisions of this Treaty;

(d) consider changes in technology or other new circumstances which have a bearing on the provisions of this Treaty; and

(e) consider possible amendments to provisions governing underground nuclear explosions for peaceful purposes.

2. The Parties through consultation shall establish, and may amend as appropriate, Regulations for the Joint Consultative Commission governing procedures, composition and other relevant matters.

ARTICLE VI

1. The Parties will develop co-operation on the basis of mutual benefit, equality, and reciprocity in various areas related to carrying out underground nuclear explosions for peaceful purposes.

2. The Joint Consultative Commission will facilitate this co-operation by considering specific areas and forms of co-operation which shall be determined by agreement between the Parties in accordance with their constitutional procedures.

3. The Parties will appropriately inform the International Atomic Energy Agency of results of their co-operation in the field of underground nuclear explosions for peaceful purposes.

ARTICLE VII

1. Each Party shall continue to promote the development of the international agreement or agreements and procedures provided for in Article V of the Treaty on the Non-Proliferation of Nuclear Weapons, and shall provide appropriate assistance to the International Atomic Energy Agency in this regard.

2. Each Party undertakes not to carry out, participate or assist in the carrying out of any explosion in the territory of another State unless that State agrees to the implementation in its territory of the international observation and procedures contemplated by Article V of the Treaty on the Non-Proliferation of Nuclear Weapons and the provisions of Article IV of and the Protocol to this Treaty, including the provision by that State of the assistance necessary for such implementation and of the privileges and immunities specified in the Protocol.

ARTICLE VIII

1. This Treaty shall remain in force for a period of five years, and it shall be extended for successive five-year periods unless either Party notifies the other of its termination no later than six months prior to its expiration. Before the expiration of this period the Parties may, as necessary, hold consultations to consider the situation relevant to the substance of this Treaty. However, under no circumstances shall either Party be entitled to terminate this Treaty while the Treaty on the Limitation of Underground Nuclear Weapon Tests remains in force.

2. Termination of the Treaty on the Limitation of Underground Nuclear Weapon Tests shall entitle either Party to withdraw from this Treaty at any time.

3. Each Party may propose amendments to this Treaty. Amendments shall enter into force on the day of the exchange of instruments of ratification of such amendments.

APPENDIX 1

ARTICLE IX

1. This Treaty including the Protocol which forms an integral part hereof, shall be subject to ratification in accordance with the constitutional procedures of each Party. This Treaty shall enter into force on the day of the exchange of instruments of ratification which exchange shall take place simultaneously with the exchange of instruments of ratification of the Treaty on the Limitation of Underground Nuclear Weapon Tests.
2. This Treaty shall be registered pursuant to Article 102 of the Charter of the United Nations.

Done at Washington and Moscow, on 28 May 1976, in duplicate, in the English and Russian languages, both texts being equally authentic.

PROTOCOL TO THE TREATY BETWEEN THE UNITED STATES OF AMERICA AND THE UNION OF SOVIET SOCIALIST REPUBLICS ON UNDERGROUND NUCLEAR EXPLOSIONS FOR PEACEFUL PURPOSES

The United States of America and the Union of Soviet Socialist Republics, hereinafter referred to as the Parties,

Having agreed to the provisions in the Treaty on Underground Nuclear Explosions for Peaceful Purposes, hereinafter referred to as the Treaty,

Have agreed as follows:

ARTICLE I

1. No individual explosion shall take place at a distance, in metres, from the ground surface which is less than 30 times the 3.4 root of its planned yield in kilotons.
2. Any group explosion with a planned aggregate yield exceeding 500 kilotons shall not include more than five individual explosions, each of which has a planned yield not exceeding 50 kilotons.

ARTICLE II

1. For each explosion, the Party carrying out the explosion shall provide the other Party:

   (a) not later than 90 days before the beginning of emplacement of the explosives when the planned aggregate yield of the explosion does not exceed 100 kilotons, or not later than 180 days before the beginning of emplacement of the explosives when the planned aggregate yield of the explosion exceeds 100 kilotons, with the following information to the extent and degree of precision available when it is conveyed:

   (1) the purpose of the planned explosion;
   (2) the location of the explosion expressed in geographical co-ordinates with

a precision of four or less kilometres, planned date and aggregate yield of the explosion;

(3) the type or types of rock in which the explosion will be carried out, including the degree of liquid saturation of the rock at the point of emplacement of each explosive; and

(4) a description of specific technological features of the project, of which the explosion is a part, that could influence the determination of its yield and confirmation of purpose; and

(b) not later than 60 days before the beginning of emplacement of the explosives the information specified in subparagraph 1(a) of this article to the full extent and with the precision indicated in that subparagraph.

2. For each explosion with a planned aggregate yield exceeding 50 kilotons, the Party carrying out the explosion shall provide the other Party, not later than 60 days before the beginning of emplacement of the explosives, with the following information:

(a) the number of explosives, the planned yield of each explosive, the location of each explosive to be used in a group explosion relative to all other explosives in the group with a precision of 100 or less metres, the depth of emplacement of each explosive with a precision of one metre and the time intervals between individual explosions in any group explosion with a precision of one-tenth second; and

(b) a description of specific features of geological structure or other local conditions that could influence the determination of the yield.

3. For each explosion with a planned aggregate yield exceeding 75 kilotons, the Party carrying out the explosion shall provide the other Party, not later than 60 days before the beginning of emplacement of the explosives, with a description of the geological and geophysical characteristics of the site of each explosion which could influence determination of the yield, which shall include: the depth of the water table; a stratigraphic column above each emplacement point; the position of each emplacement point relative to nearby geological and other features which influenced the design of the project of which the explosion is a part; and the physical parameters of the rock, including density, seismic velocity, porosity, degree of liquid saturation, and rock strength, within the sphere centred on each emplacement point and having a radius, in metres, equal to 30 times the cube root of the planned yield in kilotons of the explosive emplaced at that point.

4. For each explosion with a planned aggregate yield exceeding 100 kilotons, the Party carrying out the explosion shall provide the other Party, not later than 60 days before the beginning of emplacement of the explosives, with:

(a) information on locations and purposes of facilities and installations which are associated with the conduct of the explosion;

(b) information regarding the planned date of the beginning of emplacement of each explosive; and

(c) a topographic plan in local co-ordinates of the areas specified in paragraph 7 of Article IV, at a scale of 1:24,000 or 1:25,000 with a contour interval of 10 metres or less.

5. For application of an explosion to alleviate the consequences of an emergency situation involving an unforeseen combination of circumstances which calls for immediate action for which it would not be practicable to observe the timing requirements of paragraphs 1, 2 and 3 of this article, the following conditions shall be met:

(a) the Party deciding to carry out an explosion for such purposes shall inform the other Party of that decision immediately after it has been made and describe such circumstances;

(b) the planned aggregate yield of an explosion for such purpose shall not exceed 100 kilotons; and

(c) the Party carrying out an explosion for such purpose shall provide to the other Party the information specified in paragraph 1 of this article, and the information specified in paragraphs 2 and 3 of this article if applicable, after the decision to conduct the explosion is taken, but not later than 30 days before the beginning of emplacement of the explosives.

6. For each explosion, the Party carrying out the explosion shall inform the other Party, not later than two days before the explosion, of the planned time of detonation of each explosive with a precision of one second.

7. Prior to the explosion, the Party carrying out the explosion shall provide the other Party with timely notification of changes in the information provided in accordance with this article.

8. The explosion shall not be carried out earlier than 90 days after notification of any change in the information provided in accordance with this article which requires more extensive verification procedures than those required on the basis of the original information, unless an earlier time for carrying out the explosions is agreed between the Parties.

9. Not later than 90 days after each explosion the Party carrying out the explosion shall provide the other Party with the following information:

(a) the actual time of the explosion with a precision of one-tenth second and its aggregate yield;

(b) when the planned aggregate yield of a group explosion exceeds 50 kilotons, the actual time of the first individual explosion with a precision of one-tenth second, the time interval between individual explosions with a precision of one millisecond and the yield of each individual explosion; and

(c) confirmation of other information provided in accordance with paragraphs 1, 2, 3 and 4 of this article and explanation of any changes or corrections based on the results of the explosion.

10. At any time, but not later than one year after the explosion, the other Party may request the Party carrying out the explosion to clarify any item of the information provided in accordance with this article. Such clarification shall be provided as soon as practicable, but not later than 30 days after the request is made.

ARTICLE III

1. For the purposes of this Protocol:

(a) "designated personnel" means those nationals of the other Party identified to the Party carrying out an explosions as the persons who will exercise the rights and functions provided for in the Treaty and this Protocol; and

(b) "emplacement hole" means the entire interior of any drill-hole, shaft, adit or tunnel in which an explosive and associated cables and other equipment are to be installed.

2. For any explosion with a planned aggregate yield exceeding 100 kilotons but not exceeding 150 kilotons if the Parties, in consultation based on information provided in accordance with Article II and other information that may be introduced by either Party, deem it appropriate for the confirmation of the yield of the explo-

sion, and for any explosion with a planned aggregate yield exceeding 150 kilotons, the Party carrying out the explosion shall allow designated personnel within the areas and at the locations described in Article V to exercise the following rights and functions:

(a) confirmation that the local circumstances, including facilities and installations associated with the project, are consistent with the stated peaceful purposes;

(b) confirmation of the validity of the geological and geophysical information provided in accordance with Article II through the following procedures:

(1) examination by designated personnel of research and measurement data of the Party carrying out the explosion and of rock core or rock fragments removed from each emplacement hole, and of any logs and drill core from existing exploratory holes which shall be provided to designated personnel upon their arrival at the site of the explosion;

(2) examination by designated personnel of rock core or rock fragments as they become available in accordance with the procedures specified in subparagraph 2(b)(3) of this article; and

(3) observation by designated personnel of implementation by the Party carrying out the explosion of one of the following four procedures, unless this right is waived by the other Party:

(i) construction of that portion of each emplacement hole starting from a point nearest the entrance of the emplacement hole which is at a distance, in metres, from the nearest emplacement point equal to 30 times the cube root of the planned yield in kilotons of the explosive to be emplaced at that point and continuing to the completion of the emplacement hole; or

(ii) construction of that portion of each emplacement hole starting from a point nearest the entrance of the emplacement hole which is at a distance, in metres, from the nearest emplacement point equal to six times the cube root of the planned yield in kilotons of the explosive to be emplaced at that point and continuing to the completion of the emplacement hole as well as the removal of rock core or rock fragments from the wall of an existing exploratory hole, which is substantially parallel with and at no point more than 100 metres from the emplacement hole, at locations specified by designated personnel which lie within a distance, in metres, from the same horizon as each emplacement point of 30 times the cube root of the planned yield in kilotons of the explosive to be emplaced at that point; or

(iii) removal of rock core or rock fragments from the wall of each emplacement hole at locations specified by designated personnel which lie within a distance, in metres, from each emplacement point of 30 times the cube root of the planned yield in kilotons of the explosive to be emplaced at each such point; or

(iv) construction of one or more new exploratory holes so that for each emplacement hole there will be a new exploratory hole to the same depth as that of the emplacement of the explosive, substantially parallel with and at no point more than 100 metres from each emplacement hole, from which rock cores would be removed at locations specified by designated personnel which lie within a distance, in metres, from the same horizon as each emplacement point of 30 times the cube root of the planned yield in kilotons of the explosive to be emplaced at each such point;

(c) observation of the emplacement of each explosive, confirmation of the depth of its emplacement and observation of the stemming of each emplacement hole;

(d) unobstructed visual observation of the area of the entrance to each emplacement hole at any time from the time of emplacement of each explosive until all personnel have been withdrawn from the site for the detonation of the explosion; and

(e) observation of each explosion.

3. Designated personnel, using equipment provided in accordance with paragraph 1 of Article IV, shall have the right, for any explosion with a planned aggregate yield exceeding 150 kilotons, to determine the yield of each individual explosion in a group explosion in accordance with the provisions of Article VI.

4. Designated personnel, when using their equipment in accordance with paragraph 1 of Article IV, shall have the right, for any explosion with a planned aggregate yield exceeding 500 kilotons, to emplace, install and operate under the observation and with the assistance of personnel of the Party carrying out the explosion, if such assistance is requested by designated personnel, a local seismic network in accordance with the provisions of paragraph 7 or Article IV. Radio links may be used for the transmission of data and control signals between the seismic stations and the control signals between the seismic stations and the control centre. Frequencies, maximum power output of radio transmitters, directivity of antennas and times of operation of the local seismic network radio transmitters before the explosion shall be agreed between the Parties in accordance with Article X and time of operation after the explosion shall conform to the time specified in paragraph 7 of Article IV.

5. Designated personnel shall have the right to:

(a) acquire photographs under the following conditions:

(1) the Party carrying out the explosion shall identify to the other Party those personnel of the Party carrying out the explosion who shall take photographs as requested by designated personnel;

(2) photographs shall be taken by personnel of the Party carrying out the explosion in the presence of designated personnel and at the time requested by designated personnel for taking such photographs. Designated personnel shall determine whether these photographs are in conformity with their requests and, if not, additional photographs shall be taken immediately;

(3) photographs shall be taken with cameras provided by the other Party having built-in, rapid developing capability and a copy of each photograph shall be provided at the completion of the development process to both Parties;

(4) cameras provided by designated personnel shall be kept in agreed secure storage when not in use; and

(5) the request for photographs can be made, at any time, of the following:

(i) exterior views of facilities and installations associated with the conduct of the explosion as described in subparagraph 4(a) of Article II;

(ii) geological samples used for confirmation of geological and geophysical information, as provided for in subparagraph 2(b) of this article and the equipment utilized in the acquisistion of such samples;

(iii) emplacement and installation of equipment and associated cables used by designated personnel for yield determination;

(iv) emplacement and installation of the local seismic network used by designated personnel;

(v) emplacement of the explosives and the stemming of the emplacement hole; and

(vi) containers, facilities and installations for storage and operation of equipment used by designated personnel;

(b) photographs of visual displays and records produced by the equipment used by designated personnel and photographs within the control centres taken by cameras which are component parts of such equipment; and

(c) receive at the request of designated personnel and with the agreement of the Party carrying out the explosion supplementary photographs taken by the Party carrying out the explosion.

ARTICLE IV

1. Designated personnel in exercising their rights and functions may choose to use the following equipment of either Party, of which choice the Party carrying out the explosion shall be informed not later than 150 days before the beginning of emplacement of the explosives:

(a) electrical equipment for yield determination and equipment for a local seismic network as described in paragraphs 3, 4 and 7 of this article; and

(b) geologist's field tools and kits and equipment for recording of field notes.

2. Designated personnel shall have the right in exercising their rights and functions to utilize the following additional equipment which shall be provided by the Party carrying out the explosion, under procedures to be established in accordance with Article X to ensure that the equipment meets the specifications of the other Party: portable short-range communication equipment, field glasses, optical equipment for surveying and other items which may be specified by the other Party. A description of such equipment and operating instructions shall be provided to the other Party not later than 90 days before the beginning of emplacement of the explosives in connexion with which such equipment is to be used.

3. A complete set of electrical equipment for yield determination shall consist of:

(a) sensing elements and associated cables for transmission of electrical power, control signals and data;

(b) equipment of the control centre, electrical power supplies and cables for transmission of electrical power, control signals and data; and

(c) measuring and calibration instruments, maintenance equipment and spare parts necessary for ensuring the functioning of sensing elements, cables and equipment of the control centre.

4. A complete set of equipment for the local seismic network shall consist of:

(a) seismic stations each of which contains a seismic instrument, electrical power supply and associated cables and radio equipment for receiving and transmission of control signals and data or equipment for recording control signals and data;

(b) equipment of the control centre and electrical power supplies; and

(c) measuring and calibration instruments, maintenance equipment and spare parts necessary for ensuring the functioning of the complete network.

5. In case designated personnel, in accordance with paragraph 1 of this article, choose to use equipment of the Party carrying out the explosion for yield determination or for a local seismic network, a description of such equipment and installation and operating instructions shall be provided to the other Party not later than 90 days before the beginning of emplacement of the explosives in connexion with which such equipment is to be used. Personnel of the Party carrying out the explosion shall emplace, install and operate the equipment in the presence of designated personnel. After the explosion, designated personnel shall receive duplicate

copies of the recorded data. Equipment for yield determination shall be emplaced in accordance with Article VI. Equipment for a local seismic network shall be emplaced in accordance with paragraph 7 of this article.

6. In case designated personnel, in accordance with paragraph 1 of this article, choose to use their own equipment for yield determination and their own equipment for a local seismic network, the following procedures shall apply:

(a) the Party carrying out the explosion shall be provided by the other Party with the equipment and information specified in subparagraphs (a) (1) and (a) (2) of this paragraph not later than 150 days prior to the beginning of emplacement of the explosives in connexion with which such equipment is to be used in order to permit the Party carrying out the explosion to familiarize itself with such equipment, if such equipment and information has not been previously provided, which equipment shall be returned to the other Party not later than 90 days before the beginning of emplacement of the explosives. The equipment and information to be provided are:

(1) one complete set of electrical equipment for yield determination as described in paragraph 3 of this article, electrical and mechanical design information, specifications and installation and operating instructions concerning this equipment; and

(2) one complete set of equipment for the local seismic network described in paragraph 4 of this article, including one seismic station, electrical and mechanical design information, specifications and installation and operating instructions concerning this equipment;

(b) not later than 35 days prior to the beginning of emplacement of the explosives in connexion with which the following equipment is to be used, two complete sets of electrical equipment for yield determination as described in paragraph 3 of this article and specific installation instructions for the emplacement of the sensing elements based on information provided in accordance with subparagraph 2(a) of Article VI and two complete sets of equipment for the local seismic network as described in paragraph 4 of this article, which sets of equipment shall have the same components and technical characteristics as the corresponding equipment specified in subparagraph 6(a) of this article, shall be delivered in sealed containers to the port of entry;

(c) the Party carrying out the explosion shall choose one of each of the two sets of equipment described above which shall be used by designated personnel in connexion with the explosion;

(d) the set or sets of equipment not chosen for use in connexion with the explosion shall be at the disposal of the Party carrying out the explosion for a period that may be as long as 30 days after the explosion at which time such equipment shall be returned to the other Party;

(e) the set or sets of equipment chosen for use shall be transported by the Party carrying out the explosion in the sealed containers in which this equipment arrived, after seals of the Party carrying out the explosion have been affixed to them, to the site of the explosion, so that this equipment is delivered to designated personnel for emplacement, installation and operation not later than 20 days before the beginning of emplacement of the explosives. This equipment shall remain in the custody of designated personnel in accordance with paragraph 7 of Article V or in agreed secure storage. Personnel of the Party carrying out the explosion shall have the right to observe the use of this equipment by designated personnel during the time the equipment is at the site of the explosion. Before the beginning of emplacement of the explosives, designated personnel shall demonstrate to personnel of the Party carrying out the explosion that this equipment is in working order;

(f) each set of equipment shall include two sets of components for recording data and associated calibration equipment. Both of these sets of components in the equipment chosen for use shall simultaneously record data. After the explosion, and after duplicate copies of all data have been obtained by designated personnel and the Party carrying out the explosion, one of each of the two sets of components for recording data and associated calibration equipment shall be selected, by an agreed process of chance, to be retained by designated personnel. Designated personnel shall pack and seal such components for recording data and associated calibration equipment which shall accompany them from the site of the explosion to the port of exit; and

(g) all remaining equipment may be retained by the Party carrying out the explosion for a period that may be as long as 30 days, after which time this equipment shall be returned to the other Party.

7. For any explosion with a planned aggregate yield exceeding 500 kilotons, a local seismic network, the number of stations of which shall be determined by designated personnel but shall not exceed the number of explosives in the group plus five, shall be emplaced, installed and operated at agreed sites of emplacement within an area circumscribed by circles of 15 kilometres in radius centered on points on the surface of the earth above the points of emplacement of the explosives during a period beginning not later than 20 days before the beginning of emplacement of the explosives and continuing after the explosion not later than three days unless otherwise agreed between the Parties.

8. The Party carrying out the explosion shall have the right to examine in the presence of designated personnel all equipment, instruments and tools of designated personnel specified in subparagraph 1 (b) of this article.

9. The Joint Consultative Commission will consider proposals that either Party may put forward for the joint development of standardized equipment for verification purposes.

ARTICLE V

1. Except as limited by the provisions of paragraph 5 of this article, designated personnel in the exercise of their rights and functions shall have access along agreed routes:

(a) for an explosion with a planned aggregate yield exceeding 100 kilotons in accordance with paragraph 2 of Article III:

(1) to the locations of facilities and installations associated with the conduct of the explosion provided in accordance with subparagraph 4(a) of Article II; and

(2) to the locations of activities described in paragraph 2 of Article III; and

(b) for any explosion with a planned aggregate yield exceeding 150 kilotons, in addition to the access described in subparagraph 1(a) of this article:

(1) to other locations within the area circumscribed by circles of 10 kilometers in radius centered on points on the surface of the earth above the points of emplacement of the explosives in order to confirm that the local circumstances are consistent with the stated peaceful purposes;

(2) to the locations of the components of the electrical equipment for yield determination to be used for recording data when, by agreement between the

Parties, such equipment is located outside the area described in subparagraph 1(b)(1) of this article; and

(3) to the sites of emplacement of the equipment of the local seismic network provided for in paragraph 7 of Article IV.

2. The Party carrying out the explosion shall notify the other Party of the procedure it has chosen from among those specified in subparagraph 2(b)(3) of Article III not later than 30 days before beginning the implementation of such procedure. Designated personnel shall have the right to be present at the site of the explosion to exercise their rights and functions in the areas and at the locations described in paragraph 1 of this article for a period of time beginning two days before the beginning of the implementation of the procedure and continuing for a period of three days after the completion of this procedure.

3. Except as specified in paragraph 4 of this article, designated personnel shall have the right to be present in the areas and at the locations described in paragraph 1 of this article:

(a) for an explosion with a planned aggregate yield exceeding 100 kilotons but not exceeding 150 kilotons, in accordance with paragraph 2 of Article III, at any time beginning five days before the beginning of emplacement of the explosives and continuing after the explosion and after safe access to evacuated areas has been established according to standards determined by the Party carrying out the explosion for a period of two days; and

(b) for any explosion with a planned aggregate yield exceeding 150 kilotons, at any time beginning 20 days before the beginning of emplacement of the explosives and continuing after the explosion and after safe access to evacuated areas has been established according to standards determined by the Party carrying out the explosion for a period of:

(1) five days in the case of an explosion with a planned aggregate yield exceeding 150 kilotons but not exceeding 500 kilotons; or

(2) eight days in the case of an explosion with a planned aggregate yield exceeding 500 kilotons.

4. Designated personnel shall not have the right to be present in those areas from which all personnel have been evacuated in connexion with carrying out an explosion but shall have the right to re-enter those areas at the same time as personnel of the Party carrying out the explosion.

5. Designated personnel shall not have or seek access by physical, visual, or technical means to the interior of the canister containing an explosive, to documentary or other information descriptive of the design of an explosive nor to equipment for control and firing of explosives. The Party carrying out the explosion shall not locate documentary or other information descriptive of the design of an explosive in such ways as to impede the designated personnel in the exercise of their rights and functions.

6. The number of designated personnel present at the site of an explosion shall not exceed:

(a) for the excercise of their rights and functions in connexion with the confirmation of the geological and geophysical information in accordance with the provisions of sub-paragraph 2(b) and applicable provisions of paragraph 5 of Article III—the number of emplacement holes plus three;

(b) for the excercise of their rights and functions in connexion with confirming that the local circumstances are consistent with the information provided and with the stated peaceful purposes in accordance with the provisions in sub-paragraphs 2(a), 2(c), 2(d) and 2(e) and applicable provisions of paragraph 5 of Article III—the number of explosives plus two;

(c) for the exercise of their rights and functions in connexion with confirming that the local circumstances are consistent with the information provided and with the stated peaceful purposes in accordance with the provisions in sub-paragraphs 2(a), 2(c), 2(d) and 2(e) and applicable provisions of paragraph 5 of Article III and in connexion with the use of electrical equipment for determination of the yield in accordance with paragraph 3 of Article III—the number of explosives plus seven; and

(d) for the exercise of their rights and functions in connexion with confirming that the local circumstances are consistent with the information provided and with the stated peaceful purposes in accordance with the provisions in sub-paragraphs 2(a), 2(c), 2(d) and 2(e) and applicable provisions of paragraph 5 of Article III and in connexion with the use of electrical equipment for determination of the yield in accordance with paragraph 3 of Article III and with the use of the local seismic network in accordance with paragraph 4 of Article III—the number of explosives plus 10.

7. The Party carrying out the explosion shall have the right to assign its personnel to accompany designated personnel while the latter exercise their rights and functions.

8. The Party carrying out an explosion shall assure for designated personnel telecommunications with their authorities, transportation and other services appropriate to their presence and to the exercise of their rights and functions at the site of the explosion.

9. The expenses incurred for the transportation of designated personnel and their equipment to and from the site of the explosion, telecommunications provided for in paragraph 8 of this article, their living and working quarters, subsistence and all other personal expenses shall be the responsibility of the Party other than the Party carrying out the explosion.

10. Designated personnel shall consult with the Party carrying out the explosion in order to co-ordinate the planned programme and schedule of activities of designated personnel with the programme of the Party carrying out the explosion for the conduct of the project so as to ensure that designated personnel are able to conduct their activities in an orderly and timely way that is compatible with the implementation of the project. Procedures for such consultations shall be established in accordance with Article X.

ARTICLE VI

For any explosion with a planned aggregate yield exceeding 150 kilotons, determination of the yield of each explosive used shall be carried out in accordance with the following provisions:

1. Determination of the yield of each individual explosion in the group shall be based on measurements of the velocity of propagation, as a function of time, of the hydrodynamic shock wave generated by the explosion, taken by means of electrical equipment described in paragraph 3 of Article IV.

2. The Party carrying out the explosion shall provide the other Party with the following information:

(a) not later than 60 days before the beginning of emplacement of the explosives, the length of each canister in which the explosive will be contained in the corresponding emplacement hole, the dimensions of the tube or other device used to emplace the canister and the cross-sectional dimensions of the emplacement hole to a distance, in metres, from the emplacement point of 10 times the cube root of its yield in kilotons;

(b) not later than 60 days before the beginning of emplacement of the explosives, a description of materials, including their densities, to be used to stem each emplacement hole; and

(c) not later than 30 days before the beginning of emplacement of the explosives, for each emplacement hole of a group explosion, the local co-ordinates of the point of emplacement of the explosive, the entrance of the emplacement hole, the point of the emplacement hole most distant from the entrance, the location of the emplacement hole at each 200 metres distance from the entrance and the configuration of any known voids larger than one cubic metre located within the distance, in metres, of 10 times the cube root of the planned yield in kilotons measured from the bottom of the canister containing the explosive. The error in these co-ordinates shall not exceed 1 per cent of the distance between the emplacement hole and the nearest other emplacement hole or 1 per cent of the distance between the point of measurement and the entrance of the emplacement hole, whichever is smaller, but in no case shall the error be required to be less than one metre.

3. The Party carrying out the explosion shall emplace for each explosive that portion of the electrical equipment for yield determination described in sub-paragraph 3(a) of Article IV, supplied in accordance with paragraph 1 of Article IV, in the same emplacement hole as the explosive in accordance with the installation instructions supplied under the provisions of paragraph 5 or 6 of Article IV. Such emplacement shall be carried out under the observation of designated personnel. Other equipment specified in sub-paragraph 3(b) of Article IV shall be emplaced and installed:

(a) by designated personnel under the observation and with the assistance of personnel of the Party carrying out the explosion, if such assistance is requested by designated personnel; or

(b) in accordance with paragraph 5 of Article IV.

4. That portion of the electrical equipment for yield determination described in sub-paragraph 3(a) of Article IV that is to be emplaced in each emplacement hole shall be located so that the end of the electrical equipment which is farthest from the entrance to the emplacement hole is at a distance, in metres, from the bottom of the canister containing the explosive equal to 3.5 times the cube root of the planned yield in kilotons of the explosive when the planned yield is less than 20 kilotons and three times the cube root of the planned yield in kilotons of the explosive when the planned yield is 20 kilotons or more. Canisters longer than 10 metres containing the explosive shall only be utilized if there is prior agreement between the Parties establishing provisions for their use. The Party carrying out the explosion shall provide the other Party with data on the distribution of density inside any other canister in the emplacement hole with a transverse cross-sectional area exceeding 10 square centimetres located within a distance, in metres, of 10 times the cube root of the planned yield in kilotons of the explosion from the bottom of the canister containing the explosive. The Party carrying out the explosion shall provide the other Party with access to confirm such data on density distribution within any such canister.

5. The Party carrying out an explosion shall fill each emplacement hole, including all pipes and tubes contained therein which have at any transverse section an aggregate cross-sectional area exceeding 10 square centimetres in the region containing the electrical equipment for yield determination and to a distance, in metres, of six times the cube root of the planned yield in kilotons of the explosive from the explosive emplacement point, with material having a density not less than seven-tenths of the average density of the surrounding rock, and from that point to a distance of not less than 60 metres from the explosive emplacement point with material having a density greater than one gram per cubic centimetre.

6. Designated personnel shall have the right to:

(a) confirm information provided in accordance with sub-paragraph 2(a) of this article;

(b) confirm information provided in accordance with sub-paragraph 2(b) of this article and be provided, upon request, with a sample of each batch of stemming material as that material is put into the emplacement hole; and

(c) confirm the information provided in accordance with sub-paragraph 2(c) of this article by having access to the data acquired and by observing, upon their request, the making of measurements.

7. For those explosives which are emplaced in separate emplacement holes, the emplacement shall be such that the distance D, in metres, between any explosive and any portion of the electrical equipment for determination of the yield of any other explosive in the group shall be not less than 10 times the cube root of the planned yield in kilotons of the larger explosive of such a pair of explosives. Individual explosions shall be separated by time intervals, in milliseconds, not greater than one-sixth the amount by which the distance D, in metres, exceeds 10 times the cube root of the planned yield in kilotons of the larger explosive of such a pair of explosives.

8. For those explosives in a group which are emplaced in a common emplacement hole, the distance, in metres, between each explosive and any other explosive in that emplacement hole shall be not less than 10 times the cube root of the planned yield in kilotons of the larger explosive of such a pair of explosives, and the explosives shall be detonated in sequential order, beginning with the explosive farthest from the entrance to the emplacement hole, with the individual detonations separated by time intervals, in milliseconds, of not less than one times the cube root of the planned yield in kilotons of the largest explosive in this emplacement hole.

ARTICLE VII

1. Designated personnel with their personal baggage and their equipment as provided in Article IV shall be permitted to enter the territory of the Party carrying out the explosion at an entry port to be agreed upon by the Parties, to remain in the territory of the Party carrying out the explosion for the purpose of fulfilling their rights and functions provided for in the Treaty and this Protocol, and to depart from an exit port to be agreed upon by the Parties.

2. At all times while designated personnel are in the territory of the Party carrying out the explosion, their persons, property, personal baggage, archives and documents as well as their temporary official and living quarters shall be accorded the same privileges and immunities as provided in Articles 22, 23, 24, 29, 30, 31, 34 and 36 of the Vienna Convention on Diplomatic Relations of 1961 to the persons, property, personal baggage, archives and documents of diplomatic agents as well as to the

premises of diplomatic missions and private residences of diplomatic agents.

3. Without prejudice to their privileges and immunities it shall be the duty of designated personnel to respect the laws and regulations of the State in whose territory the explosion is to be carried out insofar as they do not impede in any way whatsoever the proper exercising of their rights and functions provided for by the Treaty and this Protocol.

ARTICLE VIII

The Party carrying out an explosion shall have sole and exclusive control over and full responsibility for the conduct of the explosion.

ARTICLE IX

1. Nothing in the Treaty and this Protocol shall affect proprietary rights in information made available under the Treaty and this Protocol and in information which may be disclosed in preparation for and carrying out of explosions; however, claims to such proprietary rights shall not impede implementation of the provisions of the Treaty and this Protocol.

2. Public release of the information provided in accordance with Article II or publication of material using such information, as well as public release of the results of observation and measurements obtained by designated personnel, may take place only by agreement with the Party carrying out an explosion; however, the other Party shall have the right to issue statements after the explosion that do not divulge information in which the Party carrying out the explosion has rights which are referred to in paragraph 1 of this article.

ARTICLE X

The Joint Consultative Commission shall establish procedures through which the Parties will, as appropriate, consult with each other for the purpose of ensuring efficient implementation of this Protocol.

Done at Washington and Moscow, on 28 May 1976.

AGREED STATEMENT

The Parties to the Treaty Between the United States of America and the Union of Soviet Socialist Republics on Underground Nuclear Explosions for Peaceful Purposes, hereinafter referred to as the Treaty, agree that under sub-paragraph 2(c) of Article III of the Treaty:

(a) Development testing of nuclear explosives does not constitute a "peaceful application" and any such development tests shall be carried out only within the boundaries of nuclear weapon test sites specified in accordance with the Treaty Between the United States of America and the Union of Soviet Socialist Republics on the Limitation of Underground Nuclear Weapon Tests,

(b) Associating test facilities, instrumentation or procedures related only to testing of nuclear weapons or their effects with any explosion carried out in accordance with the Treaty does not constitute a "peaceful application".

28 May 1976

APPENDIX 2
ANNOUNCED AND PRESUMED NUCLEAR EXPLOSIONS IN 1963–1976

CONTENTS

People's Republic of China 387
India 387
France 388
United Kingdom 389
United States of America 390
Union of Soviet Socialist Republics 400

SOURCES OF INFORMATION

The following sources have been used for the data given in Appendix 2.

Data on the atmospheric explosions in China are mainly based on official announcements by either China or the US. For the underground explosions also seismological data have been used.

The data on the Indian explosion are based on official Indian announcements and seismological data.

The information on the French explosions is mainly based on press reports and, in some cases, on official French announcements. For some of the underground explosions and for the most powerful atmospheric explosions seismological data are available.

The information on the UK explosions is based on official US and UK announcements and seismological data.

For those US explosions for which names are given, the information has been obtained from official US announcements. For all other US explosions seismological data have been used.

The location and origin time of all USSR explosions are based on seismological data. The announced yields given in figures for six PNE explosions have been obtained from USSR reports, whereas the other are taken from announcements by US authorities.

Yield estimates for the US and USSR explosions are based on reported seismological data. The procedure used to obtain these estimates is discussed in Chapter 11.

NOTATIONS USED

Origin time   GMT given in hours, minutes and seconds.

Epicenter   Latitude and longitude given in degrees. When preliminary co-ordinates only are available, this is indicated by an asterisk.

Medium   In the case of Chinese and French explosions the letter A indicates *atmospheric* and U *underground* explosion. Explosions by all other countries are underground. Explosion media for US explosions are those officially announced.

Depth   Reported depth of the device below ground.

Yield L = less than 20 kt.
$L-I$ = between 20 kt and 200 kt.
I = between 200 kt and 1000 kt.

APPENDIX 2

TABLE A2.1. Announced and presumed nuclear explosions in China in 1963–1976.

Date	Origin time	Epicenter Latitude	Epicenter Longitude	Medium	Depth (m)	Yield (kt) Announced	Yield (kt) Estimated	Name
641016	70000	41.00 N	89.00 E*	A		~20		
650514	20000	41.00 N	89.00 E*	A			L-I	
660509	80000	41.00 N	89.00 E*	A			L-I	
661027		41.00 N	89.00 E*	A			L-I	
661228		41.00 N	89.00 E*	A			L-I	
670617	190800	40.73 N	89.56 E	A		~3000		
671224		41.00 N	89.00 E*	A			L-I	
681227	73000	41.00 N	89.00 E*	A		~3000		
690922	161459	41.39 N	88.30 E	U			L-I	
690929	84026	40.72 N	89.30 E	A		~3000		
701014	72959	40.93 N	89.43 E*	A		~3000		
711118	60000	41.00 N	89.00 E*	A		~20		
720107	70000	41.00 N	89.00 E*	A			L-I	
720318	60000	41.00 N	89.00 E*	A				
730627	35951	40.56 N	89.53 E	A			I	
740617	55949	39.50 N	89.40 E	A			L	
751027	10000	41.40 N	88.40 E	U			L	
760123	60000	41.00 N	89.00 E*	A			L	
760925		41.00 N	89.00 E*	U				
761017	50000	41.00 N	89.00 E*	A		~3000		
761117	60000	41.00 N	89.00 E*	A				

★ Preliminary coordinates.

TABLE A2.2. The announced nuclear explosion in India on May 18, 1974.

Date	Origin time	Epicenter Latitude	Epicenter Longitude	Medium	Depth (m)	Yield (kt) Announced	Yield (kt) Estimated	Name
740518	23455	26.95 N	71.70 E					

TABLE A2.3. Announced and presumed nuclear explosions conducted by France in 1963–1976.

Date	Origin time	Epicenter Latitude	Epicenter Longitude	Medium	Depth (m)	Yield (kt) Announced	Yield (kt) Estimated	Name
631026	130000	24.00 N	5.00 E*	U				RUBIS
640214	110000	24.05 N	5.05 E	U				OPALE
640615	134000	24.07 N	5.03 E	U				TOPAZE
641128	103000	24.04 N	5.04 E	U				TURQUOISE
650227	113000	24.06 N	5.03 E	U				SAPHIR
650530	110000	24.05 N	5.05 E	U				JADE
651001	100000	24.06 N	5.03 E	U				CORINDON
651201	103000	24.04 N	5.05 E	U				TOURMALINE
660216	110000	24.04 N	5.04 E	U				GRENAT
660702	153400	20.00 S	140.00 W*	A				ALDEBARAN
660719	150500	20.00 S	140.00 W	A				TAMOURE
660911	173000	20.00 S	140.00 W	A		120		BETELGEUSE
660924	170000	20.00 S	140.00 W	A		150		RIGEL
661004	210000	20.00 S	140.00 W*	A				SIRIUS
670605	190000	20.00 S	140.00 W*	A				
670627	193000	20.00 S	140.00 W*	A				
670702	173000	20.00 S	140.00 W*	A				
680707	220000	20.00 S	140.00 W	A		500		
680715	190000	20.00 S	140.00 W*	A				
680803	210000	20.00 S	140.00 W*	A		2500		CANOPUS
680824	182959	22.20 S	138.80 W*	A		1000		PROCYON
680908	185959	21.78 S	139.16 W*	A				ANDROMEDE
700515	180000	20.00 S	140.00 W**	A				CASSIOPEE
700522		20.00 S	140.00 W*	A		1000		DRAGON
700624	175959	22.24 S	138.79 W*	A				ERIDAN
700624	183000	20.00 S	140.00 W*	A				LICORNE
700703	182959	21.80 S	139.20 W	A		1000		PEGASE
700727	190000	20.00 S	140.00 W*	A				ORION
700802	190000	20.00 S	140.00 W*	A				TOUCAN
700806	190000	20.00 S	140.00 W	A		H		
710605	191500	20.00 S	140.00 W	A		H		
710612	191457	23.80 S	137.20 W	A		H		ENCELADE

* Preliminary coordinates.

(Contd.)

Date	Origin time	Latitude	Longitude		Medium	Depth	Yield	Name
710704	213000	20.00 S	140.00 W	*	A			
710808	183000	20.00 S	140.00 W	*	A			
710814	185959	21.88 S	138.95 W	*	A			
720625		20.00 S	140.00 W	*	A			
720630		20.00 S	140.00 W	*	A			
720729		20.00 S	140.00 W	*	A			
730721	180000	20.00 S	140.00 W	*	A			
730728	230300	20.00 S	140.00 W	*	A			
730819		20.00 S	140.00 W	*	A			
730825		20.00 S	140.00 W	*	A			
730826		20.00 S	140.00 W	*	A			
740616		20.00 S	140.00 W	*	A			
740707		20.00 S	140.00 W	*	A			
740717		20.00 S	140.00 W	*	A			
740726		20.00 S	140.00 W	*	A			
740815		20.00 S	140.00 W	*	A			
740825		20.00 S	140.00 W	*	A			
740915		20.00 S	140.00 W	*	A			
750605	4800	20.00 S	140.00 W	*	U	622	1000	ACHILLES
751126		20.00 S	140.00 W	*	U			HECTOR
760402		20.00 S	140.00 W	*	U			PATROCLE
760711	2955	22.67 S	138.61 W		U			
760723		20.00 S	140.00 W	*	U			
761208		20.00 S	140.00 W	*	U			

TABLE A2.4. Announced and presumed underground nuclear explosions by the UK in 1963–1976.

Date	Origin time	Epicenter		Medium	Depth (m)	Yield (kt)		Name
		Latitude	Longitude			Announced	Estimated	
640717	171830	37.02 N	116.03 W	ALLUVIUM	271	L		CORMORANT
650910	171200	37.08 N	116.02 W	TUFF	455	L-I	15	CHARCOAL
740523		37.00 N	116.00 W			L-I		FALLON
760826	143000	37.00 N	116.00 W			L-I		

★ Preliminary coordinates.

TABLE A2.5. Announced and presumed underground nuclear explosions in the US in 1963–1976.

Date	Origin time	Epicenter Latitude	Epicenter Longitude	Medium	Depth (m)	Yield (kt) Announced	Yield (kt) Estimated	Name
630812	234500	37.04 N	116.02 W	ALLUVIUM	302			PEKAN
630815	130000	37.15 N	116.08 W	ALLUVIUM	225			SATSOP
630823	132000	37.12 N	116.03 W	TUFF	254			KOHOCTON
630913	135300	37.16 N	116.08 W	ALLUVIUM	225			AHTANUM
630913	170000	37.06 N	116.02 W	TUFF	714	235		BILBY
631011	140000	37.04 N	116.02 W	ALLUVIUM	261			GRUNION
631011	210000	37.12 N	116.03 W	ALLUVIUM	149			TORNILLO
631026	170000	37.20 N	116.23 W	TUFF	544			CLEARWATER
631026	170000	39.20 N	118.30 W	GRANITE	367	I		SHOAL
631114	160000	37.04 N	116.02 W	ALLUVIUM	260	12		ANCHOVY
631115	150000	37.13 N	116.05 W	ALLUVIUM	165			MUSTANG
631122	173000	37.12 N	116.04 W	ALLUVIUM	301			GREYS
631204	163830	37.04 N	116.03 W	ALLUVIUM	262			SARDINE
631212	160200	37.13 N	116.04 W	ALLUVIUM	164			EAGLE
640116	160000	37.14 N	116.05 W	TUFF	491	L-I		FORE
640123	160000	37.13 N	116.04 W	TUFF	264		19	OCONTO
640220	153000	37.15 N	116.04 W	TUFF	492	L-I		KLICKITAT
640313	160200	37.05 N	116.01 W	ALLUVIUM	114		24	PIKE
640414	144000	37.13 N	116.03 W	TUFF	203			HOOK
640425	143000	37.04 N	116.02 W	ALLUVIUM	149			STURGEON
640429	201000	37.15 N	116.05 W	ALLUVIUM	507	L-I		TURF
640429	204000	37.04 N	116.03 W	ALLUVIUM	261		100	PIPE FISH
640514	144000	37.12 N	116.04 W	ALLUVIUM	163		15	BACKSWING
640515	161500	37.04 N	116.01 W	ALLUVIUM	241			MINNOW
640611	164500	37.15 N	116.08 W	ALLUVIUM	262			ACE
640625	133300	37.11 N	116.03 W	TUFF	205			FADE
640630	133500	37.17 N	116.06 W	ALLUVIUM	258	L-I	9	DUB
640716	131500	37.18 N	116.04 W	TUFF	389			BYE
640819	160000	37.16 N	116.08 W	ALLUVIUM	166			ALVA
640822	221700	37.06 N	116.01 W	TUFF	448		18	CANVASBACK
640828	170600	37.07 N	116.02 W	TUFF	363			HADDOCK
640904	181500	37.02 N	116.02 W	ALLUVIUM	261		12	GUANAY

(Contd.)

APPENDIX 2

ID	Coord1	Coord2	Dir	Coord3	Rock	Val1	Val2	Val3	Name
641002	200300	37.08	N	116.01	TUFF	452	L		AUK
641009	140000	37.15	N	116.08	ALLUVIUM	404	38	12	PAR
641016	155930	37.04	N	116.02	ALLUVIUM	258	L	30	BARBEL
641022	1600000	31.14	N	89.57	SALT	828	5.3		SALMON
641031	170459	37.17	N	116.03	TUFF	380	L		FOREST
641105	1500000	37.11	N	116.07	DOLOMITE	402	12	9	HANDCAR
641205	211500	37.05	N	116.05	TUFF	403	L-I	10	CREPE
641216	2000000	37.03	N	116.01	ALLUVIUM	180	1.2		PARROT
641216	201000	37.18	N	116.07	TUFF	151	2.7		MUDPACK
641218	1935000	37.08	N	116.34	BASALT	27	0.1		SULKY
650114	1600000	37.12	N	116.02	TUFF	215	L		WOOL
650204	1530000	37.13	N	116.06	ALLUVIUM	232	L		CASHMERE
650212	151029	37.16	N	116.08	ALLUVIUM	224	L		ALPACA
650216	1730000	37.05	N	116.02	ALLUVIUM	296	10		MERLIN
650218	161847	36.82	N	115.95	TUFF	179	L-I		WISHBONE
650303	1913000	37.05	N	116.04	TUFF	749	L-I	65	WAGTAIL
650326	1534008	37.15	N	116.04	ALLUVIUM	537		35	CUP
650405	2100000	37.03	N	116.02	ALLUVIUM	447	L		KESTREL
650414	1314000	37.28	N	116.52	RHYOLITE	85	4.3		PALANQUIN
650421	2200000	37.01	N	116.20	TUFF	304	L	8	GUMDROP
650421	214400	37.02	N	115.99	*		L		MUSCOVY
650507	154711	37.14	N	116.07	ALLUVIUM	190	L		TEE
650512	181500	37.24	N	116.43	TUFF	696	0.8		BUTEO
650514	173236	37.06	N	116.01	TUFF	427	1.2		SCAUP
650514		37.00	N	116.03			L-I	60	CAMBRIC
650521	130852	37.12	N	116.03	TUFF	281	L	18	TWEED
650611	194500	37.04	N	115.96	ALLUVIUM	180	L-I		PETREL
650616	163000	36.82	N	116.06	QUARTZITE	195	L		DILUTED WATERS
650617	170000	37.22	N	116.03	TUFF	109	L		TINY TOT
650723	170000	37.10	N	116.04	ALLUVIUM	531	L-I		BRONZE
650806	172330	37.02	N	116.07	ALLUVIUM	321	L		MAUVE
650827	135113	37.14	N	116.02	ALLUVIUM	172	L	12	CENTAUR
650907	200823	37.02	N	116.01	TUFF	301	L		SCREAMER
650917	150823	37.14	N	116.03	ANDESITE	219	L		ELKHART
651029	210000	51.44	E	179.18	ALLUVIUM	701	80	100	LONG SHOT
651112	180000	37.05	N	116.02	TUFF	241	L		SEPIA
651203	151302	37.16	N	116.05		681	L-I		CORDUROY

★ Preliminary coordinates

(Contd.)

TABLE A2.5 (Contd.)

Date	Origin time	Epicenter Latitude	Epicenter Longitude	Medium	Depth (m)	Yield (kt) Announced	Yield (kt) Estimated	Name
651216	153918	37.14 N	116.06 W	ALLUVIUM	260	L		EMERSON
651216	191500	37.07 N	116.03 W	TUFF	500	L-I	36	BUFF
660113	153743	37.12 N	116.03 W	TUFF	183	L		MAXWELL
660118	183500	37.09 N	116.02 W	TUFF	561	L-I	32	LAMPBLACK
660121	182800	37.03 N	116.02 W	ALLUVIUM	333	L		DOVEKIE
660203	181737	37.13 N	116.07 W	ALLUVIUM	270	L		PLAID 2
660224	155507	37.27 N	116.43 W	TUFF	672	16	7	REX
660305	181500	37.17 N	116.21 W	TUFF	405	L		RED HOT
660307	184100	37.04 N	116.03 W	ALLUVIUM	195	L		FINFOOT
660312	180413	37.14 N	116.05 W	ALLUVIUM	397	L		CLYMER
660318	190000	37.01 N	116.01 W	TUFF	333	L		PURPLE
660324	145528	37.11 N	116.03 W	TUFF	150	L		TEMPLAR
660401	184000	37.10 N	116.02 W	TUFF	561	L		LIME
660406	135717	37.14 N	116.14 W	TUFF	225	L	5	STUTZ
660407	222730	37.02 N	115.99 W	TUFF	226	L		TOMATO
660414	141343	37.24 N	116.43 W	RHYOLITE	544	65	31	DURYEA
660425	183800	36.89 N	115.94 W	TUFF	295	L	4	PIN STRIPE
660504	133217	37.14 N	116.14 W	ALLUVIUM	197	L		TRAVELER
660505	140000	37.05 N	116.04 W	ALLUVIUM	305	13	8	CYCLAMEN
660506	150000	37.35 N	116.32 W	RHYOLITE	665	70	50	CHARTREUSE
660512	193726	37.13 N	116.07 W	ALLUVIUM	247	L	10	TAPESTRY
660513	133000	37.09 N	116.03 W	TUFF	548	L-I	100	PIRANHA
660527	135628	37.11 N	116.06 W	TUFF	670	L-I	190	DUMONT
660602	200000	37.18 N	116.10 W	GRANITE	337	21	17	DISCUS THROWER
660603	153000	37.23 N	116.06 W	TUFF	462	56		PILE DRIVER
660610	140000	37.07 N	116.03 W	ALLUVIUM	560	L-I	140	TAN
660615	143000	37.06 N	116.04 W	TUFF	485	L		PUCE
660615	170000	37.01 N	116.20 W	DOLOMITE	327	L-I		DOUBLE PLAY
660625	180247	37.17 N	116.05 W	ALLUVIUM	455	25		KANKAKEE
660625	171300	37.15 N	116.07 W	RHYOLITE	322	300	450	VULCAN
660630	221500	37.32 N	116.30 W	ALLUVIUM	819	L		HALFBEAK
660728	153330	37.14 N	116.10 W	ALLUVIUM	152	L		SAXON

(Contd.)

APPENDIX 2

Date										
660810	131600	37.17	N	116.05	W	ALLUVIUM	193			ROVENA
660912	153001	36.88	N	115.95	W	ALLUVIUM	254			DERRINGER
660923	180000	37.02	N	116.04	W	TUFF	561		12	DAIQUIRI
660929	144530	37.17	N	116.05	W	ALLUVIUM	228			NEWARK
661105	144500	37.13	N	116.05	W	ALLUVIUM	198		4	SIMMS
661111	1200000	37.13	N	116.05	W	ALLUVIUM	238			AJAX
661118	1502000	37.04	N	116.01	W	ALLUVIUM	211			CERISE
661203	1215000	31.14	N	89.57	W	AIR SALT	828	0.4		STERLING
661213	210000	36.88	N	115.94	W	ALLUVIUM	243			NEW POINT
661220	153000	37.30	N	116.41	W	TUFF	1215	825	10	GREELEY
670119	1645000	37.10	N	116.13	W	DOLOMITE	364	L-I	830	NASH
670120	174003	37.17	N	116.05	W	LIMESTONE	559	L-I	49	BOURBON
670208	151500	37.17	N	116.02	W	ALLUVIUM	257		29	WARD
670223	1834000	37.02	N	116.07	W	ALLUVIUM	299		10	PERSIMMON
670223	185000	37.13	N	116.05	W	TUFF	731	L-I	3	AGILE
670302	150000	37.17	N	116.02	W	ALLUVIUM	271			RIVET 3
670407	150000	37.05	N	116.04	W	ALLUVIUM	240			FAWN
670421	150900	37.02	N	116.06	W	ALLUVIUM	219		7	CHOCOLATE
670427	1445000	37.14	N	115.99	W	TUFF	499	L-I		EFFENDI
670510	134000	37.08	N	116.06	W	TUFF	746	250	10	MICKEY
670520	150000	37.13	N	116.37	W	TUFF	978	150	230	COMMODORE
670523	1500002	37.27	N	116.48	W	RHYOLITE	631	71	140	SCOTCH
670526	150000	37.13	N	116.03	W	TUFF	301		47	KNICKERBOCKER
670622	131000	37.25	N	116.21	W	TUFF	375			SWITCH
670626	160000	37.20	N	116.02	W	ALLUVIUM	310		9	MIDI MIST
670629	112500	37.03	N	116.05	W	TUFF	484		8	UMBER
670727	130000	37.15	N	116.15	W			L-I	8	STANLEY
670804	1400001	37.01	N	116.05	W	TUFF	465			WASHER
670810	141000	37.16	N	116.04	W	ALLUVIUM	332		8	BORDEAUX
670818	2012300	37.01	N	116.21	W	TUFF	446		9	DOOR MIST
670831	1630000	37.18	N	116.05	W	TUFF	518	L-I	13	YARD
670907	1345000	37.15	N	116.04	W	ALLUVIUM	174	2.2		MARVEL
670921	2045000	37.17	N	116.04	W	TUFF	667	L-I	170	ZAZA
670927	170000	37.10	N	116.06	W	TUFF	714	L-I	140	LANPHER
671018	143000	37.12	N	116.03	W	ALLUVIUM	302			SAZERAC
671025	143000	37.03	N	116.04	W	TUFF	670		7	COBBLER
671108	150000	37.09								

(Contd.)

TABLE A2.5 *(Contd.)*

| Date | Origin time | Epicenter | | Medium | Depth (m) | Yield (kt) | | Name |
		Latitude	Longitude			Announced	Estimated	
671210	193000	36.68 N	107.21 W	SHALE	1293	29		GASBUGGY
671215	150000	37.04 N	116.00 W	TUFF	333	L	2	STILT
680118	163000	37.15 N	116.07 W	ALLUVIUM	247	L	8	HUPMOBILE
680119	150000	37.16 N	116.05 W	ALLUVIUM	443	L-I	12	STACCATO
680119	181500	38.63 N	116.21 W	TUFF	975	I	1200	FAULTLESS
680126	160000	37.28 N	116.51 W	RHYOLITE	51	2,3		CABRIOLET
680131	153001	36.89 N	116.12 W					
680221	153000	37.12 N	116.05 W	TUFF	645	L-I	200	KNOX
680229	170830	37.18 N	116.21 W	TUFF	410	L	20	DORSAL FIN
680312	170400	37.01 N	116.37 W	BASALT	41	5.4		BUGGY 1
680314	151900	37.05 N	116.01 W	ALLUVIUM	209	1.4		POMMARD
680322	150000	37.33 N	116.31 W	TUFF	668	L-I	160	STINGER
680325	184427	36.87 N	115.93 W	ALLUVIUM	264	L	10	MILK SHAKE
680410	140000	37.15 N	116.08 W	ALLUVIUM	381	L	20	NOOR
680418	140500	37.15 N	116.04 W	TUFF	492	L-I	25	SHUFFLE
680423	170130	37.34 N	116.38 W	TUFF ASH	224	L-I	6	SCROLL
680426	150000	37.29 N	116.46 W	RHYOLITE	1158	1200		BOXCAR
680503	160001	37.00 N	115.99 W					
680517	130000	37.12 N	116.06 W	TUFF	472	L-I	15	CLARKSMOBILE
680606	213000	37.16 N	116.04 W	ALLUVIUM	189	L		TUB
680615	140000	37.26 N	116.31 W	TUFF	683	L-I	300	RICKEY
680628	122200	37.24 N	116.48 W	TUFF	607	L-I	58	CHATEAUGAY
680730	130000	37.13 N	116.08 W	ALLUVIUM	381	L-I	10	TANYA
680827	163000	36.88 N	115.93 W	ALLUVIUM	242	L		DIANA MOON
680829	224500	37.25 N	116.35 W	TUFF	729	L-I	260	SLED
680906	140000	37.14 N	116.05 W	TUFF	582	L-I	110	NOGGIN
680912	140000	37.03 N	116.01 W	TUFF	332	L		KNIFE A
680924	140000	37.12 N	116.13 W	TUFF	468	L-I	13	STODDARD
681003	170500	37.20 N	116.21 W	TUFF	333	L	10	HUDSON SEAL
681010	142900	37.03 N	115.99 W	TUFF	333	L	3	KNIFE C
681031	183004	36.87 N	116.27 W		301			

(Contd.)

APPENDIX 2

6811104	151500	37.13	N	116.09	W	TUFF	603		22	CREW
6811115	153005	37.00	N	116.31	W		363		8	KNIFE B
6811115	154500	37.03	N	116.03	W	ALLUVIUM	308		12	MING VASE
6811122	180000	37.01	N	116.21	W	TUFF	439		3	TINDERBOX
6811122	161000	37.14	N	116.04	W	TUFF	106	35		SCHOONER
6811208	160000	37.34	N	116.57	W	TUFF	265			TYG
6811212	151000	37.12	N	116.08	W	ALLUVIUM			20	
6812212	152001	37.01	N	116.11	W	TUFF	1402	1100	1000	BENHAM
6811219	163000	37.23	N	116.47	W	ALLUVIUM	247		3	PACKARD
6901115	190000	37.15	N	116.07	W	TUFF	518	L-I	40	WINESKIN
6901115	193000	37.21	N	116.22	W	ALLUVIUM	454	L-I	40	VISE
6901130	150000	37.05	N	116.03	W	TUFF	411		15	CYPRESS
6902212	161821	37.17	N	116.21	W	ALLUVIUM	304	L-I^a	110	BARSAC
6902320	181200	37.02	N	116.03	W	ALLUVIUM	465	L-I	35	COFFER
6904321	143000	37.13	N	116.09	W	TUFF	557	L-I		BLENTON
6904430	170000	37.08	N	116.01	W	TUFF	560			THISTLE
6904430	170000	37.09	N	116.01	W	TUFF	599			PURSE
6905507	134500	37.28	N	116.50	W	TUFF	515		180	TORRIDO
6905527	141500	37.07	N	115.99	W	TUFF	303		22	TAPPER
6906612	140000	37.01	N	116.03	W	TUFF	410	L-I	12	ILDRIM
6907716	130230	37.12	N	116.05	W	TUFF	548	H	6	HUTCH
6908814	145500	37.14	N	116.09	W	ALLUVIUM	239		300	SPIDER
6908827	143500	37.16	N	116.06	W	ALLUVIUM	239			PLIERS
6908827	134500	37.02	N	116.04	W	SANDSTONE	2575	40		RULISON
6909910	210000	39.41	N	107.95	W	ALLUVIUM	264	<1000	10	MINUTE STEAK
6909912	180220	36.88	N	115.93	W	TUFF	1158	~1000	700	JORUM
6909916	143000	37.31	N	116.46	W	PIL.LAVA	1219			MILROW
6911002	220600	51.42	E	179.18	W	TUFF	617	11	82	PIPKIN
6911008	143000	37.26	N	116.44	W	TUFF	260	L-I		CRUET
6911029	193000	37.12	N	116.13	W	TUFF	312	110		POD
6911029	200000	37.13	N	116.14	W	TUFF	625	L-I	140	CALABASH
6911029	220151	37.14	N	116.06	W	TUFF	394	L-I	17	PICCALILLI
6911121	145200	37.03	N	116.21	W	TUFF	419		16	DIESEL TRAIN
6911205	170000	37.18	N	116.00	W	TUFF	551	L-I	61	GRAPE A
6911217	150000	37.08	N	116.02	W	ALLUVIUM	378	L-I	30	LOVAGE
6911218	190000	37.12	N	116.03	W	TUFF	457	L-I	28	TERRINE

^{a} Less than 100 kt.

(Contd.)

TABLE A2.5 (Contd.)

Date	Origin time	Epicenter Latitude	Epicenter Longitude	Medium	Depth (m)	Yield (kt) Announced	Yield (kt) Estimated	Name
700123	163000	37.14 N	116.04 W	TUFF	266	L		FOB
700130	170000	37.03 N	116.03 W	ALLUVIUM	304	L	20	AJO
700204	170000	37.10 N	116.04 W	TUFF	554	L-I	120	GRAPE B
700205	150000	37.16 N	116.04 W	TUFF	442	25	8	LABIS
700211	191500	37.20 N	116.20 W	TUFF	399	L	9	DIANA MIST
700225	142838	37.04 N	116.00 W	TUFF	408	L-I	25	CUMARIN
700226	153000	37.12 N	116.06 W	ALLUVIUM	392	L-I	100	YANNIGAN
700306	142401	37.12 N	116.06 W	TUFF	289	9.0	100	CYATHUS
700306	150000	37.14 N	116.04 W	TUFF	250	L		ARABIS
700319	140330	37.09 N	116.02 W	ALLUVIUM	301	L	6	JAL
700323	230500	37.09 N	116.02 W	TUFF	560	L-I	93	SHAPER
700326	190000	37.30 N	116.53 W	TUFF	1206	>1000	1900	HANDLEY
700421	143000	37.05 N	115.99 W	TUFF	343	L-I	6	SNUBBER
700421	150000	37.12 N	116.08 W	TUFF	399	L	8	CAN
700501	141300	37.06 N	116.03 W	TUFF	390	L-I	1	BEEBALM
700501	144000	37.13 N	116.03 W	TUFF	265	L	6	HOD
700505	153000	37.22 N	116.18 W	TUFF	405	L	28	MINT LEAF
700512	140000	37.01 N	116.20 W	TUFF	253	L		DIAMOND DUST
700515	133000	37.16 N	116.04 W	TUFF	443	L-I	39	CORNICE
700521	140000	37.03 N	115.99 W	TUFF	240	L	1	MANZANAS
700521	141500	37.07 N	116.01 W	TUFF	481	L-I	20	MORRONES
700526	141600	37.18 N	116.21 W	TUFF	422	L	9	HUDSON MOON
700526	150000	37.11 N	116.06 W	TUFF	531	105	110	FLASK
700528	120001	37.18 N	116.06 W					
700626	130000	37.11 N	116.09 W	ALLUVIUM	309	L-I		ARNICA
701014	143000	37.07 N	116.00 W	TUFF	560	L-I	94	TIJERAS
701028	143001	37.27 N	115.98 W					
701105	150000	37.03 N	116.01 W	TUFF	393	L-I	11	ABEYTAS
701119	150001	36.99 N	116.05 W					
701216	160000	37.10 N	116.01 W	TUFF	485	L-I		ARTESIA
701216	160000	37.14 N	116.03 W	TUFF	294	L		CREAM
701217	160500	37.13 N	116.08 W	TUFF	662	220	170	CARPETBAG

(Contd.)

APPENDIX 2

701218	153000	37.17 N	116.10 W	TUFF	277	10	32	BANEBERRY
710616	145000	37.03 N	116.01 W	ALLUVIUM	303	L	18	EMBUDO
710623	153000	37.02 N	116.02 W	TUFF	455	L-I	10	LAGUNA
710624	140000	37.15 N	116.07 W	TUFF	519	L-I	40	HAREBELL
710629	183000	37.18 N	116.22 W					
710701	140000	37.11 N	116.20 W	TUFF	266	L	100	DIAMOND MINE
710708	140000	37.06 N	116.04 W	TUFF	529	80		MINIATA
710818	140000	37.06 N	116.04 W	TUFF	527	L-I	66	ALGODONES
710922	140000	37.07 N	115.97 W					
710929	140000	37.01 N	116.01 W	TUFF	378	L		PEDERNAL
711008	140000	37.11 N	116.04 W	TUFF	378	L	7	CATHAY
711014	143002	37.32 N	116.14 W					
711106	220000	51.47 N	179.11 E	BASALT	1791	<5000		CANNIKIN
711124	201500	36.88 N	115.93 W	ALLUVIUM	264	L		DIAGONAL LINE
711130	154501	37.16 N	116.15 W					
711214	210959	37.12 N	116.09 W	ALLUVIUM	330	L-I	24	CHAENACTIS
720203	214459	36.98 N	115.81 W					
720330	210001	36.97 N	116.05 W					
720419	163200	37.12 N	116.08 W	ALLUVIUM	326	L	19	LONGCHAMPS
720502	191500	37.21 N	116.21 W	TUFF	377	L		MISTY NORTH
720511	140002	37.25 N	116.11 W					
720517	141000	37.12 N	116.09 W	ALLUVIUM	322	L	8	ZINNIA
720519	170000	37.06 N	116.04 W	TUFF	537	L	7	MONERO
720628	163001	37.10 N	116.21 W					
720629	183003	37.21 N	116.18 W	TUFF	424	L	21	DIAMOND SCULLS
720720	171600	37.02 N	116.03 W					
720725	133001	37.08 N	116.04 W	TUFF	560	L-I	130	OSCURO
720921	153000	37.12 N	116.09 W	ALLUVIUM	295	15	15	DELPHINIUM
720926	143000	37.25 N	116.32 W					
721109	151504	37.14 N	116.08 W					
721221	201500	37.10 N	116.03 W	TUFF	688	L-I	27	FLAX
730308	161000	37.00 N	116.03 W	TUFF	569	L-I	67	MIERA
730425	222500	37.12 N	116.06 W	ALLUVIUM	453	L-I	21	ANGUS
730426	171500	39.79 N	108.37 W	TUFF	564	85	120	STARWORT
730517	160000	37.18 N	116.09 W	SANDSTONE	1900	90		RIO BLANCO
730524	133001	37.18 N	116.21 W					
730605	170000	37.18 N		TUFF	391	L	26	DIDO QUEEN

(Contd.)

TABLE A2.5 (Contd.)

| Date | Origin time | Epicenter | | Medium | Depth (m) | Yield (kt) | | Name |
		Latitude	Longitude			Announced	Estimated	
730606	130000	37.24 N	116.35 W	TUFF	1064	I	570	ALMENDRO
730628	191512	37.15 N	116.09 W	ALLUVIUM	466	L-I	60	PORTULACA
731012	170000	37.20 N	116.20 W	TUFF	413	L	9	HUSKY ACE
731128	153000	36.99 N	115.93 W			L		BERNAL
731212	190007	37.06 N	116.57 W					LATIR
740227	170000	37.10 N	116.05 W			L-I	150	MING BLADE
740619	160000	37.20 N	116.19 W			L	20	ESCABOSA
740710	160000	37.07 N	116.03 W			L-I	170	PUYE
740814	140000	37.02 N	116.04 W			L	40	PORTMANTEAU
740830	150000	37.15 N	116.08 W			L-I	200	
740925	135959	36.99 N	115.89 W					STANYAN
740926	150500	37.13 N	116.07 W			L-I	100	HYBLA FAIR
741028		37.00 N	116.00 W	★		L		
741216	173004	37.11 N	116.32 W				4	TOPGALLANT
750228	151500	37.11 N	116.06 W			L-I	185	CABRILLO
750307	150000	37.13 N	116.08 W			L-I	120	DINING CAR
750405	194500	37.19 N	116.21 W			L-I	20	EDAH
750424	141000	37.12 N	116.09 W			L-I	9	OBAR
750430	150000	37.11 N	116.03 W			L-I	41	TYBO
750514	140000	37.22 N	116.47 W		765	I	380	STILTON
750603	142000	37.34 N	116.52 W		731	L-I	275	MIZZEN
750603	144000	37.09 N	116.04 W		637	L-I	160	MAST
750619	130000	37.35 N	116.32 W		912	I	520	CAMEMBERT
750626	123000	37.28 N	116.37 W		1311	I	750	MARSH
750906	170000	37.02 N	116.03 W		427	L	15	HUSKY PUP
751024	171126	37.22 N	116.18 W		134	L	15	KASSERI
751028	143000	37.29 N	116.41 W		1265	I	1200	INLET
751120	150000	37.22 N	116.37 W		817	I	500	LEYDEN
751126	153000	37.12 N	116.02 W		320	L	5	CHIBERTA
751220	200000	37.13 N	116.06 W		716	L-I	160	MUENSTER
760103	191500	37.30 N	116.33 W		1451	I	600	KEELSON
760204	142000	37.07 N	116.03 W		640	L-I	200	

★ Preliminary coordinates.

(Contd.)

APPENDIX 2

760204	144000	37.11 N	116.04 W		655	L-I	ESRON
760212	144500	37.27 N	116.49 W		1219	I$_b$	FONTINA
760214	113000	37.24 N	116.42 W		1167	I$_b$	CHESHIRE
760309	140000	37.31 N	116.36 W		869	I$_c$	ESTUARY
760314	123000	37.31 N	116.47 W		1273	I$_b$	COLBY
760317	141500	37.26 N	116.31 W		879	I$_b$	POOL
760317	144500	37.11 N	116.05 W		780	I$_b$	STRAIT
760512	195000	37.21 N	116.21 W			L-I	
760727	203000	37.00 N	116.00 W	*	150		MIGHTY EPIC
761123	151500	37.00 N	116.00 W	*	900		
761208	145000	37.00 N	116.00 W	*	350	L	
761221		37.00 N	116.00 W	*	350	L	
761228	180000	37.00 N	116.00 W	*	900	L-I	
					500		
					200		

^{b} 200–500 kt. ^{c} 500–1000 kt.
★ Preliminary coordinates.

TABLE A2.6. Announced and presumed underground nuclear explosions in the USSR in 1963–1976.

Date	Origin time	Epicenter		Medium	Depth (m)	Yield (kt)		Name
		Latitude	Longitude			Announced	Estimated	
640315	75958	49.70 N	78.00 E				49	
640516	60058	49.90 N	78.30 E				44	
640719	55959	49.90 N	78.10 E				29	
640918	75955	72.90 N	55.20 E				2	
641025	75959	73.50 N	53.70 E				14	
641116	55957	49.70 N	78.00 E				49	
650115	55959	49.89 N	78.97 E			125	110	
650303	61457	49.82 N	78.07 E				34	
650511	63958	49.79 N	77.92 E				6	
650617	34458	49.97 N	78.07 E				21	
650729	30502	50.40 N	77.90 E					
650917	35958	49.81 N	78.05 E				15	
651008	55959	49.89 N	78.05 E				34	
651121	45758	49.77 N	78.06 E				47	
651224	45958	49.88 N	78.04 E				7	
660213	45758	49.82 N	78.13 E				270	
660320	54958	49.70 N	78.00 E				170	
660421	35758	49.81 N	78.05 E				28	
660422	25804	47.90 N	47.70 E			L	4	
660507	35758	49.74 N	77.90 E				36	
660629	65958	49.93 N	78.01 E				24	
660721	35758	49.70 N	78.00 E				29	
660805	35758	49.90 N	78.00 E				4	
660819	35301	50.40 N	77.90 E				4	
660907	35158	49.90 N	78.00 E					
660930	55953	38.80 N	64.50 E			30	65	
661019	35758	49.75 N	78.03 E				770	
661027	55758	73.44 N	54.75 E					
661203	50204	49.72 N	77.90 E				120	
661218	45758	49.93 N	77.73 E			I	210	
670226	35758	49.78 N	78.12 E				21	
670325	55759	49.77 N	78.08 E					

(Contd.)

APPENDIX 2   401

670420	40758	49.74	N	78.12	E	L-I	58
670528	40758	49.81	N	78.11	E		32
670629	25658	49.87	N	78.10	E		27
670715	32657	49.83	N	78.11	E		27
670804	65758	49.82	N	78.05	E		23
670916	40358	50.01	N	77.82	E		22
670922	50358	50.03	N	77.61	E		16
671006	70002	57.69	N	65.27	E		67
671017	50358	49.82	N	78.10	E		210
671021	45958	73.37	N	54.81	E		33
671030	60358	49.84	N	78.11	E		3
671122	40357	49.90	N	77.30	E		22
671208	60357	49.84	N	78.22	E		10
680107	34658	49.81	N	78.02	E		7
680424	103557	49.83	N	78.08	E		
680521	35912	38.92	N	65.16	E	47	18
680611	30558	49.84	N	78.16	E	L	35
680619	50557	49.96	N	79.09	E		46
680701	40202	47.92	N	47.95	E		23
680712	120757	49.67	N	78.12	E		4
680820	40558	50.00	N	78.00	E		35
680905	40557	49.76	N	78.14	E		110
680929	34258	49.77	N	78.19	E		310
681107	100205	73.40	N	54.86	E		4
681109	25358	49.79	N	78.04	E		14
681218	50157	49.72	N	78.06	E		47
690307	82658	49.81	N	78.15	E		18
690516	40257	49.77	N	78.15	E	8	25
690531	50157	49.98	N	77.73	E	8	22
690704	24657	49.75	N	78.19	E		38
690723	24658	49.87	N	78.32	E	L-I	11
690902	45957	57.41	N	54.86	E	I	11
690908	45956	57.36	N	55.11	E		8
690911	40157	49.70	N	78.11	E		78
690926	65956	45.89	N	42.47	E		21
691001	40258	49.81	N	78.21	E		340
691014	70006	73.40	N	54.81	E		

(Contd.)

TABLE A2.6 *(Contd.)*

Date	Origin time	Epicenter Latitude	Epicenter Longitude	Medium	Depth (m)	Yield (kt) Announced	Yield (kt) Estimated	Name
691130	33257	49.92 N	79.00 E			I	160	
691206	70257	43.83 N	54.78 E			L-I	100	
691228	34658	50.00 N	77.82 E				72	
691229	40158	49.73 N	78.15 E				1	
700127	70258	49.80 N	78.21 E			L-I	52	
700327	50257	49.76 N	78.01 E				10	
700625	45952	52.20 N	55.69 E				5	
700628	15758	49.83 N	78.25 E			L-I	120	
700721	30257	49.95 N	77.75 E			L-I	29	
700724	35657	49.80 N	78.17 E			L-I	21	
700906	40257	49.77 N	78.09 E			L-I	46	
701014	55957	73.31 N	55.15 E		~6000	L-I	2100	
701104	60257	49.97 N	77.79 E			L-I	34	
701212	70057	43.85 N	54.77 E			I	190	
701217	70057	49.73 N	78.13 E			L-I	35	
701223	70057	43.83 N	54.85 E			I	240	
710322	43258	49.74 N	78.18 E			L-I	86	
710323	65956	61.29 N	56.47 E		45		51	
710425	33258	49.82 N	78.09 E			L-I	140	
710525	40258	49.80 N	78.21 E			L-I	11	
710606	40257	49.98 N	77.77 E			L-I	39	
710619	40358	50.01 N	77.74 E			L-I	36	
710630	35657	49.97 N	79.05 E			L-I	25	
710702	170002	67.66 N	62.00 E				7	
710710	165959	64.17 N	55.18 E			L-I	27	
710919	110007	57.78 N	41.10 E				4	
710927	55955	73.39 N	55.10 E				770	
711004	100002	61.61 N	47.12 E			L-I	11	
711009	100002	50.00 N	77.70 E			L-I	24	
711021	60257	49.99 N	77.65 E			L-I	44	
711022	50000	51.57 N	54.54 E			L-I	34	
711129	60257	49.76 N	78.13 E			L-I	34	

(Contd.)

APPENDIX 2

Date	Number	Lat (°N)	Lon (°E)	Yield
711215	75259	49.98	77.90	3
711222	65956	47.87	48.22	210
711230	62058	49.75	78.13	90
720210	50257	49.99	78.89	43
720310	45657	49.75	78.18	33
720328	42157	49.73	78.19	15
720411	60005	37.37	62.00	7
720607	12757	49.76	78.17	34
720706	10258	49.72	77.98	1
720709	65958	49.78	35.40	6
720714	145949	50.00	46.40	0.2 (~1000)
720816	31657	49.76	78.15	15
720820	25958	49.46	48.18	87
720826	34657	49.99	77.78	35
720828	55957	73.34	55.08	690
720902	85658	49.96	77.73	7
720904	70004	67.69	33.44	21
720921	90001	52.13	51.99	88
721003	85958	46.85	45.01	350
721102	126658	49.91	78.84	111
721124	90008	52.78	51.07	20
721210	95958	51.84	64.15	700
721210	42658	49.85	78.10	620
721228	42713	50.11	78.81	3
730216	50258	51.70	77.20	48
730419	43258	49.83	78.23	27
730723	126658	50.01	77.72	28
730815	122258	49.99	78.06	420
730828	159958	42.71	78.85	28
730912	25958	50.55	67.41	14
730919	65954	73.30	68.39	2700
730927	259957	45.63	55.16	110
730930	659958	70.76	67.85	210
731026	45957	51.61	53.87	22
731026	42658	49.76	54.58	19
731026	55958	53.66	78.20	7

(Contd.)

TABLE A2.6 (Contd.)

Date	Origin time	Epicenter Latitude	Epicenter Longitude	Medium	Depth (m)	Yield (kt) Announced	Yield (kt) Estimated	Name
731027	65957	70.78 N	54.18 E			I	3200	
731214	74657	50.04 N	79.01 E				150	
740130	45658	49.89 N	77.99 E				2	
740130	45702	49.83 N	78.08 E			L-I	25	
740416	55302	49.99 N	78.82 E				3	
740516	30257	49.74 N	78.15 E			L-I	23	
740531	32657	49.95 N	78.84 E			L-I	140	
740625	35658	49.80 N	78.11 E				2	
740708	60002	53.80 N	55.20 E					
740710	25657	49.79 N	78.14 E				16	
740722	13221	70.68 N	53.54 E					
740814	145958	68.91 N	75.90 E			L-I	45	
740829	95956	73.37 N	55.09 E			L-I	870	
740829	150000	67.23 N	62.12 E				20	
740913	30258	49.82 N	78.09 E				15	
741016	63257	49.97 N	78.97 E			L-I	43	
741102	45957	70.82 N	54.06 E				1600	
741207	55957	49.91 N	77.65 E				2	
741216	62302	49.75 N	78.06 E				8	
741227	64102	49.82 N	78.12 E				6	
741227	54657	49.96 N	79.05 E			L-I	51	
750220	53258	49.82 N	78.08 E				77	
750311	54258	49.79 N	78.25 E				30	
750425	45957	47.50 N	47.50 E					
750427	53657	49.99 N	78.98 E				60	
750630	32658	49.76 N	78.00 E				35	
750630	32657	50.00 N	79.00 E				4	
750807	35658	49.81 N	78.24 E			L-I	14	
750823	85958	73.37 N	54.64 E				550	
750929	105958	69.59 N	90.40 E				6	
751005	42700	50.00 N	78.00 E*					
751018	85956	70.84 N	53.69 E				1400	

* Preliminary coordinates.

(Contd.)

APPENDIX 2

Date				L-I	
751021	115957	73.35 N	55.08 E		700
751029	44658	49.98 N	78.97 E		900
751213	45657	49.80 N	78.20 E		100
751225	51657	50.04 N	78.90 E		904
760115	44658	49.87 N	78.25 E		100
760421	45758	49.82 N	78.20 E		110
760421	50257	49.93 N	78.82 E		20
760519	25658	49.86 N	78.01 E		
760609	30258	50.02 N	79.08 E		
760704	25658	49.91 N	78.95 E		25
760723	23258	49.79 N	78.05 E		900
760729	45958	47.78 N	48.12 E *		10
760804	25700	50.00 N	78.00 E *		
760828	25700	50.00 N	79.00 E *		
760929	30000	73.00 N	55.00 E *		
761020	80000	73.00 N	55.00 E *		
761030	45700	50.00 N	78.00 E *		
761123	50300	50.00 N	79.00 E *		
761207	45700	50.00 N	78.00 E *		
761230	35700	50.00 N	78.00 E *		

* Preliminary coordinates.

REFERENCES

AEC, 1973. *Atomic Energy Programs.* Vol. 1. Operating and developmental functions. U.S. Atomic Energy Commission.

AEC, 1974. *Annual Report to Congress.* U.S. Atomic Energy Commission, Washington, DC., 250 pp.

Aggarwal, Y.P., Sykes, L.R., Simpson, D.W. & Richards, P.G., 1975. Spatial and temporal variations of t_s/t_p and in P-wave residuals at Blue Mountain Lake, New York: Application to earthquake prediction. *J. Geophys. Res.*, 80: 718–732.

Aki, K., 1972. Scaling law of earthquake source time function. *Geophys. J. Roy. Astron. Soc.*, 31:3–25.

Aki, K., 1973. Scattering of P waves under the Montana Lasa. *J. Geophys. Res.*, 78:1334–1347.

Aki, K. & Tsai, Y.B., 1972. Mechanism of Love-wave excitation by explosive sources. *J. Geophys. Res.*, 77:1452–1475.

Allen, C.R., Bonilla, M.G., Brace, W.F., Bullock, M., Clough, R.W., Hamilton, R.M., Hofheinz, R., Kisslinger, C., Knopoff, L., Park, M., Press, F., Raleigh, C.G., & Sykes, L.R., 1975. Earthquake research in China. *Trans. Amer. Geophys. Union*, 56:838–881.

Ambraseys, N.N., 1971. Value of historical records of earthquakes. *Nature*, 232:375–379.

Anderson, T.W. & Bahadur, R., 1962. Classification into two multivariate normal distributions with different covariance matrices. *Ann. Math. Stat.*, 33:422–431.

Anglin, F.M., 1971. Detection capabilities of the Yellowknife seismic array and regional seismicity. *Bull. Seism. Soc. Amer.*, 61:993–1008.

Anglin, F.M., 1972. Discrimination of earthquakes and explosions using short period seismic array data. *Nature*, 233:51–52.

Anglin, F.M. & Israelson, H., 1973. Seismological discrimination of earthquakes and explosions using multistation short period data. *Bull. Seism. Soc. Amer.*, 63:321–323.

Aptikayev, F.F., Gurbunova, I.V., Dokutjaev, M.M., Melovatskij, B.V., Nersesov, I.L., Rautian, T.G., Romaskov, A.N., Rulev, B.G., Fomitjev, A.G., Chalturin, V.I., & Charin, D.A., 1967. The results of scientific observations during the Medeo explosion. (In Russian.) *An. KazSSR Vestnik*, 5:30–40.

Archambeau, C.G., 1976. *Identification of seismic events: Discrimination of underground explosions and earthquakes.* To be published by Elsevier.

ARPA, 1973. *Progress and problems in seismic verification research.* Defense Advanced Research Projects Agency, Arlington, Report No. T10 73–3.

REFERENCES

Austegard, A., 1974. *Seismic events in the USSR in 1972.* National Defense Research Institute, Stockholm, Report C 20018–T1.

Bakun, W.H. & Johnson, L.R., 1970. Short period spectral discriminants for explosions. *Geophys. J. Roy. Astron. Soc.*, 22:139–152.

Bakun, W.H. & Johnson, L.R., 1973. The deconvolution of teleseismic P waves from explosions Milrow and Cannikin. *Geophys. J. Roy. Astron. Soc.*, 34:321–342.

Bamford, S.A.D., 1972. Evidence for a low velocity zone in the crust beneath the Western British Isles. *Geophys. J. Roy. Astron. Soc.*, 30:101–105.

Barazangi, M. & Isacks, B., 1971. Lateral variation of seismic wave attenuation. *J. Geophys. Res.*, 76:8493–8516.

Basham, P.W., 1969. Canadian magnitudes of earthquakes and nuclear explosions in South Western North America. *Geophys. J. Roy. Astron. Soc.*, 17:1–13.

Basham, P.W., 1971. A new magnitude formula for short period continental Rayleigh waves. *Geophys. J. Roy. Astron. Soc.*, 23:255–260.

Basham, P.W. & Anglin, F.M., 1973. Multiple discriminant screening procedure for test ban verification. *Nature*, 246:474–475.

Basham, P.W. & Ellis, R.M., 1969. The composition of P codas using magnetic tape seismograms. *Bull. Seism. Soc. Amer.*, 59:473–486.

Basham, P.W. & Horner, R.B., 1973. Seismic magnitudes of underground nuclear explosions. *Bull. Seism. Soc. Amer.*, 63:105–132.

Basham, P.W. & Whitham, K., 1970. *Seismological detection and identification of underground nuclear explosions.* Seism. Div., Earth Physics. Branch, Department of Energy Mines and Resources, Ottawa.

Beauchamp, K.G., 1973. *Signal processing using analog and digital techniques.* George Allen & Unwin Ltd., London, 547 pp.

Berg, J.W., Trembly, L.D. & Laun, P.R., 1964. Primary ground displacements and seismic energy near the Gnome explosion. *Bull. Seism. Soc. Amer.*, 54:1115–1126.

Berteussen, K. & Husebye, E.S., 1972. *Predicted and observed seismic event detectability of the NORSAR array.* NTNF/NORSAR, Kjeller, Technical Report No. 42.

Billington, S. & Isacks, B.L., 1975. Identification of fault planes associated with deep earthquakes. *Geophys. Res. Lett.*, 2:62–66.

Blamey, C. & Gibbs, P.G., 1968. The epicentres and origin times of some large explosions. *Geophys. J. Roy. Astron. Soc.*, 16:1–7.

Block, B., 1970. *Report on a new broad band vertical accelerometer.* In: Copies of papers presented at Woods Hole Conference on seismic discrimination. Laboratories of Teledyne Geotech, Alexandria, Virginia.

Bolt, B.A., 1960. The revision of earthquake epicenters, focal depths, and origin-time using a high-speed computer. *Geophys. J. Roy. Astron. Soc.*, 3:434–440.

Bolt, B.A., 1970. Earthquake location for small networks using the generalized inverse matrix. *Bull. Seism. Soc. Amer.*, 60:1823–1828.

Bolt, B.A. (Ed.), 1972a. *Methods in computational physics.* Vol. 11, Seismology: Surface waves and earth oscillations. Academic Press, New York, 309 pp.

Bolt. B.A. (Ed.), 1972b. *Methods in computational physics*. Vol. 12, Seismology: Body waves and sources. Academic Press, New York, 391 pp.

Bolt, B.A., 1974. Earthquake studies in the People's Republic of China. *Trans. Amer. Geophys. Union*, 55:108—117.

Bolt, B.A., 1976. *Nucelar explosions and earthquakes. The parted veil*. W.H. Freeman and Co., San Fransisco, 309 pp.

Booker, A. & Mitronovas, W., 1964. An application of statistical discrimination to classify seismic events. *Bull. Seism. Soc. Amer.*, 54:961—971.

Bormann, P., 1972. Identification of teleseismic events in the records of Moxa station. *Gerlands Beitr. Geophys.*, 81:105—116.

Bradner, H. & Dodds, J.G., 1964. Comparative seismic noise on the ocean bottom and on land. *J. Geophys. Res.*, 69:4339—4348.

Bradner, H., Dodds, J.G. & Foulks, R.E., 1965. Investigation of microseism sources with ocean bottom seismometers. *Geophysics,* 30:511—526.

Brazee, R.J., 1969. Further reporting on the distribution of earthquakes with respect to magnitude (m_b). *Earthquake Notes*, XL:49—51.

Briscoe, H.W. & Fleck, P.L., 1965. Data recording and processing for the experimental large aperture seismic array. *Proc. IEEE*, 53:1852—1859.

Briscoe, H.W. & Sheppard, R.M., 1966. *A study of the capability of a LASA to aid the identification of a seismic source*. Massachusetts Institute of Techonology, Lincoln Laboratory, Technical Note 1966—38.

Broding, R.A., Bentley-Lewellyn, N.J. & Hearn, D.P., 1964. Study of three-dimensional seismic detection system. *Geophysics*, 29:221—249.

Bullen, K.E., 1963. *An introduction to the theory of seismology*. Cambridge University Press, London, 381 pp.

Bungum, H., 1972. *Array stations as a tool for microseismic research*. NTNF/ NORSAR, Kjeller, Technical Report No. 46.

Bungum, H. & Husebye, E.S., 1974, Analysis of the operational capabilities for detection and location of seismic events at NORSAR. *Bull. Seism. Soc. Amer.*, 64:637—656.

Bungum, H. & Ringdahl, F., 1974. *Diurnal variation of seismic noise and its effect on detectability*. NTNF/NORSAR, Kjeller, Scientific Report No. 5.

Bungum, H. & Tjøstheim, D., 1976. Discrimination between Eurasian earthquakes and underground explosions using the $m_b:M_s$ method and short period autoregressive parameters. *Geophys. J. Roy. Astron. Soc.*, 45:371—392.

Bungum, H., Rygg, E. & Bruland, L., 1971a. Short period seismic noise structure at the Norwegian seismic array. *Bull. Seism. Soc. Amer.*, 61:357—373.

Bungum, H., Husebye, E.S. & Ringdahl, F., 1971b. The NORSAR array and preliminary results of data analysis. *Geophys. J. Roy. Astron. Soc.*, 25:115—126.

Burch, R.F., 1968. *A comparison of the short period noise at four UKAEA type arrays and an estimate of their detection capabilities*. United Kingdom Atomic Energy Authority, AWRE, Aldermaston, Report No. O 79/68.

Burke, M.D., Kanasewich, E.R., Malinsky, J.D. & Mantalbetti, J.F., 1970. A wide band digital seismograph system. *Bull. Seism. Soc. Amer.*, 60:1417–1426.

Burridge, R. & Knopoff, L. 1964. Body force equivalents for seismic dislocations. *Bull. Seism. Soc. Amer.*, 54:1875–1888.

Burton, P.W., 1975a. Estimations of Q_γ^{-1} from seismic Rayleigh waves. *Geophys. J. Roy. Astron. Soc.*, 36:167–189.

Burton, P.W. 1975b. *An array of broad band seismographs: Some initial results.* In XIVth General Assembly of the European Seismological Commission, Trieste, 16–22 September 1974. Akademie der Wissenschaften der DDR, Berlin, pp. 145–155.

Burton, P.W. & Kennett, B.L.N., 1972. Upper mantle zone of low Q. *Nature Physical Science*, 238:87–90.

Butkovich, T.R., 1968. *Frozen earth materials as seismic decoupling mediums.* Lawrence Radiation Laboratory, UCRL–50486.

Capon, J., 1969a. Investigation of long period noise at the large aperture seismic array. *J. Geophys. Res.*, 74:3182–3194.

Capon, J., 1969b. High resolution frequency wave number spectral analysis. *Proc. IEEE*, 57:1408–1418.

Capon, J., 1970. Analysis of Rayleigh-wave multipath propagation at LASA. *Bull. Seism. Soc. Amer.*, 60:1701–1732.

Capon, J., 1971. *Analysis of long period microseismic noise at NORSAR.* In: Seismic Discrimination, Semiannual Technical Summary Report to the Advanced Research Projects Agency 1 July–31 December 1971. Massachusetts Institute of Technology, Lincoln Laboratory p. 17.

Capon, J., 1973. Analysis of microseismic noise at LASA, NORSAR and ALPA. *Geophys. J. Roy. Astron. Soc.*, 35:39–54.

Capon, J. & Evernden, J., 1971. Detection of interfering Rayleigh-waves at LASA. *Bull. Seism. Soc. Amer.*, 61:807–849.

Capon, J. & Greenfield, R.J., 1967. *Matched filtering of long period Rayleigh-waves.* In: Seismic Discrimination, Semiannual Technical Summary Report to the Advanced Research Projects Agency 1 January–30 June 1967. Massachusetts Institute of Technology, Lincoln Laboratory, pp. 21–23.

Capon, J., Greenfield, R.J. & Lacoss, R.T., 1967. *Design of seismic arrays for efficient on-line beamforming.* Massachusetts Institute of Technology, Lincoln Laboratory, Technical Note 1967–26.

Capon, J., Greenfield, R.J., Kolker, R.J. & Lacoss, R.T., 1968. Short period signal processing results for the large aperture seismic arrays. *Geophysics*, 33:452–472.

Carder, D.S. & Mickey, W.V., 1960. *Seismic ground effects from coupled and decoupled shots in salt.* Coast and Geodetic Survey, US Dept. of Commerce.

Carpenter, E.W., 1965. A historical review of seismic array development. *Proc. IEEE*, 53:1816–1821.

Carpenter, E.W., Marshall, P.D. & Douglas, A., 1967. The amplitude distance curve for short period teleseismic P-waves. *Geophys. J. Roy. Astron. Soc.*, 13:61–70.

CCD/296, 1970. United Kingdom. *Working paper on Verification of a Comprehensive Test Ban Treaty.* Disarmament Conference Document.

CCD/305, 1970. Canada. *Working paper on Seismological Capabilities in Detecting and Identifying Underground Nuclear Explosions.* Disarmament Conference Document.

CCD/330, 1971. United States. *Working paper containing remarks of Dr. Stephan Lukasik, Director of the US Advanced Research Projects Agency, regarding research on seismic detection, location and identification of earthquakes and explosions, presented at Informal Meeting on 30 June.* Disarmament Conference Document.

CCD/348, 1971. Sweden. *Working paper with suggestions as to possible provisions of a Treaty Banning Underground Nuclear Weapon Tests.* (revised version of the Swedish Working paper ENDC/242, April 1, 1969). Disarmament Conference Document.

CCD/363, 1972. United Kingdom. *Working paper on seismic yields of underground explosions—Estimating yields of underground explosions from amplitudes of seismic signals.* Disarmament Conference Document.

CCD/376, 1972. Canada, Japan and Sweden. *Working paper on measures to improve tripartite cooperation among Canada, Japan and Sweden in the detection, location and identification of underground nuclear explosions by seismological means.* Disarmament Conference Document.

CCD/380, 1972. Canada and Sweden. *Working paper on an experiment in international cooperation: shortperiod seismological discrimination of shallow earthquakes and underground nuclear explosions.* Disarmament Conference Document.

CCD/386, 1972. United Kingdom. *Working paper on seismic data handling and analysis for a comprehensive test ban.* Disarmament Conference Document.

CCD/388, 1972. United States of America. *A review of current progress and problems in seismic verification.* Disarmament Conference Document.

CCD/397, 1973. Sweden. *Working paper with points to be considered by experts on the verification of ban on underground nuclear explosions.* Disarmament Conference Document.

CCD/398, 1973. Sweden. *Working paper presenting the ways in which verification has been dealt with in various arms control and disarmament treaties and proposals.* Disarmament Conference Document.

CCD/399, 1973. Japan. *Working paper on problems in determining the body wave magnitude.* Disarmament Conference Document.

CCD/401, 1973. United Kingdom. *Working paper on a review of the United Kingdom seismological research and development programme.* Disarmament Conference Document.

CCD/402, 1973. United Kingdom. *Working paper on the estimation of depth of seismic events.* Disarmament Conference Document.

CCD/404, 1973. United States of America. *Working paper. A program of research related to problems in seismic verification.* Disarmament Conference Document.

CCD/405, 1973. Sweden. *Working paper reviewing recent Swedish scientific work on the verification of a ban on underground nuclear explosions.* Disarmament Conference Document.

CCD/406, 1973. Canada. *Working paper on the verification of a comprehensive test ban by seismological means.* Disarmament Conference Document.

CCD/407, 1973. United States of America. *Comments on CCD/399, concerning magnitude determinations.* Disarmament Conference Document.

CCD/440, 1974. United Kingdom. *Working paper on a development in discriminating between seismic sources.* Disarmament Conference Document.

CCD/454, 1975. Japan. *Working paper containing views of a Japanese expert. Arms control implications of peaceful nuclear explosions (PNE).* Disarmament Conference Document.

CCD/455, 1975. *Letter dated 24 June 1975 from the Director General of the Internal Atomic Energy Agency to the Secretary-General of the United Nations concerning the studies on the peaceful applications of nuclear explosions, their utility and feasibility, including legal helth and satefy aspects. Annex II. Nuclear Explosions for peaceful purposes. I. Technical committee on peaceful nuclear explosions. (20—24 January 1975).* Disarmament Conference Document.

CCD/456, 1975. United States of America. *Working paper on arms control implications of nuclear explosions for peaceful purposes (PNE).* Disarmament Conference Document.

CCD/457, 1975. Canada, Japan and Sweden. *Working paper reporting the summary proceedings of an informal scientific conference held 14—19 April 1975 to promote Canadian—Japanese—Swedish cooperation in detection, location and identification of underground nuclear explosion by seismological means.* Disarmament Conference Document.

CCD/459, 1975. United Kingdom. *Working paper on safeguards against the employment of multiple explosions to simulate earthquakes.* Disarmament Conference Document.

CCD/482, 1976. Sweden. *Working paper on international cooperative measures to monitor a comprehensive test ban treaty.* Disarmament Conference Document.

CCD/484, 1976. *Letter dated 8 April 1976 from the Charge d'affaires A.I. of Norway to the special representative of the Secretary-General to the Conference of the Committee on Disarmament transmitting a working paper on some new results in seismic discrimination.* Disarmament Conference Document.

CCD/491, 1976. United States of America. *Current status of research in seismic verification.* Disarmament Conference Document.

CCD/492, 1976. United Kingdom. *Text of a statement on a comprehensive test ban made by Mr. Fakely at an informal meeting of the CCD on Tuesday, 20 April, 1976.* Disarmament Conference Document.

CCD/513, 1977. *First progress report to the Conference of the Committee on Disarmament by the Ad Hoc Group of Scientific Experts to consider international co-operative measures to detect and to identify seismic events.* Disarmament Conference Document.

CCD/523, 1977. Union of Soviet Socialist Republics. *Draft treaty on the complete and general prohibition of nuclear weapon tests.* Disarmament Conference Document.

CCD/526, 1977. Sweden. *Draft treaty banning nuclear weapon test explosions in all environments.* Disarmament Conference Document.

CCD/PV.546, 1971. Disarmament Conference Document.

CCD/PV.553, 1972. Disarmament Conference Document.

CCD/PV.599, 1973. Disarmament Conference Document.

CCD/PV.610, 1973. Disarmament Conference Document.

CCD/PV.625, 1973. Disarmament Conference Document.

CCD/PV.637, 1974. Disarmament Conference Document.

CCD/PV.638, 1974. Disarmament Conference Document.

CCD/PV.647, 1974. Disarmament Conference Document.

CCD/PV.653, 1974. Disarmament Conference Document.

CCD/PV.664, 1975. Disarmament Conference Document.

CCD/PV.688, 1976. Disarmament Conference Document.

CCD/PV.689, 1976. Disarmament Conference Document.

CCD/PV.692, 1976. Disarmament Conference Document.

CCD/PV.704, 1976. Disarmament Conference Document.

Chedd, G., 1973. Plowshare death rattle at Rio Blanco. *New Scientist*, 57:544–545.

Cherry, J.T. & Petersen, F.L., 1970. *Numerical simulation of stress wave propagation from underground nuclear explosions.* In: Peaceful Nuclear Explosions, Proceedings of a Panel Vienna, 2–6 March 1970, IAEA, Vienna, pp. 241–325.

Chinnery, M.A. & North, R.G., 1974. *Frequency-magnitude curves and M_s-m_b relationship.* In: Seismic Discrimination, Semiannual Technical Summary Report to the Advanced Research Projects Agency 1 January–30 June 1974. Massachusetts Institute of Technology, Lincoln Laboratory, pp. 14–15.

Christofferson, A. & Husebye, E.S., 1974. Least squares signal estimation techniques in analysis of seismic array recorded P-waves. *Geohpys. J. Roy. Astron. Soc.*, 38:525–552.

Christofferson, A. & Jansson, B., 1969. *Maximum likelihood estimation of an unknown signal in multiple Gaussian noise with known covariance matrix.* Institute of Statistics, Uppsala, Sweden, Seminar Report.

Coffer, H.F. & Higgins, G.H., 1968. *Nuclear explosions for oil and gas stimulation and shale oil recovery.* Proceedings of the Southwestern Legal Foundation. Exploration and Economics of the Petroleum Industry, Volume 6.

Cohen, T.J., 1970. Source-depth determinations using spectral, pseudo-auto correlation and cepstral analysis. *Geophys. J. Roy. Astron. Soc.*, 20:223–231.

Covington, P.A., 1974. *Seismograph station abbreviations and coordinates.* United States Dept. of the Interior, Geological Survey.

Crampin, S., 1970. A method for the location of near seismic events using travel times along ray paths. *Geophys. J. Roy. Astron. Soc.*, 21:535–539.

Curtis, J.W., 1973. A magnitude domain study of the seismicity of Papua, New Guinea and The Solomon Islands. *Bull. Seism. Soc. Amer.*, 63:787–806.

REFERENCES

Dahlman, O., 1969. *Short period seismic noise in Western Central Sweden.* National Defense Research Institute, Stockholm, Report C 4388—20.

Dahlman, O., 1974. Seismic source and transmission functions from underground nuclear explosions. *Bull. Seism. Soc. Amer.*, 64:1275—1293.

Dahlman, O. & Elvers, E., 1977. *Grouping estimated explosion yields.* (In preparation.)

Dahlman, O., Israelson, H., Hörnström, G., Lindh, B., Nedgård. I., Nordgren, L. & Slunga, R., 1971. *Hagfors Observatory 1970, Annual Report.* National Defense Research Institute, Stockholm, Report A 4501—26.

Dahlman, O., Israelson, H., Austegard, A. & Hörnström, G., 1974. Definition and identification of seismic events in the USSR in 1971. *Bull. Seism. Soc. Amer.*, 64:607—636.

Davies, D., 1973. Monitoring underground explosions. *Nature*, 241:19—24.

Davies, D. & Frasier, C.W., 1970. *A Chinese puzzle.* In: Seismic Discrimination, Semiannual Technical Summary Report to the Advanced Research Projects Agency 1 July—31 December 1970. Massachusetts Institute of Technology, Lincoln Laboratory, pp. 17—19.

Davies, D. & Julian, B., 1972. A study of short period P-wave signals from Longshot. *Geophys. J. Roy. Astron. Soc.*, 29:184—202.

Davies, D. & Sheppard, R.M., 1972. Lateral heterogeneity in the earth's mantle. *Nature*, 239:318—323.

Davies, G.L., 1961. *Magnetic tape instrumentation.* McGraw-Hill, New York, 263 pp.

Dean, W.C., 1972. Detection threshold of the LASA/SAAC system. *Geophys. J. Roy. Astron. Soc.*, 31:271—278.

Dean, W.C., 1975. *Operational experience with the LASA/SAAC system.* In: K.G. Beauchamp (Ed.), Exploitation of seismograph networks. Noordhoff, Leiden, pp. 149—166.

De Geer, L.E. & Persson, G., 1975. Some aspects of preventing clandestine development of nuclear weapons in connection with peaceful nuclear explosions. *FOA Reports*, 9 (No. 5), 4 pp.

De Noyer, J., 1963. *Identification of earthquakes and underground nuclear explosions by seismic networks.* Institute for Defence Analyses Research and Engineering Support Div., Research paper P—19.

Dickinson, H. & Tamarkin, P., 1965. Systems for the detection and identification of nuclear explosions in the atmosphere and in space. *Proc. IEEE*, 53:1921—1934.

Douglas, A., 1967. Joint epicentre determination. *Nature*, 215:47—48.

Douglas, A., 1975. *Modelling as an aid to interpreting P-wave seismograms.* In: K.G. Beauchamp (Ed.), Exploitation of seismograph networks. Noordhoff, Leiden, pp. 409—439.

Douglas, A. & Lilwall, R.C., 1972. Methods of estimating travel times and epicentres. *Geophys. J. Roy. Astron. Soc.*, 30:187—197.

Douglas, A., Hudson, J.A. & Kembhavi, V.K., 1971. The analysis of surface wave spectra using a reciprocity theorem for surface waves. *Geophys. J. Roy. Astron. Soc.*, 23:207–223.

Douglas, A., Corbishley, D.J., Blamey, C. & Marshall, P.D., 1972a. Estimating the firing depth of underground explosions. *Nature*, 237:26–28.

Douglas, A., Hudson, J.A. & Blamey, C., 1972b. A quantitative evaluation of seismic signals at teleseismic distances. III. Computed P and Rayleigh wave seismograms. *Geophys. J. Roy. Astron. Soc.*, 28:385–410.

Douglas, A., Marshall, P.D., Gibbs, P.G., Young, J.B. & Blamey, C., 1973. P signal complexity reexamined. *Geophys. J. Roy. Astron. Soc.*, 33:195–221.

Douglas, A., Young, J.B. & Hudson, J.A., 1974a. Complex P-wave seismograms from simple earthquake sources. *Geophys. J. Roy. Astron. Soc.*, 37:141–150.

Douglas, A., Hudson, J.A., Marshall, P.D. & Young, J.B., 1974b. Earthquakes that look like explosions. *Geophys. J. Roy. Astron. Soc.*, 36:227–233.

Douze, E.J., 1964. Signal and noise in deep wells. *Geophysics*, 29:721–732.

Douze, E.J., 1967. Short period seismic noise. *Bull. Seism. Soc. Amer.*, 57:55–81.

Dumas, H., 1971. *Unattended seismological observations.* In: Hearings before the Subcommittee on Research, Development and Radiation of the Joint Committee on Atomic Energy, Congress of the United States, Ninety-second Congress. First Session on extent of present capabilities for detecting and determining nature of underground events. Oct. 27 and 28. US Government Printing Office. Washington, D.C., pp. 118–122.

Dziewonski, A.M. & Gilbert, F. 1974. Temporal variation of the seismic moment tensor and the evidence of precursive compression for two deep earthquakes. *Nature*, 247:185–188.

Eaton, J.P., O'Neill, M.E. & Murdock, J.N., 1970. Aftershocks of the 1966 Parkfield-Cholame, California, earthquake: A detailed study. *Bull. Seism. Soc. Amer.*, 60:1151–1197.

Elvers, E., 1974. Seismic event identification by negative evidence. *Bull. Seism. Soc. Amer.*, 64:1671–1684.

Elvers, E., 1975. *Seismic event identification by the $m_b(M_s)$-method.* National Defense Research Institute, Stockholm, Report C 20070–T1.

ENDC/PV.224, 1965. Disarmament Conference Document.

ENDC/232, 1968. United Kingdom. *Working paper on the comprehensive test ban treaty.* Disarmament Conference Document.

Enescu, D., Georgescu, A., Jianu, D. & Zamarca, I., 1973. Theoretical model for the process of underground explosions. Contributions to the problem of the separation of large explosions from earthquakes. *Bull. Seism. Soc. Amer.*, 63:765–786.

Engdahl, E.R., 1972. Seismic effects of the MILROW and CANNIKIN nuclear explosions. *Bull. Seism. Soc. Amer.*, 62:1411–1424.

Engdahl, E.R., 1973. Relocation of intermediate depth earthquakes in the Central Aleutians by seismic ray tracing. *Nature*, 245:23–25.

Engelder, J.T., 1974. Microscopic wear grooves on slickensides: Indicator of paleoseismicity. *J. Geophys. Res.*, 79:4387–4392.

REFERENCES

Ericsson, U., 1969. *Hagfors Seismological Observatory in Sweden.* National Defense Research Institute, Stockholm, Report C 4400–20.

Ericsson, U., 1970. Event identification for test ban control. *Bull. Seism. Soc. Amer.*, 60:1521–1546.

Ericsson, U., 1971a. *Event identification by m(M) observations from networks.* National Defense Research Institute, Stockholm, Report C 4480–A1.

Ericsson, U., 1971b. *Seismometric estimates of underground nuclear explosion yields.* National Defense Research Institute, Stockholm, Report C 4464–26.

Ericsson, U., 1971c. A linear model for the yield dependence of magnitudes measured by a seismographic network. *Geophys. J. Roy. Astron. Soc.*, 25:49–69.

Ericsson, U., 1973. *Multivariate and strength independent discrimination between underground explosions and earthquakes.* Paper presented at the IASPEI meeting in Lima.

Eriksen, B., 1972. *Undersökningar av luftburet radioaktivt material härrörande från en underjordisk kärnladdningsexplosion i USSR den 23 mars 1971.* (In Swedish.) National Defense Research Institute, Stockholm, Report C 4502–A1.

Evans, D.M., 1966. The Denver area earthquakes and Rocky Mountain arsenal disposal well. *Mountain Geologist*, 3:23–26.

Evernden, J.F., 1967. Magnitude determination at regional and near-regional distances in the US. *Bull. Seism. Soc. Amer.*, 57:591–639.

Evernden, J.F., 1969a. Precision of epicentres located by small numbers of worldwide stations. *Bull. Seism. Soc. Amer.*, 59:1365–1398.

Evernden, J., 1969b. Identification of earthquakes and explosions by use of teleseismic data. *J. Geophys. Res.*, 74:3828–3856.

Evernden, J., 1970. Study of regional seismicity and associated problems. *Bull. Seism. Soc. Amer.*, 60:393–446.

Evernden, J.F., 1971a. Location capability of various seismic networks. *Bull. Seism. Soc. Amer.*, 61:241–273.

Evernden, J.F., 1971b. Variation of Rayleigh wave amplitude with distance. *Bull. Seism. Soc. Amer.*, 61:231–240.

Evernden, J., 1975. Further studies on seismic discrimination. *Bull. Seism. Soc. Amer.*, 65:359–392.

Ewing, W.M., Jardetsky, W.S. & Press, F., 1957. *Elastic waves in layered media.* McGraw–Hill, New York, 380 pp.

Fenix and Scission Inc., 1970. *Project Payett, Final Summary Report on the Feasibility of Constructing a Large Underground Chamber for Clandestine Nuclear Testing.* Fenix and Scission Inc., Engineers and Constructors, Tulsa, Oklahoma, USA.

Ferguson, T.S., 1967. *Mathematical statistics. A decision theoretical approach.* Academic Press, New York, 396 pp.

Filson, J.R., 1969. *Short period seismic spectrum at NORSAR.* Seismic Discrimination, Semiannual Technical Summary Report to the Advanced Research Projects Agency 1 January–30 June 1969. Massachusetts Institute of Technology, Lincoln Laboratory, pp. 1–5.

Filson, J.R., 1973. *Estimating the effect of Asian earthquake codas on the explosion detection capability of LASA.* Massachusetts Institute of Technology, Lincoln Laboratory, Technical Note 1973—29.

Filson, J.R., 1974. *Long period results from the International Seismic Month.* Massachusetts Institute of Technology, Lincoln Laboratory, Technical Note 1974—14.

Filson, J.R. & Bungum, H., 1972. Initial discrimination results from the Norwegian seismic array. *Geophys. J. Roy. Astron. Soc.*, 31:315—328.

Filson. J.R. & Frasier, C.W., 1972. Multisite estimation of explosive source parameters. *J. Geophys. Res.*, 77:2045—2061.

Fix, J., 1972. Ambient earth motion in the period range from 0.01—2560 s. *Bull. Seism. Soc. Amer.*, 62:1753—1760.

Flinn, E.A., 1965. Confidence regions and error determinations for seismic event location. *Rev. Geophys.* 3:157—185.

Forbes, C.B., Obenchain, R. & Swain, R.J., 1965. The LASA sensing system design, installation and operation. *Proc. IEEE*, 53:1834—1843.

Foster, J.S., 1971. *Statement.* In: Hearings before the Subcommittee on Research, Development, and Radiation of the Joint Committee on Atomic Energy. Congress of the United States. Ninety-Second Congress. First session on extent of present capabilities for detecting and determining nature of underground events. Oct. 27 and 28. US Government Printing Office, Washington D.C., pp. 3—16.

Foster, J.S., 1973. *Statement.* In: Hearing before the sub-committee on Arms Control, International Law and Organization of the Committee on Foreign Relations. United States Senate, Ninety-third Congress. First session on S. Res. 67. US Government Printing Office, Washington D.C., pp. 91.

Frannti, G.E., 1963. The nature of high frequency earth noise spectra. *Geophysics*, 28:547—562.

Frasier, C.W., 1972. Observations of pP in the short-period phases of NTS explosions recorded in Norway. *Geophys. J. Roy. Astron. Soc.*, 31:99—109.

Frasier, C.W., 1974. *Single-channel event detector in real time.* In: Seismic Discrimination, Semiannual Technical Summary to the Advanced Research Projects Agency 1 January—30 June 1974. Massachusetts Institute of Technology, Lincoln Laboratory, pp. 51.

Frasier, C.W. & Filson, J.R., 1972. A direct measurement of the earth's short period attenuation along a teleseismic ray. *J. Geophys. Res.*, 77:3782—3787.

Gandhi, I., 1974. *Underground nuclear explosion experiment.* Prime Minister's statement. Press Information Bureau, Government of India, New Dehli.

George, T.A., 1963. *Statement.* In: Hearings before the Joint Committee on Atomic Energy. Congress of the United States. Eighty-eighth Congress. First Session on Developments in technical capabilities for detecting and identifying nuclear weapons tests. US Government Printing Office, Washington D.C., pp. 238—260.

Gibbs, J.F., Healy, J.H., Raleigh, G.B. & Coakley, J., 1973. Seismicity in the Rangely, Colorado Area: 1962—1970. *Bull. Seism. Soc. Amer.*, 63:1557—1570.

Gjøystdahl, H. & Husebye, E.S., 1972. *A comparison of performance between pre-*

diction error and bandpass filters. NTNF/NORSAR, Kjeller, Technical Report No. 43.

Gjøystdahl, H., Husebye, E.S. & Rieber-Mohn, D., 1973. One-array and two-array location capabilities. *Bull. Seism. Soc. Amer.*, 63:549—569.

Glasstone, S. (Ed.), 1964. *The effects of nuclear weapons.* US Atomic Energy Commission, Washington D.C., 730 pp.

Gough, D.I., 1975. *Induced seismicity.* Document SC—76/CONF. 224/COL. 22, United Nations Educational, Scientific and Cultural Organization.

Grafenberg, 1974. *The seismological stations in the Federal Republic of Germany.* Central Seismological Observatory, Grafenberg, Erlangen.

Green, D.M. & Swets, J.A., 1966. *Signal detection theory and psychophysics.* Wiley, New York, 455 pp.

Green, Jr., P.E., Frosch, R.A. & Romney, C.F., 1965. Principles of an experimental large aperture seismic array. *Proc. IEEE*, 53:1821—1833.

Greensfelder, R.W., 1956. The $P_g - P_n$-method of determining depth of focus with applications to Nevada earthquakes. *Bull. Seism. Soc. Amer.*, 55:391—405.

Gutenberg, B., 1945. Amplitudes of surface waves and the magnitudes of shallow earthquakes. *Bull. Seism. Soc. Amer.*, 35:57—69.

Gutenberg, B. & Andrews, F., 1956. *Bibliography of microseisms.* 2nd Ed., Seismological Laboratory, California Institute of Technology, Pasadena, 134 pp.

Gutenberg, B. & Richter, C.F., 1954. *Seismicity of the earth and associated phenomena.* Princeton Univ. Press, Princeton, N.J., 310 pp.

Gyldén, N. & Holm, L.W., 1974. Risker för kärnladdningsframställning i det fördolda. (In Swedish.) National Defense Research Institute, Stockholm, Report C 4567—T3.

Hamilton, R.M., Smith, B.E., Fisher, F.G. & Papanek, P.J., 1972. Earthquakes causes by underground nuclear explosions on Pahute Mesa, Nevada Test Site. *Bull. Seism. Soc. Amer.*, 62:1319—1341.

Hasegawa, H.S., 1971. Analysis of teleseismic signals from underground nuclear explosions originating in four geological environments. *Geophys. J. Roy. Astron. Soc.*, 24:365—381.

Haskell, N.A., 1967. Analytic approximation for the elastic radiation from a contained underground explosion. *J. Geophys. Res.*, 72:2583—2587.

Haubrich, R.A. & McCamy, K., 1969. Microseisms: Costal and pelagic sources. *Rev. Geophys. Space Phys.*, 7:539—572.

Haubrich, R.A., Munk, W.H. & Snodgrass, F.E., 1963. Comparative spectra of microseisms and swell. *Bull. Seism. Soc. Amer.*, 53:27—37.

Herbst, R.F., Werth, G.C. & Springer, D.L., 1961. Use of large cavities to reduce seismic waves from underground explosions. *J. Geophys. Res.*, 66:959—978.

Herrin, E., Arnold, E.P., Bolt, B.A., Clawson, G.E., Engdahl, E.R., Freedman, H.W., Gordon, D.W., Hales, A.L., Lobdell, J.L., Nuttli, O., Romney, C., Taggart, J. & Tucker, W., 1968. 1968 Seismological tables for P-phases. *Bull. Seism. Soc. Amer.*, 58:1193—1241.

Hjortenberg, E., 1967. *Bibliography of microseisms 1955–1964.* Geodaetisk Instituts Skrifter, 3. Raekke, Bind 38, Copenhagen.

Hoagland, A.S., 1963. *Digital magnetic recording.* Wiley, New York, 154 pp.

Holzer, A., 1971. *Statement.* In: Hearings before the Subcommittee on Research, development and radiation of the Joint Committee on Atomic Energy. Congress of the United States. Ninety-second Congress. First session on extent of present capabilities for detecting and determining nature of underground events. Oct. 27 and 28. US Government Printing Office, Washington D.C., pp. 98–114.

Husebye, E.S., Dahle, A. & Berteussen, 1974. Bias analysis of NORSAR and ISC reported seismic event m_b magnitudes. *J. Geophys. Res.*, 79:2967–2978.

IAEA, 1970. *Peaceful nuclear explosions. Phenomenology and Status report, 1970.* Proceedings of a Panel on the Peaceful uses of nuclear explosions organized by the International Atomic Energy Agency and held in Vienna 2–6 March 1970. IAEA, Vienna, 454 pp.

IAEA, 1971. *Peaceful nuclear explosions. II. Their practical applications.* Proceedings of a Panel on the practical applications of the peaceful uses of nuclear explosions organized by the International Atomic Energy Agency and held in Vienna 18–22 January 1971. IAEA, Vienna, 355 pp.

IAEA, 1974. *Peaceful nuclear explosions. III. Applications, Characteristics and Effects.* Proceedings of a Panel organized by the International Atomic Energy Agency and held in Vienna 27 November–1 December 1972. IAEA Vienna, 488 pp.

IAEA, 1975. *Peaceful nuclear explosions. IV.* Proceedings of a technical committe on the peaceful uses of nuclear explosions organized by the International Atomic Energy Agency and held in Vienna 20–24 January 1975. IAEA, Vienna, 479 pp.

IAEA, 1976. *Basic material for a further study of legal aspects of nuclear explosions for peaceful purposes.* IAEA Gov/COM.23/13, 30 June 1976, Vienna.

Ichikawa, M. & Mochizuki, E., 1971. Traveltime tables for local earthquakes in and near Japan. (In Japanese.) *Papers in Meteorology and Geophysics*, XXII, pp. 229–290.

Israelson, H., 1971. Spectral content of teleseismic P-waves recorded at the Hagfors Observatory. *Geophys. J. Roy. Astron. Soc.*, 25:89–95.

Israelson, H., 1972. *Seismic identification using short period Hagfors data.* In: Proceedings from the seminar on seismology and seismic arrays, Oslo, 22–25 November 1971. NTNF/NORSAR, Kjeller, pp. 61–78.

Israelson, H., 1975. *Short period identification.* In: K.G. Beauchamp (Ed.), Exploitation of Seismograph Networks. Noordhoff, Leiden, pp. 441–459.

Israelson, H. & Wägner, H., 1973. *Short period seismic criteria obtained at the Hagfors Observatory from North American events.* National Defense Research Institute, Stockholm, Report D 4250–AR73.

Israelson, H. & Yamamoto, M., 1974. *Seismic magnitudes from explosions and earthquakes obtained at the Hagfors and Matsushiro Observatories.* National Defense Research Institute, Stockholm, Report C 20010–T1.

Israelson, H., Slunga, R. & Dahlman, O., 1974. Aftershocks caused by the Novaya Zemlya Explosion on October 27, 1973. *Nature*, 237:450–452.

Israelson, H., Basham, P., Elvers, E., Slunga, R, & Dahlman, O., 1976. *Seismic events in the US in 1972.* (In preparation.)

Jackson, P.L., 1970. Seismic ray simulation for a spherical earth. *Bull. Seism. Soc. Amer.*, 60:1021–1026.

Jackson, D.D., & Anderson, D.L., 1970. Physical mechanisms of seismic wave attenuation. *Rev. Geophys. Space Phys.*, 8:1–63.

Jacobs, K.H., 1970. Three-dimensional seismic ray tracing in a laterally heterogenous spherical earth. *J. Geophys. Res.*, 75:6675–6689.

Jacobs, K.H., 1972. Global tectonic implications of anomalous seismic P travel times from the nuclear explosion Longshot. *J. Geophys. Res.*, 77:2556–2573.

Jeffreys, H., 1970. *The earth.* Cambridge University Press, Cambridge 525 pp.

Jefferys, H. & Bullen, K.E., 1967. *Seismological tables.* British Association for the Advancement of Science, Gray Milne Trust. Reprinted by Smith and Ritchie, Edinburgh, 50 pp.

Jeppsson, I., 1975a. *Evasion by hiding in earthquake.* National Defense Research Institute, Stockholm, Report C 20043–T1.

Jeppsson, I., 1975b. *Evasion by hiding nuclear explosions in earthquakes from the Aleutian Islands, the Kuril Islands and Kamchatka.* National Defense Research Institute, Stockholm, Report C 20073–T1.

Joint Committee on Atomic Energy, 1963. *Hearings before the Joint Committee on Atomic Energy.* Congress of the United States. Eighty-eighth Congress. First session on developments in technical capabilities for detecting and identifying nuclear weapons tests. US Government Printing Office, Washington D.C., 518 pp.

Jordan, T.H., 1975. The continental tectosphere. *Rev. Geophys. Space Phys.*, 13:1–12.

Jordan, T.H. & Lynn, W.S., 1974. A velocity anomaly in the lower mantle. *J. Geophys. Res.*, 79:2679–2685.

Julian, B.R., 1970a. *Ray tracing in arbitrarily heterogeneous media.* Massachusetts Institute of Technology, Lincoln Laboratory, Technical Note 1970–45.

Julian, B., 1970b. *Seismic rays in heterogeneous media.* In: Seismic Discrimination, Semiannual Technical Summary Report to the Advanced Research Projects Agency 1 July–31 December 1970. Massachusetts Institute of Technology, Lincoln Laboratory, pp. 21–23.

Julian, B.R., 1973a. *Predicting Rayleigh-wave propagation behaviour from geological and geophysical data.* In: Seismic Discrimination. Semiannual Technical Summary Report to the Advanced Research Projects Agency 1 January–30 June 1973. Massachusetts Institute of Technology, Lincoln Laboratory, pp. 27.

Julian, B., 1973b. *Extension of standard event location procedures.* In: Seismic discrimination, Semiannual Technical Summary Report to the Advanced Research Projects Agency 1 January–30 June 1973. Massachusetts Institute of Technology, Lincoln Laboratory, pp. 4–9.

Julian, B.R. & Sengupta, M.K., 1974. Seismic travel time evidence for lateral inhomogeneity in the deep mantle. *Nature*, 242:443–447.

Kaila, K.L., 1970. Decay rate of P-wave amplitudes from nuclear explosions and the magnitude relations in the epicentral distance range 1° to 98°. *Bull. Seism. Soc. Amer.*, 60:447—460.

Kaila, K.L. & Sarkar, D., 1975. *P-wave amplitude variation with epicentral distance and the magnitude relations. Bull. Seism. Soc. Amer.*, 65:915—926.

Kaila, K.L., Gaur, V.K. & Narain, H., 1972. Quantitative seismicity maps of India. *Bull. Seism. Soc. Amer.*, 62:1119—1133.

Kanamori, H., 1967. Attenuation of P-waves in the upper and lower mantle. *Bull. Earthquake Res. Inst.*, Tokyo, 45:299—312.

Kedrov, O.K., 1971. Dynamic features of the recordings of longitudinal waves from distant earthquakes. *Izvestiya, Physics of the Solid Earth* (English edition), No. 11:762—770.

Kedrovski, O.L., 1970. *The application of contained nucelar explosions in industry.* (In Russian with English abstract.) In: Peaceful Nucelar Explosions Phenomenology and Status report, 1970. Proceedings of a Panel on the Peaceful uses of nuclear explosions organized by the International Atomic Energy Agency and held in Vienna, 2—6 March 1970. IAEA, Vienna, pp. 163—185.

Kedrovski, O.L., Manikov, K.V., Leonov, E.A., Romadin, N.M., Dorodnov, V.F. & Nikiforev, G.A., 1975. *The use of contained nuclear explosions to create underground reservoirs, and experience of operating these for gas condensate storage.* (In Russian with English abstract.) In: Peaceful nuclear explosions IV. Proceedings of a technical committee on the Peaceful uses of nuclear explosions organized by the International Atomic Energy Agency and held in Vienna 20—24 January 1975. IAEA, Vienna, pp. 399—420.

Keilis-Borok, V.L., Mebel, S.S., Pyatetskii-Shapiro, I.I., Vartanova, L.Yu & Zhelankina, T.S., 1972. *Computer determination of earthquake focal depth.* In: V.L. Keilis-Borok (Ed.), Computational Seismology. (Translated from Russian.) Consultants Bureau, New York, pp. 16—24.

Kelleher, J.A., 1970. Space-time seismicity of the Alaska-Aleutian seismic zone. *J. Geophys. Res.*, 75:5745—5756.

Kelleher, J. & Savino, J., 1975. Distribution of seismicity before large strike slip and thrust-type earthquakes. *J. Geophys. Res.*, 80:260—271.

Kelleher, J., Sykes, L. & Oliver, J., 1973. Possible criteria for predicting earthquake locations and their application to major plate boundaries of the Pacific and the Caribbean. *J. Geophys. Res.*, 78:2547—2585.

Kelly, E.J., 1964. *The representation of seismic waves in frequency—wave number space.* Massachusetts Institute of Technology, Lincoln Laboratory, Group Report 1964—15.

Kelly, E.J., 1968. *A study of two short period discriminants.* Massachusetts Institute of Technology, Lincoln Laboratory, Technical Note 1968—8.

Kendall, M.G. & Stuart, A., 1963. *The advanced theory of statistics.* Vol. 3. Charles Griffin & Company Ltd., London, 552 pp.

Kennedy, R.F., 1969. *13 Days. The Cuban missile crisis.* Pan Books Ltd., London, 190 pp.

Kennett, B.L.N., 1973. The interaction of seismic waves with horizontal velocity contrasts. *Geophys. J. Roy. Astron. Soc.*, 33:431—450.

REFERENCES

King, C., Abu-Zenn, A. & Murdock, J.N., 1974. Teleseismic source parameters of the Longshot, Milrow and Cannikin. *J. Geophys. Res.*, 79:721–718.

Kirstein, P.T., 1975. *UK experiences with the ARPA computer netwoek*. In: K.G. Beauchamp (Ed.), Exploitation of Seismograph Networks. Noordhoff, Leiden, pp. 55–80.

Kisslinger, C. & Wyss, M., 1975. Earthquake prediction. *Rev. Geophys. Space Phys.*, 13:298–300.

Knopoff, L. & Randall, M.J., 1970. The compensated linear-vector dipole; A possible mechanism for deep earthquakes. *J. Geophys. Res.*, 75:4957–4964.

Kogeus, K., 1968. A synthesis of short period P-wave records from distant explosion sources. *Bull. Seism. Soc. Amer.*, 58:663–680.

Kolar, O.C. & Pruvost, N.I., 1975. Earthquake simulation by nuclear explosions. *Nature*, 253:242–245.

Korhonen, H., 1971. *Types of storm microseism spectra at Oulu*. Dept. of Geophysics, Univ. of Oulu, Contribution No. 15.

Korhonen, H. & Kukkonen, E., 1974. *Spectral composition of short period noise at Oulu seismograph station*. Dept. of Geophysics, University of Oulu, Contribution No. 42.

Kosminskaya, I.P., Puzyrev, N.N. & Alekseyev, A.S., 1972. Explosion seismology, its past, present and future. *Tectonophysics*, 13:309–323.

Koyama, S., Sotobuyashi, T. & Suzuki, T., 1966. Highly fractional nuclear debris resulting from the venting of a Soviet underground nuclear test. *Nature*, 209:239–240.

Lacoss, R.T., 1965. *Geometry and patterns of large aperture seismic arrays*. Massachusetts Institute of Technology, Lincoln Laboratory, Technical Note 1965–64.

Lacoss, R.T., 1969. *A large population LASA discrimination experiment*. Massachusetts Institute of Technology, Lincoln Laboratory, Technical Note 1969–24.

Lacoss, R.T., 1971. *Discussion of a suite of long period signals recorded in Norway from Asian events*. In: Seismic discrimination, Semiannual Technical Summary Report to the Advanced Research Projects Agency 1 July–31 December 1971. Massachusetts Institute of Technology, Lincoln Laboratory, pp. 19–22.

Lacoss, R.T. & Kuster, G.T., 1970. *Processing a partially coherent large seismic array for discrimination*. Massachusetts Institute of Technology, Lincoln Laboratory. Technical Note 1970–30.

Lacoss, R.T. & Lande, L., 1969. *LASA-NORSAR Depth Phase Study*. In: Seismic discrimination, Semiannual Technical Summary Report to the Advanced Research Projects Agency 1 July–31 December 1969. Massachusetts Institute of Technology, Lincoln Laboratory, pp. 4–5.

Lacoss, R.T., Kelly, E.J. & Toksöz, M.N., 1969. Estimation of seismic noise structure using arrays. *Geophysics*, 34:21–38.

Lacoss, R.T., Needham, R.E. & Julian, B.R., 1974. *International seismic month event list*. Massachusetts Institute of Technology, Lincoln Laboratory, Technical Note 1974–14.

Lamb, G.L., 1960. *Some seismic effects of underground explosions in cavities.* Los Alamos Scientific Laboratory of the University of California, LA–2405.

Lambert, D.G. & Becker, E.S., 1973. *Evaluation of the detection and discrimination capabilities of the very long period experiments (VLPE) single stations, VLPE network and the VLPE-ALPA-NORSAR combined network.* Texas Instruments Inc., Dallas, Texas, Special Report No. 6, Extended array evaluation program.

Lambert, D.G., von Seggern, D.H., Alexander, S.S. & Galat, T.A., 1969. *The longshot experiment. Vol. II, Comprehensive analysis.* Seismic Data Lab., Alexandria, Virginia, Report No. 234.

Landers, T.E., 1972. Some interesting central Asian events on the M_s:m_b diagram. *Geophys. J. Roy. Astron. Soc.*, 31:329–339.

Landers, T.E., 1974. *Improved depth-phase identification via maximum entropy cepstral analysis.* In: Seismic discrimination, Semiannual Technical Summary Report to the Advanced Research Projects Agency 1 January–30 June 1974. Massachusetts Institute of Technology, Lincoln Laboratory, pp. 26.

Larsson, T., 1974. *Kompendium i A-stridsmedel.* (In Swedish.) National Defense Research Institute, Stockholm, 199 pp.

Lee, W.H. & Taylor, P.T., 1966. Global analysis of seismic refraction measurements. *Geophys. J. Roy. Astron. Soc.*, 11:389–414.

Lee, W.H., Eaton, M.S. & Brabb, E.E., 1971. The earthquake sequence near Danville, California, 1970. *Bull. Seism. Soc. Amer.*, 61:1711–1794.

Lee, W.H., Meagher, K.L., Bennet, R.E. & Matamoros, E.E., 1972. *Catalogue of earthquakes along the San Andreas fault system in Central California for the year 1971.* National Center for Earthquake Research, Menlo Park, Open-file Report.

Le Pichon, X., Francheteau, J. & Bonnin, J., 1973. *Plate tectonics.* Elsevier, Amsterdam, 300 pp.

Lin, M. & Filson, J.R., 1974. *More upper mantle reflections.* In: Seismic discrimination, Semiannual Technical Summary Report to the Advanced Research Projects Agency 1 January–30 June 1974. Massachusetts Institute of Technology, Lincoln Laboratory, pp. 38.

Lomnitz, C., 1974. *Global tectonics and earthquake risk.* Elsevier, Amsterdam, 320 pp.

Long, F.A., 1963. *Statement.* In: Hearings before the Joint Committee on Atomic Energy Congress of the United States. Eighty-eighth Congress. First Session on Developments in technical capabilities for detecting and identifying nuclear weapons tests. US Government Printing Office, Washington D.C., pp. 410–431.

Love, A.E.H., 1944. *A treatise on the mathematical theory of elasticity.* Dover Publications, New York, 643 pp.

Lukasik, S.J., 1971. *Statement.* In: Hearings before the subcommittee on Research, Development and Radiation of the Joint Committee on Atomic Energy. Congress of the United States. Ninety-second Congress. First session on extent of present capabilities for detecting and determining nature of underground events. Oct. 27 and 28. US Government Printing Office, Washington D.C., pp. 17–67.

Lund, C., 1975. *The Scandinavian "Blue Road" traverse.* (In preparation.)

Mack, H., 1969. Nature of short-period P-wave signal variations at LASA. *J. Geophys. Res.*, 74:3161–3170.

Manchee, E.B., 1972. Short period seismic discrimination. *Nature*, 239:152–153.

Marshall, P.D., 1970. *Some seismic results of the Medeo explosions in the Alma-Ata region of the USSR.* United Kingdom Atomic Energy Authority, AWRE, Aldermaston, Reporr No. O 33/70.

Marshall, P.D., 1972. *Some seismic results from a world wide sample of large underground explosions.* United Kingdom Atomic Energy Authority, AWRE, Aldermaston, Report No. O 49/72.

Marshall, P.D. & Basham, P.W., 1972. Discrimination between earthquakes and underground explosions employing an improved M_s scale. *Geophys. J. Roy. Astron. Soc.*, 28:431–458.

Marshall, P.D. & Carpenter, E.W., 1966. Estimations of Q for Rayleigh waves. *Geophys. J. Roy. Astron. Soc.*, 10:549–550.

Marshall, P.D. & Hurley, R.W., 1975. Recognising simulated earthquakes. *Nature*, 259:378–380.

Marshall, P.D. & Key, P., 1973. *An analysis of seismic waves from earthquakes and explosions in the Sino Soviet area during 1966.* United Kingdom Atomic Energy Authority, AWRE, Aldermaston, Report No. O 5/73.

Marshall, P.D., Douglas, A. & Hudson, J.A., 1971. Surface waves from underground explosions. *Nature*, 234:8–9.

Marshall, P.D., Burch, R.F. & Douglas, A., 1972. How and why to record broad band seismic signals. *Nature*, 239:154–155.

Massé, R.P., Lambert, D.G. & Harkrider, D.G., 1973. Precision of the determination of focal depth from the spectral ratio of Love/Rayleigh surface waves. *Bull. Seism. Soc. Amer.*, 63:59–100.

Massinon, B. & Plantet, J.L., 1975. *A large aperture seismic network implementation in France: description and some results concerning epicentre location and upper mantle anomalies.* Laboratorie de Détection et de Géophysique, Commissariat à l'Energie Atomique, Montrouge.

McEvilly, T.V. & Peppin, W.A., 1972. Source characteristics of earthquakes, explosions and afterevents. *Geophys. J. Roy. Astron. Soc.*, 31:67–82.

Melton, B.S. & Kirkpatrick, B.M., 1970. The symmetrical triaxial seismometer—Its design for application to long period seismometry. *Bull. Seism. Soc. Amer.*, 60:717–740.

Miyamura, S. & Hori, M., 1972. Body wave magnitude at 1 Hz and 2 Hz as a short period discriminant between earthquakes and explosions. *Bull. Seism. Soc. Amer.*, 62:411–412.

Mohorovicic, A., 1909. Das Beben vom 8.X.1909. *Jb. Met. Obs. Zagreb*, 9:1–63.

Montalbetti, J.F. & Kanasewich, E.R., 1970. Enhancement of teleseismic body phases with a polarization filter. *Geophys. J. Roy. Astron. Soc.*, 21:119–129.

Mueller, G., 1973a. Seismic moment and longperiod radiation of underground nuclear explosions. *Bull. Seism. Soc. Amer.*, 63:847–858.

Mueller, S. (Ed.) 1973b. The structure of the Earth's crust based on seismic data. *Tectonophysics*, 20:1–391.

Mueller, R.A. & Murphy, B.R., 1971. Seismic characteristics of underground nuclear detonations. Part I. Seismic spectrum scaling. *Bull. Seism. Soc. Amer.*, 61:1675–1692.

Murphy, J.C. & Hewlett, R.A., 1975. Identification of earthquakes and explosions using near field seismic data. *Trans. Amer. Geophys. Union*, 56:1148.

Murphy, A., Savino, J., Rynn, J., Choy, G. & McCamy, K., 1972. Observations of long period (10–100 s) seismic noise at several worldwide locations. *J. Geophys. Res.*, 77:5042–5049.

Myrdal, A., 1974. The international control of disarmament. *Scientific American*, 231, No.4:21–33.

Myrdal, A., 1976. *The Game of Disarmament.* Pantheon, New York, 443 pp.

Needham, R. E., 1975. *Worldwide Detection Capability of a Prototype Network of Seismograph Stations.* Massachusetts Institute of Technology, Lincoln Laboratory, Technical Note 1975–42.

Nevada Operations Office, 1976. *Announced United States Test Statistics.* Energy Research and Development Administration, Las Vegas, Nevada.

Noponen, I., 1973. *Compressional wave power spectrum from seismic sources.* University of Helsinki.

Noponen, I., 1974. Seismic ray direction anomalies caused by deep structure in Fennoscandia. *Bull. Seism. Soc. Amer.*, 64:1931–1941.

Noponen, I., 1975. *Compressional wave power spectrum from seismic sources.* Final Scientific Report. University of Helsinki.

Nordyke, M.D., 1970. *Peaceful uses of nuclear explosions.* In: Peaceful nuclear explosions. Phenomenology and Status report, 1970. Proceedings of a Panel on the Peaceful uses of nuclear explosions organized by the international Atomic Energy Agency and held in Vienna 2–6 March 1970. IAEA, Vienna, pp. 49–107.

Nordyke, M.D., 1974. *A review of Soviet data on the peaceful uses of nuclear explosions.* Lawrence Livermore Laboratory, Livermore, UCRL–51414.

North, R.G., 1974. *Focal depth from surface waves.* Seismic discrimination, Semi-annual Technical Summary Report to the Advanced Research Projects Agency 1 January–30 June 1974. Massachusetts Institute of Technology, Lincoln Laboratory, pp. 26–28.

NPT, 1975. NPT/CONF/35/I. *Review Conference of the Parties to the Treaty on the non-proliferation of nuclear weapons.* Final document.

Nuttli, O.W. & Kim, S.G., 1975. Surface-wave magnitudes of Eurasian earthquakes and explosions. *Bull. Seism. Soc. Amer.*, 65:693–710.

Orhaug, T. & Dyring, E., 1973. *Stormakternas vakande rymdögon.* (In Swedish.) National Defense Research Institute, Stockholm, Report B 2045–E1.

Panofsky, W., 1973. *Statement.* In: Hearing before the Subcommittee on Arms Control, International Law and Organization of the Committee on Foreign Relations. United States Senate. Ninety-third Congress. First session on S. Res. 67. US Government Printing Office, Washington D.C., pp. 110–119.

REFERENCES

Pasechnik, I.P., 1968. *Statement.* In: Seismic methods for monitoring underground explosions. International Institute for Peace and Conflict Research, Stockholm, pp. 83.

Pasechnik, I.P., 1972. *Characteristics of seismic waves in nuclear explosions and earthquakes.* (Translated from Russian.) Foreign Technology Division WP–AFB, Ohio, 318 pp.

Pattersson, D.W., 1966. Nuclear decoupling, full and partial. *J. Geophys. Res.*, 71:3427–3436.

Peacock, K.L. & Treitel, S., 1969. Predictive deconvolution. Theory and practice. *Geophysics*, 34:155–169.

Peppin, W.A. & McEvilly, T.V., 1974. Discrimination among small magnitude events on Nevada Test Site. *Geophys. J. Roy. Astron. Soc.*, 37:227–243.

Pfluke, J.H. & Murdock, J.N., 1971. A theoretical model of the NORSAR seismic noise field. *Geophys. J. Roy. Astron. Soc.*, 25:17–24.

Pirhonen, S., 1976. *Preliminary observations from the Jyväskylä seismic station.* Paper presented at the Seventh Nordic Seminar on Seismology in Copenhagen 24–26 May, 1976.

Pisarenko, V.G. & Poplavskii, A.A., 1972. *A statistical method for determining earthquake focal depth from the record of one seismic station.* In: V.L. Keilis-Borok (Ed.), Computational Seismology. (Translated from Russian.) Consultants Bureau, New York, pp. 202–213.

Pomeroy, P.W., 1970. *Correlation of infrasonic microbarometric disturbances and long period seismic phenomena.* The University of Michigan, Arlington, Virginia, Report 02637–1–p.

Powell, P. & Fries, D., 1966. *Handbook: World-Wide Standard Seismograph Network.* Institute of Science and Technology, The University of Michigan, Ann Arbor.

Power, D.V., 1972. Ground motion prediction method for simultaneously detonated multiple underground nuclear explosives. *Nuclear Technology*, 16:437–443.

Press, F. & Siever, R., 1974. *Earth.* W.H. Freeman & Co., San Fransisco, 945 pp.

Randall, M.J., 1973. The spectral theory of seismic sources. *Bull. Seism. Soc. Amer.*, 63:1133–1144.

Randall, M.J. & Knopoff, L., 1970. The mechanism at the focus of deep earthquakes. *J. Geophys. Res.*, 75:4965–4976.

Richter, C.F., 1935. An instrumental earthquake magnitude scale. *Bull. Seism. Soc. Amer.*, 25:1–32.

Richter, C.F., 1958. *Elementary seismology.* W.H. Freeman & Co., San Francisco, 768 pp.

Rieber-Mohn, D., 1975. *Aspects of ARPANET usage at NORSAR.* In: K.G. Beauchamp (Ed.), Exploitation of Seismograph Networks. Noordhoff, Leiden, pp. 81–88.

Rikitake, T., 1975. Earthquake Precursors. *Bull. Seism. Soc. Amer.*, 65:1133–1162.

Ringdahl, R., Husebye, E.S. & Dahle, A., 1971. *Event detection problems using a partially coherent seismic array.* NTNF/NORSAR, Kjeller, Technical Report No. 45.

Robinson, E.A., 1967. *Statistical communication and detection with special reference to digital data processing of radar and seismic signals.* Charles Griffin and Co. Ltd., London, 362 pp.

Rocard, Y., 1962. *Various characteristics and some possible causes of seismic ground noise.* In: Problems in seismic background noise. University of Michigan, Vesiac Advisory Report 4410−32−X, pp. 85−93.

Rodean, H.C., 1971a. *Nuclear explosion seismology.* US AEC, Oak Ridge, 156 pp.

Rodean, H.C., 1971b. *Cavity decoupling of nuclear explosions.* Lawrence Radiation Laboratory, Livermore, UCRL−51097.

Romney, C., 1971. *Seismic system improvement.* In: Hearings before the Subcommittee on Research, Development, and Radiation of the Joint Committee on Atomic Energy. Congress of the United States, Ninety-second Congress. First Session on extent of present capabilities for detecting and determining nature of underground events. Oct. 27 and 28. US Government Printing Office, Washington D.C., pp. 85−98.

Ryall, A., 1970. *Seismic identification at short distances.* In: Copies of papers presented at Woods Hole Conference on Seismic Discrimination. Laboratories of Teledyne, Geotech, Alexandria, Virginia.

Sammon, J.W., Foley, D. & Proctor, A., 1970. *Considerations of dimensionality versus sample size.* In: Proc. 1970 IEEE symposium on adaptive processes, decision and control. The University of Texas at Austin, December 7−9, IX:2.1−2.9.

Savage, J.C., 1966. Radiation from a realistic model of faulting. *Bull. Seism. Soc. Amer.*, 56:577−592.

Savage, J.C., 1972. Relation of corner frequency to fault dimensions. *J. Geophys. Res.*, 77:3788−3805.

Savarensky, E.F., Sololev, S.L. & Kharin, D.A., 1962. *Atlas of the seismicity of the USSR.* (In Russian.) Akad. Nauk SSSR, Moscow, 388 pp.

Savarensky, E.F., Proskurjakova, T.A. & Voronina, E.V., 1967. *On microseisms phase velocities and the direction to the exitation source.* In: Papers presented at the 9th Assembly of the European Seismological Commission. Akademisk Forlag, Copenhagen, pp. 347−356.

Savino, J., Sykes, L.R., Liebermann, R.C. & Molnar, P., 1971. Excitation of seismic surface waves with periods of 15 to 70 seconds for earthquakes and explosions. *J. Geophys. Res.*, 76:8003−8020.

Savino, J., McCamy, K. & Hade, G., 1972a. Structures in earth noise beyond twenty seconds−A window for earthquakes. *Bull. Seism. Soc. Amer.*, 62:141−176.

Savino, J., Murphy, A., Rynn, J., Tatham, R., Sykes, L.R., Choy, G. & McCamy, K., 1972b. Results from the high gain long period seismograph experiment. *Geophys. J. Roy. Astron. Soc.*, 31:179−203.

Sax, L.R., 1968. Stationarity of seismic noise. *Geophysics*, 33:668−674.

Scholz, C.H., 1968. The frequency−magnitude relation of microfracturing in rock and its relation to earthquakes. *Bull. Seism. Soc. Amer.*, 58:399−416.

Scoville, H., 1973. *Statement.* In: Hearing before the Subcommittee on Arms Control, International Law and Organization of the Committee on Foreign Relations.

United States Senate. Ninety-third Congress. First session on S. Res. 67. US Government Printing Office, Washington D.C., pp. 119–128.

Sheppard, R.M., 1968. *Comparison of LASA observations with Kuriles ocean bottom net observatories*. In: Seismic Discrimination, Semiannual Technical Summary Report to the Advanced Research Projects Agency 1 January–30 June, 1968. Massachusetts Institute of Technology, Lincoln Laboratory, pp. 11–15.

Shimsoni, M. & Smith, S.W., 1964. Seismic signal enhancement with three component detectors. *Geophysics*, 29:664–671.

Shishkevish, C., 1975. *Soviet Seismographic Stations and Seismic Instruments*. Rand Corporation, Santa Monica, Cal., Report R-1647-ARPA.

Shlien, S. & Toksöz, N., 1973. Automatic event detection and location capabilities of large aperture seismic arrays. *Bull. Seism. Soc. Amer.*, 63:1275–1288.

Shor, Jr., G.G, & Raitt, R.W., 1969. *Explosion seismic refraction studies of the crust and upper mantle in the Pacific and Indian Oceans*. In: The earth's crust and upper mantle. Amer. Geophys. Union, Washington, D.C., pp. 225–230.

Shumway, R. & Blandford, R., 1970. *Simulation of discrimination analysis*. In: Copy of papers presented at Woods Hole Conference on Seismic discrimination. Laboratories of Teledyne Geotech, Alexandria, Virginia.

Singer, S., 1965. The Vela satellite program for detection of high-altitude nuclear detonations. *Proc. IEEE*, 53:1935–1940.

Singh, D., Rastogi, B.K. & Gupta, H.K., 1975. Surface wave radiation pattern and source parameters of Koyna earthquake of Dec. 10, 1967. *Bull. Seism. Soc. Amer.*, 65:711–731.

SIPRI, 1968. *Seismic methods for monitoring underground explosions*. Stockholm International Peace Research Institute, Stockholm, 130 pp.

SIPRI, 1969. *SIPRI Yearbook of world armaments and disarmament 1968/69*. Almqvist & Wiksell, Stockholm, 440 pp.

SIPRI, 1970. *SIPRI Yearbook of world armaments and disarmament 1969/70*. Almqvist & Wiksell, Stockholm, 540 pp.

SIPRI, 1972. *World armaments and disarmament*. SIPRI Yearbook 1972. Almqvist & Wiksell, Stockholm, 611 pp.

SIPRI, 1973. *World armaments and disarmament*. SIPRI Yearbook 1973. Almqvist & Wiksell, Stockholm, 510 pp.

SIPRI, 1974. *World armaments and disarmament*. SIPRI Yearbook 1974. Almqvist & Wiksell, Stockholm, 526 pp.

SIPRI, 1975. *World armaments and disarmament*. SIPRI Yearbook 1975. Almqvist & Wiksell, Stockholm, 618 pp.

SIPRI, 1976. *World armament and disarmament*. SIPRI Yearbook 1976. Almqvist & Wiksell, Stockholm, 493 pp.

Slunga, R., 1973. Simultaneous location of local events and travel time modelling. *Earthquake Notes*, XLIV:14.

Slunga, R., 1975. Depth determination by use of teleseismic P-arrivals. *Geophys. J. Roy. Astron. Soc.*, 42:737–746.

Slunga, R., 1976. *NTS-explosions recorded at Hagfors, a comparison of theoretical and observed seismograms.* Paper presented at the Seventh Nordic Seminar on Seismology in Copenhagen 24–26 March 1976.

Smith, S.W., 1972. The anelasticity of the mantle. *Tectonophysics*, 13:601–622.

Smith, W.D., 1975. The application of finite element analysis to body wave propagation problems. *Geophys. J. Roy. Astron. Soc.*, 42:747–768.

Springer, D.L. & Hannon, W.J., 1973. Amplitude–yield scaling for underground nuclear explosions. *Bull. Seism. Soc. Amer.*, 63:477–500.

Springer, D.L. & Kinnaman, R.L., 1971. Seismic source summary for US underground nuclear explosions 1961–1970. *Bull. Seism. Soc. Amer.*, 61:1073–1098.

Springer, D.L. & Kinnaman, R.L., 1975. Seismic source summary for US underground nuclear explosions. *Bull. Seism. Soc. Amer.*, 65:343–349.

Springer, D., Denny, M., Healy, J. & Mickey, W., 1968. The Sterling experiment: Decoupling of seismic waves by a shot-generated cavity. *J. Geophys. Res.*, 73:5995–6012.

Strauss, A.C., 1973. *Final evaluation of the detection and discrimination capability of the Alaskan long period array.* Texas Instruments Inc., Dallas, Texas, Special Report No. 8. Extended Array Evaluation Program.

Strelitz, R., 1975. The September 5, 1970, Sea of Okhotsk earthquake: A multiple event with evidence of triggering. *Geophys. Res. Lett.*, 2:124–127.

Sykes, L.R., 1971. Aftershock zones of great earthquakes, seismicity gaps and earthquake prediction for Alaska and the Aleutians. *J. Geophys. Res.*, 76:8021–8041.

Talandier, J. & Kuster, G., 1976. Seismicity and submarine volcanic activity in French Polynesia. *J. Geophys. Res.*, 81:936–948.

Teledyne, 1970. *Alaskan long period array.* Geotech Teledyne, Alexandria, Virginia Technical Report No. 70–39.

Thatcher, W. & Hamilton, R.M., 1973. Aftershocks and source characteristics of the 1969 Coyote Mountain earthquake, San Jacinto Fault Zone, California. *Bull. Seism. Soc. Amer.*, 63:647–661.

Thirlaway, H.I.S., 1965. Earthquake or explosion? *New Scientist*, 18:311–315.

Thirlaway, H.I.S., 1975. *Exploitation of seismograph networks.* In: K.G. Beauchamp (Ed.), Exploitation of seismograph networks. Noordhoff, Leiden, pp. 5–10.

Tjöstheim, D., 1975a. Some autoregressive models for short-period seismic noise. *Bull. Seism. Soc. Amer.*, 65:677–691.

Tjöstheim, D., 1975b. Autoregressive representation of seismic P-wave signals with an application to the problem of short-period discriminants. *Geophys. J. Roy. Astron. Soc.*, 43:269–291.

Toksöz, M.N., 1970. *Crustal effects on long period chirp filters.* In: Copies of papers presented at Woods Hole Conference on Seismic discrimination, Laboratories of Teledyne Geotech, Alexandria, Virginia.

Toksöz, M.N., 1975. Lunar and planetary seismology. *Rev. Geophys. Space Phys.*, 13:306–312.

Toman, J., 1975. *Project Rio Blanco—Part II. Production test data and preliminary analysis of Top Chimney/Cavity.* In: Peaceful Nuclear Explosions IV. Proceedings of a Technical Committee on the Peaceful Uses of Nuclear Explosions organized by the International Atomic Energy Agency and held in Vienna, 20–24 January, 1975. IAEA Vienna, pp. 117–140.

Trembly, L.D. & Berg, J.W., 1966, Amplitudes and energies of primary seismic waves near the Hardhat, Haymaker, and Shoal nuclear explosions. *Bull. Seism. Soc. Amer.*, 56:643–654.

Trifunac, M.D., 1974. A three dimensional dislocation model for the San Fernando, Calif., Earthquake of Febr. 9, 1971. *Bull. Seism. Soc. Amer.*, 64:149–172.

Tsai, Y.B. & Aki, K., 1970a. Precise focal depth determination from amplitude spectra of surface waves. *J. Geophys. Res.*, 75:5729–5744.

Tsai, Y.B. & Aki, K., 1970b. Source mechanism of the Truckee, Calif., earthquake of Sept. 12, 1966. *Bull. Seism. Soc. Amer.*, 60:1199–1208.

Tsai, Y.B. & Aki, K., 1971. Amplitude spectra of surface waves from small earthquakes and underground nuclear explosions. *J. Geophys. Res.*, 76:3940–3952.

Udias, A., 1971. Source parameters of earthquakes from spectra of Rayleigh waves. *Geophys. J. Roy. Astron. Soc.*, 22:353–376.

UED Earth Sciences Division, 1966. *Long range seismic measurements. Chase IV 16 September 1965.* Seismic Data Laboratory, Alexandria, Virginia, Report No. 137.

UKAEA, 1965. *The detection and recognition of underground explosions.* United Kingdom Atomic Energy Authority.

UKAEA, 1967. *Data processing facilities and data available at the UKAEA Data Analysis Centre for Seismology.* AWRE, Aldermaston.

UN, 1954. *Official Records of the Disarmament Commission.* Supplement for April, May and June, 1954. Document DC/44 and Corr. 1.

UN, 1958. *Ibid.*, Thirteenth Session, Annexes, Agenda items 64, 70, and 72, Document A/3897.

UN, 1967. *The United Nations and Disarmament 1945–1965.* UN Publication Sales No. 67–I–9, 338 pp.

UN, 1969. *UN Document.* A/RES/2604 (XXIV).

UN, 1970. *Basic problems of disarmament.* Reports of the Secretary General. United Nations, New York, pp. 73–154.

UN, 1971. *UN Document.* A/C.1/PV.1847 (XXVI).

UN, 1974. *UN Document.* A/RES/3235 (XXIX).

UN, 1975a. *UN Document.* A/RES/3466 (XXX).

UN, 1975b. *UN Document.* A/C.1/PV.2107 (XXX).

UN, 1975c. *UN Document.* A/C.1/PV.2105 (XXX).

UN, 1975d. *UN Document.* A/C.1/PV.2073 (XXX).

USCGS, 1960. *Performance specifications of the instrumentation for a world-wide network of standardized seismographs.* The Committee on Seismological Stations. US Department of Coast and Geodetic Survey, Washington, D.C.

Van Trees, H.L., 1969. *Detection, estimation and modulation theory.* Part I. John Wiley & Sons Inc., New York, 647 pp.

Veith, K.F. & Clawson, G.E., 1972. Magnitude from short period P-wave data. *Bull. Seism. Soc. Amer.*, 62:435–452.

Vinnik, L.P. & Purchkina, N.M., 1964. A study of structure of short period microseisms. *Isz. Geophys.*, Ser. 5:688–700.

Von Seggern, D., 1972. Relative location of seismic events using surface waves. *Geophys, J. Roy. Astron. Soc.*, 26:499–513.

Ward, P., 1970. *A high gain, braod-band long period seismic experiment.* In: Copies of papers presented at Woods Hole Conference on Seismic Discrimination. Laboratories of Teledyne Geotech, Alexandria, Virginia.

Weichert, D.H., 1971. Short period spectral discriminant for earthquake and explosion differentiation. *Z. Geophys.*, 37:147–152.

Weichert, D.H., 1975a. *The role of medium aperture arrays: The Yellowknife system.* In: K.G. Beuachamp (Ed.), Exploitation of Seismograph networks. Noordhoff, Leiden, pp. 167–196.

Weichert, D.H., 1975b. Reduced false alarm rates in seismic array detection by nonlinear beamforming. *Geophys. Res. Lett.* 2:121–123.

Weichert, D.H. & Basham, P.M., 1973. Deterrence and false alarms in seismic discrimination. *Bull. Seism. Soc. Amer.*, 69:1119–1132.

Weichert, D.H., Basham, P.W. & Anglin, F.M., 1975. *Explosion identification and monitoring.* In: K.G. Beauchamp (Ed.), Exploitation of seismograph networks. Noordhoff, Leiden, pp. 383–395.

Weidner, D.J. & Aki, K., 1973. Focal depth and mechanism of Mid-Ocean ridge earthquakes. *J. Geophys. Res.*, 78:1818–1831.

Werth, G.C. & Herbst, R.F., 1963. Comparison of amplitudes of seismic waves from explosions in four mediums. *J. Geophys. Res.*, 68:1463–1475.

Whiteway, F.E., 1965. The recording and analysis of seismic body-waves using linear cross arrays. *Radio Electronic Engineer*, 29:33–46.

Willis, D.E., De Noyer, J. & Wilson, J.T., 1963. Differentiation of earthquakes and underground nuclear explosions on the basis of amplitude characteristics. *Bull. Seism. Soc. Amer.*, 53:979–987.

Willis, D.E., George, G.D., Petzl, K.G., Saltzer, C.E., Shakal, A.F., Torfin, R.D., Woodzick, T.L. & Wolosin, C., 1972. Seismological aspects of the Cannikin nuclear explosion. *Bull. Seism. Soc. Amer.*, 62:1377–1395.

Willmore, P.L., 1975. *Characteristics of seismic transducers, array transmission and recording.* In: K.G. Beauchamp (Ed.), Exploitation of seismograph networks. Noordhoff, Leiden, pp. 25–36.

Willrich, M. & Taylor, T.B., 1974. *Nuclear theft: Risks and safeguards.* Ballinger Publishing Co., Cambridge, 243 pp.

REFERENCES

Wood, R.V., Enticknap, R.G., Lin, C.S. & Martinson, R.M., 1965. Large aperture seismic array signal handling system. *Proc. IEEE*, 53:1844–1851.

Woodruff, W.R. & Guido, M.S., 1975. *Project Rio-Blanco—Part I. Nuclear operations and chimney reentry.* In: Peaceful Nuclear Explosions IV. Proceedings of a Technical Committee on the Peaceful Uses of Nucelar Explosions organized by the International Atomic Energy Agency and held in Vienna, 20–24 January 1975. Vienna, pp. 29–116.

Wyss, M., 1973. Towards a physical understanding of the earthquake frequency distribution. *Geophys. J. Roy. Astron. Soc.*, 31:341–359.

Wyss, M., Hanks, T.R. & Liebermann, R.C., 1971. Comparison of *P*-wave spectra of underground explosions and earthquakes. *J. Geophys. Res.*, 76:2716–2729.

Yamamoto, M., 1974. *Estimation of focal depth by pP and sP phases.* National Defense Research Institute, Stockholm, Report C 20027–T1.

Zander, I. & Araskog, R., 1973. *Nuclear explosions 1945–1972. Basic data.* National Defense Research Institute, Stockholm, Report A 4505–A1.

Ziolkowski, A., 1973. Prediction and suppression of long period non-propagating seismic noise. *Bull. Seism. Soc. Amer.*, 63:937–958.

ABBREVIATIONS AND SYMBOLS

ABM	Anti-Ballistic Missile	M_L	Local magnitude
ALPA	Alaskan Long-Period Array (seismological station)	M_s	Surface-wave magnitude
ARPA	Advanced Research Projects Agency of the US Department of Defense	MOS	Institute of Physics of the Earth, Moscow
		Mt	Megaton
ARPANET	Computer network supported by ARPA	NORSAR	Norwegian Seismic Array (seismological station)
ASRO	Auxiliary Seismic Research Observatory	NPT	Non-Proliferation Treaty
		NTS	Nevada Test Site
CCD	Conference of the Committee on Disarmament	P	Compressional seismic wave
CTB	Comprehensive Test Ban Treaty	PNE	Nuclear explosion for peaceful purposes
dB	Decibel	PTB	Partial Test Ban Treaty
$dT/d\Delta$	Inverse velocity, or slowness	SALT	Strategic Arms Limitation Talks
ENDC	Eighteen-Nation Committee on Disarmament	SIPRI	Stockholm International Peace Research Institute
f	Signal frequency (Hz)	SNR	Signal-to-noise ratio
GBA	Gauribidanur (seismological station in India)	SP	Short-period
		SRO	Seismological Research Observatory
HFS	Hagfors Observatory in Sweden, or a substation thereof	TMF	Third moment of frequency
		TNT	Trinitrotoluene
IAEA	International Atomic Energy Agency	TTBT	Threshold Test Ban Treaty
		UKAEA	United Kingdom Atomic Energy Authority
IC	Identification curve		
ILPA	Iranian Long-Period Array (seismological station)	USAEC	United States Atomic Energy Commission
ISM	International Seismic Month (cooperative experiment in 1972)	USGS	United States Geological Survey
JMA	Japan Meteorological Agency	VLPE	Very Long Period Experiment seismograph station
KSRS	Korean Seismic Research Station	WRA	Warramunga (seismological station in Australia)
kt	Kiloton	WWSSN	World Wide Seismological Station Network
LASA	Large Aperture Seismic Array (seismological station in Montana, US)	Y	Explosion yield
		YKA	Yellowknife (seismological station in Canada)
LP	Long-period		
LRSM	Long-Range Seismic Measurement Station (temporary seismological station in the US)	Δ	Epicentral distance
		ρ	Density
m_b	Body-wave magnitude		

INDEX

ABM Treaty, 20–21
A-bomb, 36
Afghanistan, seismic instrumentation, 131
Afiamalu, seismological station, 163
Aftershock, 81, 83, 255
Alamogordo, New Mexico, 6
Alaska, 85
Aleutian Islands, 83, 85, 89, 160, 181, 286, 319
Algeria, 49
Alice Springs, seismological station, 163
Alluvium, 95, 264, 274, 307
 – deposits, 308, 343
ALPA, seismological station, 127, 130, 169
 – detection thresholds, 160, 162
 – $m_b(M_s)$ data, 225
Analog-to-digital conversion, 115
Anelasticity, 57, 63, 241
Angular distance, 59
Antarctic Treaty, 10, 12, 13
Apollo, satellite, 330
Apparent velocity, seismic waves, 149, 155, 172
Argentina, 9, 14, 15, 17
ARPA, 131
ARPANET, 132, 352
Array station, 116, 125, 128, 147
 – detection process, 147
 – event location, 172, 183
 – three-dimensional, 149
Arrival time, P waves, 61, 200
ASRO, 131
Atmospheric explosion, seismic discriminant, 221
 – seismic signals, 135, 146

Australia, seismic instrumentation, 126, 128, 131
Autoregressive model, 239

Baguio City, seismological station, 163
Baker Lake, seismological station, 163
Bandpass filtering, 144–146
Bandwidth, 109
Baneberry, nuclear explosion, 315
Bangui, seismological station, 163
Basalt, crustal layer, 203
Beam, 150, 152
Beamforming, 117, 149
Beam pattern, 150
Belgium, 40
Berkeley, $m_b(M_s)$ data, 222, 225, 228
Binding energy, 35
Black box, 132
Blacknest, data analysis center, 189, 260
 – broadband array, 253
Body wave, 59–67
Bolivia, seismic instrumentation, 128, 131
Brasilia, seismological station, 126
 – detection threshold, 163
Bravo, nuclear explosion, 6
Brazil, 9, 15, 17
 – seismic instrumentation, 126
Broadband seismic signals, 109, 252, 313
Buchara, nuclear explosion, 300
Bulawayo, seismological station, 163
Bulgaria, 9
Burma, 9

Canada, 9
 – position on test ban, 22

433

(Canada)
— seismic instrumentation, 126, 128
— seismological network, 225, 260
Cannikin, nuclear explosion, 61, 81, 89, 288
Cavity, 292
— decoupling, 308
CCD, 8–9, 22–34
CEA, seismological network, 126
Centre Séismologique Européo-Mediterranéen, 189
Cepstrum, 199
Chain reaction, 37
Charters Towers, seismological station, 131
— detection threshold, 159
Chiengmai, seismological station, 163
Chimney, nuclear explosion, 292
China, 8, 10, 14, 16, 17, 40, 45–47
— nuclear explosions, 387
— position on test ban, 23
— seismicity, 84
Coda, seismic signal, 194, 318
Coherency, noise, 137
— signal, 153
Collapse, explosion cavity, 255
College Outpost, seismological station, 163
Collmberg, seismological station, 163
Colombia, seismic instrumentation, 131
Complexity, 206, 214, 238
Comprehensive Test Ban, 5
Compression, 98
Compressional seismic waves, 55
Conference of the Committee on Disarmament, 8–9, 22–34
Convergence, plate boundary, 53, 58, 76, 106, 190
Core, earth, 52, 57, 321
Corner frequency, 101, 252
Cowboy, decoupling experiment, 308
Crater, nuclear explosion, 274, 291, 326, 334, 335
Critical mass, 37
Crust, earth, 52
— continental, 55
— oceanic, 55
— seismic waves, 71
Crustal waves, 71–73

Cumulative detection capability, 159
Czechoslovakia, 9

Data exchange, 188
Deconvolution, 199
Decoupling, 7, 306
— factor, 307
Definition, of seismic event, 171
Depth, of seismic event, 190
— distribution, 190
— earthquakes, 77
— estimation methods, 192
— event identification, 261
Detection, 134, 143
— array station, 147
— capability, 156
— network capability, 163
— processes, 143
— single station, 144
— station capability, 156
— threshold, 159
— yield thresholds, 162
Deterrence, 213
Deuterium, 37
Digital recording, 114
Dilatation, 98
Disarmament Commission, 8
Discriminant, 213
— applicability, 211
— distribution function, 214
— efficiency, 211
— linear combination, 217
— multidimensional, 217, 239
— multistation, 256
Dispersion, surface waves, 68
Displacement potential, 94
Divergence, plate boundary, 53
Double couple, point source, 98, 208
Dynamic range, 109

Earth, internal structure, 52–58
Earthquake, aftershock, 81, 83, 326
— artificial, 83
— depth, 77, 190
— epicenter, 77
— foreshock, 82
— lunar, 82
— magnitude, 77
— prediction, 82

(Earthquake)
— seismic record, 144, 146, 151, 152, 154, 192, 194, 195, 205
— source, 97
— source displacement, 99
— space–time pattern, 80
Eastern Kazakh test area, 47, 88, 182, 183, 286
Egypt, 9, 17
Eighteen-Nation Committee on Disarmament, 8
Elasticity, 53
Elastic sphere, 43, 93
ENCD, 8
Epicenter, 76, 97
— event identification, 261
Eskdalemuir, 127
— seismic noise, 141
— monitoring experiment, 260
Ethiopia, 9
Evasion, 306
Event, definition, 171
— location, 171
Explosion, seismic record, 218
— source, 92
Ezine, seismological station, 163

False alarm, 213
Fangataufa, 49
Fault plane, 97
Federal Republic of Germany, 9
— seismic instrumentation, 126, 128
Filter, 145
Finland, seismic instrumentation, 126
Fission, 36
Focal depth, 190
Focal mechanism, 208
Focus, 97
France, 8, 9, 10, 14, 16, 17, 40, 45, 46, 47
— nuclear explosions, 388
— position on test ban, 24
— seismic instrumentation, 126
— seismicity, 85
— test sites, 85
Frequency wave number filtering, 150
Frobisher Bay, seismological station, 163
Fusion, 36

Gain-ranging amplifier, 113, 120
Gas stimulation, 293
Gauribidanur, 126
— detection threshold, 163
— monitoring experiment, 260
Geometrical spreading, 63
German Democratic Republic, 9
Grafenberg, seismological station, 126
Granite, crustal layer, 203
— shot medium, 95, 264, 274, 307
Group velocity, 68

Hagfors, seismological station, 112, 118, 119, 126, 129, 130
— detection capability, 157
— detection procedure, 155
— detection threshold, 160, 162, 163
— International Seismic Month, 157
— location error, 175
— $m_b(M_s)$ data, 255
— monitoring experiment, 260
— seismic noise, 139, 142
— short-period discriminant, 246
Handley, nuclear explosion, 236
H-bomb, 37
Hide-in-earthquake, 315
Hungary, 9
Hypocenter, 97, 200

IAEA, 15, 291, 304, 353
Identification of seismic events, 211
Identification curve, 215, 228, 249, 257
ILPA, seismological station, 126
Incoherent beamforming, 153
Incremental detection capability, 159
India, 9, 10, 16, 45–47
— nuclear explosion, 294, 328, 387
— position on test ban, 25
— seismic instrumentation, 126, 128, 130, 131
— seismicity, 87
Indonesia, 17
Institute of Physics of the Earth, Moscow, 108, 208
Interference of seismic signals, 149, 151, 159
International Geophysical Year, 13

International Seismic Month, 156, 163, 166, 167, 169, 188, 346
International Seismological Centre, 188
International seismological data exchange, 188
Iran, 9
— seismic instrumentation, 126, 131
ISM, 259
Israel, 17
Italy, 9

Japan, 9, 40
— position on test ban, 26
— seismic instrumentation, 123
Japan Meteorological Agency, 189, 203
Joint epicenter determination, 176, 180
Joint hypocenter determination, 200
Jyväskylä, seismological station, 126

Kabul, seismological station, 131
— detection threshold, 163
Kamchatka, 85, 89, 160, 319
KANAB, seismological station, 255
Kenya, seismic instrumentation, 131
Kosan Boka, seismological station, 163
KSRS, seismological station, 126
Kuril, 85, 89, 160, 319

Landsat, 330
La Paz, seismological station, 131
LASA, seismological station, 118, 127, 129
 — beamforming, 151
 — detection threshold, 160, 162, 163
 — International Seismic Month, 156
 — location error, 175, 177
 — monitoring experiment, 260
 — seismic noise, 138
Laser, 39
Lithium, 40
Lithosphere, 52, 75, 191
Local station network, 189
 — depth estimation, 202
 — event identification, 255
 — event location, 186
Location, event, 171
 — error, 173

Long-period seismic signals, 109
 — depth estimation, 204
 — detection capability, 169
 — frequency band, 137
 — noise, 137
 — seismometer, 110
Longshot, nuclear explosion, 81, 89, 181, 200, 241
Lop Nor, 49, 84, 86
Love wave, 60, 67
 — discriminant, 253
 — noise, 138
 — signal duration, 140
 — signal spectrum, 204
Low-velocity zone, 56, 69
LRSM, seismological station, 128
Lucky Dragon, 6

Magnetic-tape recording, 114
Magnitude, body wave, 66
 — distribution, 77, 158, 159, 167
 — local, 73
 — original definition, 73
 — surface wave, 69
Mantle, earth, 52
Matsushiro, seismological station, 163
m_b, body-wave magnitude, 66
$m_b(M_s)$ method, definition, 219
 — evasion, 311
 — theoretical basis, 103
m_b-yield relation, 267
Medeo, chemical explosions, 269
Mexico, 9
Microseism, 135
Milrow, nuclear explosion, 81, 89
M_L, local magnitude, 73
Moho, 55
Moment function, seismic source, 96
Mongolia, 9
Monitoring experiment, 260
Morocco, 9
Mould Bay, seismological station, 163
M_s, surface-wave magnitude, 69
M_s-yield relation, 267
Multiarray location, 185
Multipathing, 66, 195, 240
Multiple explosions, 311
Mururoa, 49

National Geophysical Data Center, 131
National means of verification, 21—22, 325
Negative evidence, 219, 234
Netherlands, 9, 40
Network, event detection capability, 163
— event location capability, 176, 185
— station, 121
Nevada Test Site, 47, 83, 90, 178, 286, 308, 334
New Zealand Seismological Observatory, 189
— seismic instrumentation, 126, 128, 130
Nigeria, 9
Noise, seismic, 134
— suppression filter, 146
Non Proliferation Treaty, 10, 12, 16, 303
— text, 360
NORSAR, 126, 129, 130
— beamforming, 151
— detection thresholds, 160, 162, 163
— International Seismic Month, 156
— location error, 175
— seismic noise, 139, 141
— short-period discriminant, 246
Notch filter, 147
Novaya Zemlya, 47, 83, 88, 241, 286
Nuclear explosion, 41
— cavity radius, 42
— shock wave, 41
— shot depth, 199
— source model, 92
— thermal radiation, 41
— yield, 263
Nurmijärvi, seismological station, 163

Ogdensburg, $m_b(M_s)$ data, 225
— monitoring experiment, 260
Oil stimulation, 293
On-site inspection, 7, 8, 12, 20, 326
OPANAL, 15, 339
Origin time, 172
Outer Space Treaty, 10, 12, 14

P, seismic wave, 59, 93, 98, 103
Pakistan, 9, 14, 17
Partial decoupling, 308
Partial hiding, 309
Partial Test Ban Treaty, 8, 10, 12, 14, 339
— text, 358
Payette, cavity experiment, 309
PcP, seismic wave, 61, 62, 193, 196, 240
Peaceful nuclear explosions, 291
Pechora—Kama canal, 300
Penas, seismological station, 163
Peru, 9
Pg, seismic wave, 71, 72
PKP, seismic wave, 61, 62
Plate tectonics, 52
Plowshare, 294
Plutonium, 36
P_n, seismic wave, 71, 73
PNE, peaceful nuclear explosions, 291
PNE Treaty, 10, 12, 18, 303, 329
— text, 368
Point source, 98
Poland, 9
Polarity, seismic signal, 98
— discriminant, 252
Polarization filter, 147, 195
Port Moresby, seismological station, 163
pP, seismic wave, 62, 192
Prediction-error filter, 146
Pulse width, 239
P wave, 59—60
— amplitude, 64, 65
— depth estimation, 200, 206
— detection, 156, 164
— theoretical, 105
— travel path, 59, 62, 106
— travel time, 61
— velocity, 55, 56

Q value, 57, 58
— attenuation, 63, 69, 105
— lateral variation, 58, 64
Quetta, seismological station, 163

Radiation pattern, seismic waves, 93, 98, 254

Radioactivity, 300, 304
Radiochemical sampling, 263, 288, 326
Rajasthan, 88
Rarotonga, seismic noise, 142
Rayleigh wave, 67, 70, 107
 — depth estimation, 204
 — detection, 156, 169
 — event location, 175
 — noise, 138
 — signal duration, 140
 — spectrum, 204
 — travel time, 68
 — velocity, 68
Ray path, seismic waves, 59, 62
Ray tracing, 202
Receiver operating characteristics, 152, 215
Recording, seismic signal, 113
Reduced displacement potential, 94
Response curve, seismometer, 112
Rg, seismic wave, 72–73
Rhyolite, 274, 307
Rio Blanco, nuclear explosion, 295, 315
Romania, 9

S, seismic wave, 59, 98, 254, 255
Saint Saulge, seismological station, 163
SALT, arms control, 20–22
Salt, shot medium, 307, 309
Sampling rate, 115
San Fernando, earthquake, 236
Satellite, 329
 — ground resolution, 330
 — orbit altitude, 330
 — photos, 330
Screening, event identification, 258
SDAC, Seismic Data Analysis Center, 130, 132
Sea Bed Treaty, 10, 12, 17, 339
Sedan, nuclear explosion, 292, 334, 335
Seismic, array station, 117
 — instrumentation, 109
 — moment, 97, 110
 — noise, 134
 — record, 60, 218
 — source, 92, 96, 97, 101, 103
Seismicity, 73–91

Seismological stations, 108
Seismometer, 110, 111
Sg, seismic wave, 71–73
Shale, shot medium, 307
Shear wave, 55
Shillong, seismological station, 163
Shiraz, seismological station, 163
Shock wave, explosion, 93, 326
Short-period seismic signals, 109, 164
 — discriminant, 236, 261, 313
 — frequency band, 137
 — noise, 140
 — seismometer, 111
 — spectrum, 207
Signal matching, 237
Signal-to-noise ratio, 144
SIPRI, 3
Skylab, 332
Slapdown, explosion, 199
Slowness, 63
S_n, seismic wave, 71
Source function, 272
 — seismic, 92
 — spectrum, 96, 101, 103
South Africa, 17
South Korea, seismic instrumentation, 126, 130
South Pole, seismological station, 163
sP, seismic wave, 62, 192
Spectrum, seismic noise, 136, 142
 — seismic signals, 107, 207, 220
 — seismic source, 94, 96, 101, 103
SRO, Seismological Research Observatory, 131, 145, 342
 — detection process, 145
Sterling, decoupling experiment, 307, 309
Stockpile test, nuclear explosion, 50, 338
Stress drop, earthquake, 99, 100
Surface-reflected wave, 192
Surface wave, 59, 67–71
S wave, 55, 58
 — discriminant, 255
 — travel path, 59
 — velocity, 55, 56
Sweden, 9
 — position on test ban, 28
 — seismic instrumentation, 126

INDEX 439

Switzerland, 17

Taiwan, seismic instrumentation, 131
Tamanarasset, seismic noise, 143
Tamper, nuclear explosion, 37
Teleseismic, 79
Thailand, seismic instrumentation, 131
Thermonuclear reaction, 38
Third moment of frequency, discriminant, 242, 243, 245, 250
Thorium, 40
Threshold Test Ban Treaty, 10, 12, 17, 171, 180, 263, 339
 — text, 365
Tibet, earthquakes, 231, 256
Tlatelolco, treaty, 10
TMF, discriminant, 242, 243, 245, 250
Transcurrence, plate boundary, 53
Transmission function, 272, 277
Travel time, seismic waves, 61, 68, 193
 — derivative, 63, 172, 200
Treaty of Tlatelolco, 10, 12, 15, 339
Trinity, nuclear explosion, 44
Tritium, 37
Tsukuba, short-period discriminant, 246
Tucson, seismological station, 163
Tuff, shot medium, 95, 274, 307
Turkey, seismic instrumentation, 131

UN, resolution on seismic data exchange, 164
Underground nuclear explosion, 35
 — depth, 199
Unita Basin, seismological station, 163
United Kingdom, 9, 10, 40, 45–47
 — nuclear explosions, 389
 — position on test ban, 32
 — seismic instrumentation, 127, 129
 — test site, 88, 89
United States, 9, 10, 40, 45–47
 — nuclear explosions, 283, 285, 287, 390
 — peaceful nuclear explosions, 296, 297
 — position on test ban, 32
 — seismic instrumentation, 123, 127
 — seismicity, 90, 191
 — test site, 88, 89

Unmanned seismic observatory, 132
Uranium, 36
 — enrichment, 39, 40
US Geological Survey, 188, 197, 259–260
USSR, 9, 10, 40, 45–47
 — nuclear explosions, 283, 285,
 — peaceful nuclear explosions, 298, 299, 302
 — position on test ban, 30
 — seismic instrumentation, 110, 123
 — seismicity, 88, 191
 — test site, 88

Vela uniform, 108, 120, 128
Velocity filtering, 154
 — seismic waves, 55, 56
Verification, national technical means, 21–22, 325
 — by challenge, 29
Visual recording, 114
VLPE, seismological station, 112, 115, 130, 342
 — detection capability, 168
 — $m_b(M_s)$ data, 225
 — monitoring experiment, 260
 — seismic noise, 139
 — signal interference, 161

Warramunga, seismological station, 126
 — monitoring experiment, 260
 — seismic noise, 141
Waveform filter, 145, 147
Wellington Seismological Observatory, 189
World Meteorological Organization, 352
WWSSN, 112, 123

Yellowknife, seismological station, 126, 129
 — beamforming, 152
 — detection threshold, 160, 162, 163
 — International Seismic Month, 157
 — location error, 175
 — monitoring experiment, 260
 — short-period discriminant, 246
Yield, 263

(Yield)
    — announced, 266
    — estimates, 272
    — estimation, 263
    — m_b relation, 267
    — M_s relation, 270

(Yield)
    — theoretical amplitude, 265
Yugoslavia, 9

Zaire, 9